ERRATUM

Atmospheric Ammonia – Detecting emission changes and environmental impacts

References within this publication, to other chapters of this book are incorrect.

For all instances of:

'xxxxx (2008) xxxxx. In: Sutton M.A., Reis. S., Baker S.M.H.: Atmospheric Ammonia – Detecting emission changes and environmental impacts, xx-xx, Springer publishers'

Please read:

'xxxxx (2009) xxxxx. In: Sutton M.A., Reis., Baker S.M.H.: Atmospheric Ammonia – Detecting emission changes and environmental impacts, xx-xx, Springer publishers'

Atmospheric Ammonia

Atmospheric Ammonia

Detecting emission changes and environmental impacts

Results of an Expert Workshop under the Convention on Long-range Transboundary Air Pollution

Edited by

Mark A. Sutton
Centre for Ecology & Hydrology,
(CEH), UK

Stefan Reis
Centre for Ecology & Hydrology,
(CEH), UK

Samantha M.H. Baker
Department for Environment Food and Rural Affairs
(Defra), UK

 Springer

Editors
Mark A. Sutton
Centre for Ecology & Hydrology
(CEH)
UK

Stefan Reis
Centre for Ecology & Hydrology
(CEH)
UK

Samantha M.H. Baker
Department for Environment
Food and Rural Affairs (Defra)
UK

ISBN 978-1-4020-9120-9   e-ISBN 978-1-4020-9121-6

Library of Congress Control Number: 2008937477

Cover figures:
Main Figure: Reindeer moss - Cladonia portentosa damaged by ammonia from an agricultural point
source. Photo by Ian Leith, CEH.
Top row figures: Pristine and damaged lichen (Cladonia portentosa) and Sphagnum moss (Sphagnum
capillifolium) exposed to ambient and enhanced ammonia concentrations at the CEH nitrogen
manipulation facility at Whim Bog, south-east Scotland. Photo by Ian Leith, CEH

Printed on acid-free paper

springer.com

Preface

This volume represents the fruits of a major international synthesis on the environmental behaviour and effects atmospheric ammonia. Specifically, it provides a contribution to the work of the Geneva Convention on Long Range Transboundary Air Pollution (CLRTAP), established under the auspices of the United Nations Economic Commission for Europe (UNECE).

Atmospheric ammonia has for some years been recognized by the Convention as being of key importance to air pollution effects on the eutrophication and acidification of ecosystems. More recently, the contribution of ammonia to the formation of secondary particulate matter has also been highlighted.

Partly as a result of the many environmental interactions, ammonia has, until now, mostly been reviewed alongside the other air pollutants treated by the Convention. This is, in many ways, exactly as it should be. However, ammonia is an unusual air pollutant in that its largest sources are rural rather than urban, with livestock agriculture being the major source. Coupled with high uncertainties in the emissions, in the measurement of atmospheric concentrations and in the environmental impacts, there was therefore a strong case to dedicate a specific review to current understanding of ammonia.

The original motivation for the present book arose in 2003, related to concerns over the existing 'critical level' for ammonia. With the Gothenburg Protocol signed in 1999, ammonia had been included in national emissions ceilings of air pollutants across the UNECE for the first time. This heightened the need to properly quantify the emissions, trends and impacts of ammonia. The empirical critical loads for nitrogen deposition had recently been reviewed, at the Bern Workshop of 2002. However, that assessment incorporated the role of ammonia only as far as deposition was concerned. By contrast, the critical level for ammonia concentrations had not been reviewed for a decade, since the Egham Workshop of 1992. By 2003, new data were emerging that suggested the need for a major re-evaluation of the ammonia critical level. To do this required that the evidence be assembled and that the Convention formally review the current values.

Hot on the heels of the critical level question was concern about the 'ammonia gap' in the Netherlands. Emission abatement measures had been implemented, but apparently atmospheric ammonia concentrations had not decreased. These issues were reviewed in the Bern Workshop (2000), but already substantial new datasets were appearing, pointing to the need for a new scientific update. These issues were coming particularly to the fore as other countries started to review their options for meeting the ammonia commitments under the Gothenburg Protocol.

The third issue to arise was that of ammonia in 'hot-spot' areas. Previously, there had been little international assessment of these impacts, which had been considered a local concern. Yet, it was rapidly becoming apparent that the regional assessments depended on how well the science could treat dispersion and deposition in source areas. In addition, since the main ammonia hot-spots occur in rural areas, a better understanding of these processes was recognized as essential to maximise overall environmental protection.

Finally, a fourth issue was added to the review. Increasingly, different models were becoming available to simulate the transboundary transport and deposition of ammonia at regional scales. Yet, there remained major uncertainties in the parametrization of these models for ammonia, as well as substantial differences when compared with measurements. It was therefore agreed that the review must address the current status of the regional models, with a particular emphasis on the key uncertainties related to ammonia.

These four themes have thus provided the core of the present review on ammonia, with special attention to the detection of emission changes and environmental impacts. The review itself was conducted in the form of a UNECE Expert Workshop (the 'Edinburgh Workshop'), with background documents prepared in advance to inform the discussions, and with targeted additional papers presented during the workshop to support development of the conclusions.

The UNECE Edinburgh Workshop was hosted by the Scottish Government at Victoria Quay, Leith (4–6 December 2006). It was jointly funded by COST 729, NitroEurope IP, Defra, the Centre for Ecology & Hydrology (CEH) and the Scottish Government. We thank each of these organizations for their respective contributions. In particular, we would like to thank Emma Giles (CEH) for her achievement in coordinating the organizational aspects of the workshop, and Geeta Wonnacott and Philip Wright for their hosting of the workshop at the Scottish Government offices.

The key conclusions of the workshop have been reported to the Convention, including new values for ammonia critical levels, a new position on the status of the 'ammonia gap', and clear recommendations for the assessment of ammonia dispersal and deposition at local and regional scales. The present volume provides the background to these conclusions, reporting the current state-of-the-art in each of these areas.

Mark A. Sutton
Stefan Reis
Samantha M.H. Baker

Contents

Contributors

Achermann, Beat
Federal Office for the Environment, Air Pollution Control Division,
Air Quality Management Section, CH-3003 Bern, Switzerland,
beat.achermann@bafu.admin.ch

Agostini, Francesco
SLU Swedish Agriculture university, Department Energy and Technology,
Ulls Väg 30A, 756 51 Uppsala, Sweden, dragostini62@hotmail.com

Alebic-Juretic, Ana
Institute of Public Health, Braće Branchetta 20, 51000 Rijeka, Croatia,
ana-alebic.juretic@ri.htnet.hr

Andersen, Helle Vibeke
National Environmental Research Institute, Aarhus University, Frederiksborgvej
399, DK-4000 Roskilde, Denmark, hva@dmu.dk

Aneja, P. Viney
Department of Marine, Earth, and Atmospheric Sciences, North Carolina State
University, Box 8208, 1125 Jordan Hall, Raleigh, NC 27695-8208, USA,
viney_aneja@ncsu.edu

ApSimon, M. Helen
Imperial College London, Centre for Environmental Policy, Faculty of Natural
Sciences, South Kensington Campus, London SW7 2AZ, United Kingdom
h.apsimon@imperial.ac.uk

Asman, A.H. Willem
International Institute for Applied Systems Analysis (IIASA), A-2361
Laxenburg, Austria, Willem.Asman@agrsci.dk

Ayres, John
Government of Canada, Department of the Environment, 351 St. Joseph Blvd,
9th Floor, Gatineau, Quebec, K1A 0H3, Canada, John.Ayres@ec.gc.ca

Baker, M.H. Samantha
Air and Environment Quality Division, Department for Environment, Food and
Rural Affairs, 4/G17 Ashdown House, 123 Victoria Street, London, SW1E 6DE,
United Kingdom, samantha.baker@defra.gsi.gov.uk

Bareham, Simon
Countryside Council for Wales, Joint Nature Conservation Committee, Maes y
Ffynnon, Ffordd Penrhos, BANGOR, Gwynedd, LL57 2DW, United Kingdom,
s.bareham@ccw.gov.uk

Bealey, J. William
Centre for Ecology & Hydrology, Bush Estate, Penicuik, Midlothian EH26 0QB,
United Kingdom, bib@ceh.ac.uk

Bergström, Robert
Swedish Meteorological and Hydrological Institute, Research Department,
SE-601 76 Norrkoping, Sweden, robert.bergstrom@smhi.se

Bittmann, Shabtai
Agriculture and Agri-Food Canada, Pacific Agri-Food Research Centre,
Box 1000, Agassiz, BC, Canada, bittmans@agr.gc.ca

Błas, Marek
Department of Meteorology and Climatology, University of Wrocław, ul.
Kosiby 6/8 51–670 Wrocław, Poland, blasm@meteo.uni.wroc.pl

Bleeker, Albert
Energy Research Centre of the Netherlands, P.O. Box 1, 1755 ZG Petten,
The Netherlands, a.bleeker@ecn.nl

Bobbink, Roland
Utrecht University, Landscape Ecology, Institute of Environmental Biology,
P.O. Box 800.84, 3508 TB Utrecht, The Netherlands, r.bobbink@bio.uu.nl

Branquinho, Christina
Centro de Ecologia e Biologia Vegetal (CEBV), Department Biologia Vegetal da
Faculdade de Ciências, Universidade de Lisboa, Campo Grande, Edifício C2,
4° Piso, 1749–016 Lisboa, Portugal, cmbranquinho@fc.ul.pt

Cape, J. Neil
Centre for Ecology & Hydrology, Bush Estate, Penicuik, Midlothian EH26 0QB,
United Kingdom, jnc@ceh.ac.uk

Cellier, Pierre
Institut National de la Recherche Agronomique (INRA) UMR Environment
and cropland (EGC), BP1, 78850 Thiverval Grignon, France,
cellier@bcgn.grignon.inra.fr

Crittenden, Peter
University of Nottingham, School of Biology, University Park, Nottingham,
NG7 2RD, United Kingdom, Peter.Crittenden@nottingham.ac.uk

Crossley, Alan
Centre for Ecology & Hydrology, Bush Estate, Penicuik, Midlothian EH26 0QB,
United Kingdom, alan@acrosstech.org.uk

Cruz, Cristina
University of Lisbon, Faculdade de Ciencias, Centro de Ecologia e Biologia
Vegetal, Campo Grande, Bloco C4, Piso 1, 1749-016 Lisboa, Portugal,
ccruz@fc.ul.pt

Dämmgen, Ulrich
Johann Heinrich von Thunen-Institute, Federal Research, Institute for Rural Areas
Forestry and Fisheries, Institute of Agricultural Climate Research, Bundesallee 50,
38116 Braunschweig, ulrich.daemmgen@daemmgen.de

Dias, Teresa
University of Lisbon, Faculdade de Ciencias, Centro de Ecologia e Biologia
Vegetal, Campo Grande, Bloco C4, Piso 1, 1749-016 Lisboa, Portugal
mtdias@fc.uk.pt

Dore, J. Anthony
Centre for Ecology & Hydrology, Bush Estate, Penicuik, Midlothian EH26 0QB,
United Kingdom, todo@ceh.ac.uk

Dragosits, Ulrike
Centre for Ecology & Hydrology, Bush Estate, Penicuik, Midlothian EH26 0QB,
United Kingdom, ud@ceh.ac.uk

Ellermann, Thomas
National Environmental Research Institute, Aarhus University, Frederiksborgvej
399, DK-4000 Roskilde, Denmark, tel@dmu.dk

Erisman, Jan Willem
Energy Research Center of the Netherlands, ECN, P.O. Box 1, 1755 ZG Petten,
The Netherlands, erisman@ecn.nl

Eurich-Menden, Brigitte
Kuratorium fur Technik und Bauwesen in der Landwirtschaft (KTBL)
e.V. Bartningstrasse 49, 64289 Darmstadt, Germany, b.eurich-menden@ktbl.de

Fagerli, Hilde
Norwegian Meteorological Institute, Research Department, P.O. Box 43, Blindern,
NO-0313, Oslo, Norway, hilde.fagerli@met.no

Fangmeier, Andreas
University of Hohenheim, Institute for Landscape and Plant Ecology, Faculty of
Agricultural Sciences, August-von-Hartmannstr 3, 70599 Stuttgart, Germany,
afangm@uni-hohenheim.de

Flechard, J. Chris
Soils, Agronomy and Spatialization (SAS) Unit, INRA, 65 rue de St-Brieuc,
35042 Rennes Cedex, France, chris.flechard@rennes.inra.fr

Fowler, David
Centre for Ecology & Hydrology, Bush Estate, Penicuik, Midlothian EH26 0QB,
United Kingdom, dfo@ceh.ac.uk

Fudala, Janina
Institute for Ecology of Industrial Areas, ul. Kossutha 6, 40-844 Katowice,
Poland, jfudala@ietu.katowice.pl

Gauger, Thomas
University of Stuttgart, Institut für Navigation, Universität Stuttgart,
Institut für Navigation, Breitscheidstraße 2, 70174 Stuttgart, Germany
thomas.gauger@fal.de

Geels, Camilla
NERI, Department of Atmospheric Environment, Frederiksborgvej 358,
P.O. Box 358, DK-4000, Roskilde, Denmark, cag@dmu.dk

Gehrig, Robert
Empa, Swiss Federal Laboratories for Materials Testing and Research,
Air Pollution/Environmental Technology, Ueberlandstrasse 129, CH-8600
Dubendorf, Switzerland, Robert.Gehrig@empa.ch

Genermont, Sophie
Institut National de la Recherche Agronomique (INRA), UMR Environment
and cropland (EGC), BP1, 78850 Thiverval Grignon, France,
sophie.genermont@grignon.inra.fr

Gillespie, Colin
Scottish Environment Protection Agency (SEPA), Air Policy Unit,
Corporate Office, Castle Business Park, Stirling, United Kingdom,
Colin.Gillespie@Sepa.org.uk

Harlen, Karen S.
Central Analytical Laboratory Illinois State Water Survey, Champaign IL, USA
kharlin@uiuc.edu

Hassouna, Mélynda
Institut National de la Recherche Agronomique (INRA), UMR SAS,
65 rue de Saint-Brieuc, CS 84215, 35042 Rennes CEDEX,
Melynda.Hassouna@rennes.inra.fr

Hertel, Ole
NERI, Department of Atmospheric Environment, Frederiksborgvej 358,
P.O. Box 358, DK-4000, Roskilde, Denmark, oh@dmu.dk

Hole, Lars Robert
Norwegian Institute of Air Research, NILU, Department Atmosphere and Climate,
Polar Envionment Centre, 9296 Tromso, Norway, lrh@nilu.no

Horvath, Laszlo
Hungarian Meteorological Service, Gilice 39, 1181 Budapest, Hungary
lbrech.l@met.hu

Hutchings, J. Nicholas
Department of Agroecology, Faculty of Agricultural Sciences, University of Aarhus, P.O. Box 50, Research Centre Foulum, 8830 Tjele, Denmark, Nick.Hutchings@agrsci.dk

James, P.W.
NHM London, Department of Botany, Cromwell Rd, London SW7 7BD, United Kingdom

Jones, Matthew
US Environment Protection Agency, USEPA, Research Triangle Park, NC 27711, USA, matjceh@yahoo.com

Kinsella, Liam
Department of Agriculture and Food, Specialist Farm Services, Environment and Evaluation Division, Agriculture House, Kildare Street, Dublin 2, Ireland, Liam.Kinsella@agriculture.gov.ie

Klimont, Zbigniew
IIASA, APD, Schlossplatz 1, Laxenburg 2361, Austria, klimont@iiasa.ac.at

Kruijt, Roy Wichink
National Institute for Public Health and Environment, A. van Leeuwenhoeklaan 9, 3720 BA Bilthoven, The Netherlands, roy.wichink.kruit@rivm.nl

Kryza, Maciej
Department of Meteorology and Climatology, University of Wrocław, ul. Kosiby 6/8 51-670 Wrocław, Poland, kryzam@meteo.uni.wroc.pl

Leith, D. Ian
Centre for Ecology & Hydrology, Bush Estate, Penicuik, Midlothian EH26 0QB, United Kingdom, idl@ceh.ac.uk

Lewandowska, Anita
University of Gdansk, Institute of Oceanography, Marine Chemistry and Environmental Protection Department, Al. Pilsudskiego 46, 81-378 Gdynia, Poland, nadsta@sat.ocean.univ.gda.pl

Løfstrøm, Per
National Environmental Research Institute, Department of Atmospheric Environment, Frederiksborgvej 399, P.O. Box 358, DK 4000, Denmark, pl@dmu.dk

Loubet, Benjamin
Institut National de la Recherche Agronomique (INRA) UMR Environment and cropland (EGC), BP1, 78850 Thiverval Grignon, France, loubet@bcgn.grignon.inra.fr

Love, Linda
Centre for Ecology & Hydrology, Bush Estate, Penicuik, Midlothian EH26 0QB, United Kingdom, llove@ceh.ac.uk

Martins-Loução, Maria Amélia
University of Lisbon, Sciences Faculty, Reitoria da Universidade de Lisboa,
Alameda da Universidade, Cidade Universitaria, 1649-004 Lisboa, Portugal,
maloucao@reitoria.ul.pt

Máguas, Cristina
Centre for Ecology and Plant Biology, Departamento de Biologia Vegetal,
Faculdade de Ciências da Universidade de Lisboa, Bloco C2, Piso 4,
1749–016 Lisbon, Portugal, cmaguas@fc.ul.pt

Misselbrook, H. Tom
IGER, Soil, Environment and Ecological Services, North Wyke, Okehampton,
EX20 2SB, United Kingdom, Tom.Misselbrook@bbsrc.ac.uk

Mitosinkova, Marta
Slovak Hydrometeorological Institute, Jeseniova street 17,833 15 Bratislava,
Slovak Republic, Marta.Mitosinkova@shmu.sk

Neftel, Albrecht
Agroscope Reckenholz Tanikon Research Station, Air Pollution/Climate
Department, Reckenholzstrasse 191, 8046 Zurich, Switzerland,
lbrecht.neftel@fal.admin.ch

Nemitz, Eiko
Centre for Ecology & Hydrology, Bush Estate, Penicuik, Midlothian EH26 0QB,
United Kingdom, en@ceh.ac.uk

Peter, Kathrin
Evaluationen, Spitalgasse 14, CH-3011 Berne, Switzerland,
kathrin.peter@evaluationen.ch

Pinder, Rob
NOAA/Atmospheric Sciences Modeling Division, in partnership with the U.S.
Environmental Protection Agency, Office: E-231G-1, 109 T.W. Alexander Drive,
Research Triangle Park, NC 27711, USA, pinder.rob@epa.gov

Pinho, Pedro
CEBV-Centro de Ecologia e Biologia Vegetal/Center for Ecology Plant Biology
CEBV, FCUL, Campo Grande, C1, Piso 3, sala 41, 1749-016 Lisboa, Portugal,
ppinho@fc.ul.pt

Pitcairn, E.R. Carole
South House, Old Church, Aberlady, EH32 0RA, United Kingdom,
thepitcairns@yahoo.co.uk

Place, J. Chris
School of Geosciences, Institute of Geography, Geography Building, Drummond
Street, Edinburgh EH8 9XP, United Kingdom, C.J.Place@ed.ac.uk

Raes, Caroline
European Commission, DG Agriculture and Rural Development, Unit F1 –
Environment, GMO and Genetic Resources, B-1049 Brussels, Belgium,
caroline.raes@ec.europa.eu

Reidy, Beat
Swiss College of Agriculture (SHL), Langgasse 85, 3052 Zollikofen, Switzerland,
beat.reidy@shl.bfh.ch

Reis, Stefan
Centre for Ecology & Hydrology, Bush Estate, Penicuik, Midlothian EH26 0QB,
United Kingdom, srei@ceh.ac.uk

Rihm, Beat
Meteotest, Fabrikstrasse 14, CH-3012 Berne, Switzerland, rihm@meteotest.ch

Robin, Paul
INRA Institut National de la Recherche Agronomique (INRA), UMR SAS,
65 rue de Saint-Brieuc, CS 84215, 35042 Rennes CEDEX, France,
Paul.Robin@rennes.inra.fr

Rosa, Ana Paula
Universidade de Lisboa, Faculdade de Ciências, Centro de Biologia Ambiental,
Campo Grande, Edifício C2, Piso 5. 1749-016 Lisboa, Portugal,
paulayana@gmail.com

Sauter, Ferd
National Institute for Public Health and Environment, LVM, P.O. Box 1,
3720 BA Bilthoven, The Netherlands, ferd.sauter@rivm.nl

Schaap, Martijn
TNO Built Environment and Geosciences, Environment, Health and Safety, P.O.
Box 342, 7300 AH Apeldoorn, The Netherlands, martijn.schaap@tno.nl

Sheppard, J. Lucy
Centre for Ecology & Hydrology, Bush Estate, Penicuik, Midlothian EH26 0QB,
United Kingdom, ljs@ceh.ac.uk

Simmons, Ivan
Centre for Ecology & Hydrology, Bush Estate, Penicuik, Midlothian EH26 0QB,
United Kingdom, ivsi@ceh.ac.uk

Smith, Ken
ADAS UK Ltd, Integrated Water and Environmental Management, ADAS
WolverhamptWoodthorne, Wolverhampton, WV6 8TQ, United Kingdom,
ken.smith@adas.co.uk

Smith, I. Rognvald
Centre for Ecology & Hydrology, Bush Estate, Penicuik, Midlothian EH26 0QB,
United Kingdom, ris@ceh.ac.uk

Sobik, Mieczysław
Department of Meteorology and Climatology, University of Wrocław, ul. Kosiby
6/8 51-670 Wrocław, Poland, sobik@meteo.uni.wroc.pl

Sponar, Michel
European Commission, B-1049 Brussels, Belgium, msponar@mobistarmail.be

Spranger, Till
German Fed.Env.Agency (UBA), Wörlitzer Platz 1, 06844 Dessau, Germany,
till.spranger@uba.de

Strizik, Michal
VSB – Technical University of Ostrava, Faculty of Safety Engineering, Labora-
tory of Risk Research and Management, Lumirova 13, CZ-700 30, OSTRAVA 3
– Vyskovice, Czech Republic, michal.strizik@vsb.cz

Sutton, A. Mark
Centre for Ecology & Hydrology, Bush Estate, Penicuik, Midlothian EH26 0QB,
United Kingdom, ms@ceh.ac.uk

Tang, Y. Sim
Centre for Ecology & Hydrology, Bush Estate, Penicuik, Midlothian EH26 0QB,
United Kingdom, yst@ceh.ac.uk

Theobald, R. Mark
Centre for Ecology & Hydrology, Bush Estate, Penicuik, Midlothian EH26 0QB,
United Kingdom, mrtheo@ceh.ac.uk

Urech, Martin
puls, Mühlemattstrasse 45, CH-3007 Berne, martin.urech@pulsbern.ch

van der Eerden, Ludger
OBRAS Centre for Art and Science, Herdade, da Marmeleira. N18 Evoramonte,
CP2 7100 300 Estremoz, Portugal, Obrasart@hotmail.com

van der Hoek, Klaas
National Institute for Public Health and Environment, P.O. Box 1, 3720 BA,
Bilthoven, The Netherlands, klaas.van.der.hoek@rivm.nl

van Dijk, Netty
Centre for Ecology & Hydrology, Bush Estate, Penicuik, Midlothian EH26 0QB,
United Kingdom, nvd@ceh.ac.uk

van Jaarsveld, A. Hans
The Netherlands Environmental Assessment Agency, P.O. Box 303, 3720 AH
Bilthoven, The Netherlands, hans.van.jaarsveld@mnp.nl

van Pul, Addo
RIVM, LVM, P.O. Box 1, 3720 BA Bilthoven, The Netherlands,
addo.van.pul@rivm.nl

Vidic, Sonja
Meteorological and Hydrological Service of Croatia, Research and Development
Department/Air Quality Research Unit, Gric 3, Zagreb, 10000, Croatia,
vidic@cirus.dhz.hr

Vieno, Massimo
University of Edinburgh, School of Geosciences, Crew Building,
The King's Building, West Mains Road, EH9 3JN, United Kingdom,
mvieno@staffmail.ed.ac.uk

Walker, John
US Environment Protection Agency, USEPA Mailroom, E305-02,
Research Triangle Park, NC 27711, USA, walker.johnt@epa.gov

Wallasch, Marcus
Umweltbundesamt, Paul-Ehrlich-Str.29, 63225, Langen, Germany
markus.wallasch@uba.de

Webb, J.
AEA, Gemini Building, Harwell Business Centre, Didcot, Oxfordshire OX11
0QR, United Kingdom, j.webb@aeat.co.uk

Whitfield, Clare
Joint Nature Conservation Committee, Peterborough, United Kingdom
Clare.Whitfield@jncc.gov.uk

Wolseley, A. Patricia
NHM London, Department of Botany, Cromwell Rd, London SW7 7BD,
United Kingdom, P.Wolseley@nhm.ac.uk

Xuejun, Liu
Centre for Ecology & Hydrology, Bush Estate, Penicuik, Midlothian EH26 0QB,
United Kingdom, XLiu@wpo.nerc.ac.uk

Zelinger, Zdenek
Academy of Sciences of the Czech Republic, J. Heyrovsky Institute of Physical
Chemistry, Dolejskova 3, 182 23 Prague 8, Czech Republic
zdenek.zelinger@jh-inst.cas.cz

Acronyms and Abbreviations

CAFE	Clean Air For Europe programme of the European Commission
CLE	Critical Level
CLO	Critical Load
CLRTAP	Convention on Long-Range Transboundary Air Pollution
Defra	Department of Environment, Food and Rural Affairs (United Kingdom)
EC	European Commission
JNCC	Joint Nature Conservation Committee (United Kingdom)
NEC(D)	National Emission Ceilings (Directive) of the European Commission
NH_3	In this document NH_3 refers to atmospheric ammonia gas
NOEC	No observable effect concentration
UBA	Umweltbundesamt (Germany)
UNECE	United Nations – Economic Commission for Europe
V_d	Deposition velocity
Whim	Whim experimental site in south-east Scotland, where moorland vegetation has been fumigated with NH_3 at low concentrations for several years
BBD	Bern Background Document. Paper prepared for the Bern Workshop (18-20 September 2000) of the UNECE Ammonia Expert Group, pp 57-84 in: Menzi H. and Achermann B. (eds.), SAEFL, Bern, 2001
WG	Working Group convened during the UNECE Edinburgh Workshop on ammonia

Chapter 1
Introduction

Mark A. Sutton

1.1 Overview

Atmospheric ammonia is emerging increasingly as the key transboundary air pollutant that will contribute to future impacts of nitrogen and acidity on terrestrial ecosystems in Europe. At the same time, ammonia is making an increasing relative contribution to particulate matter, with its associated human health risks. These features are shown clearly in Fig. 1.1 below, from the *Clean Air For Europe* (CAFE) programme.

Thus, by 2020, it is estimated that NH_3 will be the largest single contributor to each of acidification, eutrophication and secondary particulate matter in Europe. This increase is particularly a reflection of the success of European policies to reduce SO_2 and NO_x emissions. As a result, NH_3 is increasingly dominating nitrogen and acidifying inputs, while reducing ammonia emissions and the associated environmental impacts remain major challenges for the future.

This book is the result of an Expert Workshop held under the auspices of the UNECE Convention on Long Range Transboundary Air Pollution (CLRTAP). It reports an analysis focused on several key questions regarding the detection of changes and quantification of environmental impacts due to ammonia. Following the classical format of UNECE Expert Workshops, it aimed at reaching agreement on current scientific understanding and providing recommendations that would be useful to the Convention on Long Range Transboundary Air Pollution.

The Expert Workshop addressed the following key questions:

1. Do existing critical thresholds for ammonia reflect current scientific understanding and, if not, what are suitable values according to current knowledge?
2. To what extent can independent atmospheric measurements verify where regional changes in NH_3 emissions have and have not occurred?
3. How can transboundary air pollution assessment be downscaled to deal with ammonia hotspots in relation to operational modelling and monitoring?

Mark A. Sutton
Centre for Ecology & Hydrology, Bush Estate, Penicuik, Midlothian, EH26 0QB,
United Kingdom

M. Sutton, S. Reis and S.M.H. Baker (eds), *Atmospheric Ammonia,*
© Springer Science + Business Media B.V. 2009

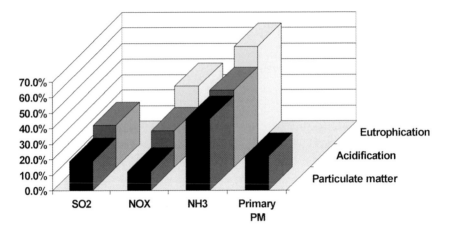

Fig. 1.1 The estimated contribution of NH₃, SO₂, NOx and primary PM to environmental problems in 2020. From the *Clean Air For Europe* (CAFE) programme

4. What are the differences between mesoscale (regional) atmospheric transport and chemistry models in relation to their formulation and results for ammonia?

These questions have been addressed previously in UNECE workshops, but in all cases the workshop conclusions can now be considered as somewhat dated, emphasizing the need for a review of the issues. Several previous UNECE Expert Workshops were particularly relevant in relation to ammonia:

- **Bad Harzburg (1988)** Critical levels workshop (origination of critical level for NH₃, Posthumus 1988)
- **Egham (1992)** Critical levels workshop (included: update review on critical level for NH₃, Ashmore and Wilson 1994; van der Eerden et al. 1991, 1994)
- **Göteborg (1992)** Workshop on the deposition of acidifying substances (included: review of reduced nitrogen dry deposition, Sutton et al. 1993)
- **Aspenaas Herrgaard (1997)** Workshop on strategies for monitoring of regional air pollution (Included: review of air monitoring for assessment of air pollution effects, Lövblad and Sutton 1997)
- **Bern (2000)** Workshop of the UNECE Ammonia Expert Group (development of UNECE Framework advisory code of good agricultural practice for reducing ammonia emissions (Menzi and Achermann 2001) and review of the link between control of ammonia emissions and atmospheric measurements of reduced nitrogen (Sutton et al. 2001)

More recently the effects of ammonia as a contribution to nitrogen deposition were reviewed in the Bern 2002 workshop (Achermann and Bobbink 2003; Bobbink et al. 2003). However, this focused on empirical critical loads for nitrogen, rather than explicitly on the role of ammonia. As part of that workshop, the importance of uncertainties in atmospheric N deposition, including NH₃ was addressed by Sutton et al. (2003). In addition, a more recent review of overall monitoring strategies in the EMEP programme, including regarding gas-particle distribution, was made at the Oslo 2004 Expert Workshop (Aas 2005).

In the following, some of the main issues associated with each of the four key questions are summarized, particularly bearing in mind developments since the previous UNECE Expert Workshops.

1.2   To What Extent Can Independent Atmospheric Measurements Verify Where Regional Changes in NH₃ Emissions Have and Have Not Occurred? (Working Group 1)

This question included the following objectives:

(a) To quantify the extent to which estimated regional changes in ammonia emissions have been reflected in measurements of ammonia and ammonium in the atmosphere

(b) To distinguish cases where the estimated changes in ammonia emission are due to altered sectoral activity or the implementation of abatement policies and thereby assess the extent to which atmospheric measurements verify the effectiveness of ammonia abatement policies

(c) To make recommendations for future air monitoring and systems for assessing the national implementation of ammonia abatement policies

The discussion on this topic was initiated in the late 1990s following the observation that expected changes in NH_3 emissions in the Netherlands and in Eastern Europe were not matched by observed reductions in NH_3 concentrations. In the Netherlands, an extensive NH_3 emissions reduction policy was implemented and it was therefore surprising that by 1997, NH_3 concentrations were no smaller than in 1993, when the policy was initiated (Erisman et al. 1998; van Jaarsveld et al. 2000). The issue became known as the "Ammonia Gap", raising questions regarding the cost effectiveness of the NH_3 abatement policy. Additionally, in eastern Europe, following the crash in agricultural livestock populations and fertilizer usage after the political changes of 1989, it was curious that available monitoring in Hungary could also not detect the expected reductions in NH_3 emissions (Horvath and Sutton 1998). Since the emissions in east Europe must have decreased, due to reduced sector activity, this raised the question of whether there were non-linearities in the link between NH_3 emissions and atmospheric concentrations and deposition. These issues were reviewed at the Bern Workshop in 2000 (Sutton et al. 2001), which noted how interactions with changing SO_2 emissions, local spatial variability, short term meteorological variability and interactions with NH_3 compensation points were among the factors explaining the difficulty to make the links.

One of the key findings of the Bern workshop, described in the Working Group Report (Menzi and Achermann 2001), was the severe lack of NH_3 monitoring data across Europe. Recommendations were therefore made regarding the need to establish robust monitoring networks, especially with the ability to speciate between NH_3 gas and NH_4^+ aerosol, a finding which was re-enforced by the Oslo Workshop (2004) on monitoring strategies (Aas 2005).

Since the Bern Workshop, major new datasets on European NH$_3$ and NH$_4^+$ monitoring and their relationship to estimated NH$_3$ emissions have become available. The focus of Working Group 1 of the present Expert Workshop was therefore to update the current scientific understanding based on these new datasets and assessments. In particular, the group asked whether there is still an "Ammonia Gap" in the Netherlands, whether such a gap exists in other countries, whether we can be confident of the effectiveness of ammonia mitigation policies, and how can we best address the relationships between emission and deposition using atmospheric modelling and improved monitoring activities. A major review of these issues was prepared by Bleeker et al. This was revised following the workshop review process and is presented in Chapter 4 of this volume (Bleeker et al. 2009, this volume)

1.3   Do Existing Critical Thresholds for Ammonia Reflect Current Scientific Understanding and if Not What Are Suitable Values According to Current Knowledge? (Working Group 2)

This question included the following objectives:

(a) To examine the case for setting new ammonia critical thresh-old(s) based on current evidence of direct impacts of ammonia on different receptors
(b) To discuss the extent to which vegetation and sensitive ecosystems appear to be differentially sensitive to ammonia versus other forms of reactive N
(c) To debate the case for establishing indicative air concentration limits for indirect effects of ammonia, which would be consistent with current critical loads for nitrogen

The critical level for NH$_3$ has stood for around 15 years since its last revision at the Egham Workshop in 1992 (Ashmore and Wilson 1994). This lack of activity was probably because critical loads of nitrogen tend to be exceeded at much lower NH$_3$ concentrations than required to exceed the NH$_3$ critical levels values set at that workshop. The values set were 3,300, 270, 23 and 8 µg m^{-3} for hourly, daily, monthly and annual values, based on the work of van der Eerden et al. (1991, 1994). This resulted in NH$_3$ receiving a lower level of attention, since its contribution was in some way hidden in the calculation of total nitrogen deposition, despite the fact that gaseous NH$_3$ is often the largest single component of atmospheric nitrogen inputs.

Among the limitations of the data available at the Egham Workshop was the fact that these only referred to very short-term studies. As a result, no attention was given to what are the long-term effects of high ammonia concentrations. Thus the annual critical level refers to the exceedance for a particular year, rather than being a long-term annual value which should not be exceeded, as is the case for the empirical critical loads. Recent evidence from the UK showed that the effects of NH$_3$ concentrations interact over longer periods than a year, pointing to the need to set a long-term annual mean critical level.

One of the questions asked at the Expert Workshop was how to name a NH_3 concentration threshold that equates to indicative exceedance of the critical load values, such as may be used in setting air quality target values. If such long-term effects of NH_3 are considered "indirect", then there is potentially a difficulty with the critical level nomenclature, if this strictly refers only to "direct" effects. On the other hand, it appears that the original meaning of "indirect" in this context was that effects are mediated via the soil. Hence, it might be argued that even long-term effects of NH_3 are direct, since most of the uptake occurs directly to the plants. In some cases, the mechanism of damage also demonstrates the role of direct effects, such as damage to sensitive lichens, which is partly mediated through changes in substrate pH rather than just nitrogen supply.

Studies have also shown that for a given N input, NH_3 can show larger effects than other nitrogen forms. This is obviously a difficulty for the critical loads approach, where all forms of N are considered to have an equal effect simply through N dose. The Expert Workshop therefore asked whether there a practical way that such differences might be included in the critical loads estimation, or whether the simpler approach of incorporating the NH_3-specific effects in the critical level would be more achievable.

These issues are considered further in a first discussion document prepared in advance of the Edinburgh Workshop, and brought up-to-date with the current questions (Sutton et al. 2009, this volume). A comprehensive review was subsequently prepared as background material for the Expert Workshop, and which has been revised to take account of the Workshop review and conclusions (Cape et al. 2009, this volume).

1.4 How Can Transboundary Air Pollution Assessment Be Downscaled To Deal with Ammonia Hotspots in Relation to Operational Modelling and Monitoring? (Working Group 3)

This question included the following objectives:

(a) To review current emission and atmospheric dispersion modelling methods for downscaling NH_3 dispersion and deposition in hot spots
(b) To examine the status of methods for effect assessment and air monitoring in NH_3 hot spots
(c) To recommend broad principles for assessment approaches in ammonia hot spots, including spatial approaches and the interactions between transboundary ammonia emission reduction targets and other policy measures

To date, most attention in modelling NH_3 dispersion and deposition has focused on the regional and European scale. Certainly, within the UNECE Convention on Long-Range Transboundary Air Pollution, little attention has been given to dealing with NH_3 in hot-spot areas. In the past, it was considered that this was a local problem and not relevant for the transboundary interest of the Convention. However, the role of hot-spots is increasingly recognized by the Task Force on Measurement and Modelling, for example, when modelling the so called city-delta, which is the urban enhancement of particulate matter concentrations above background.

In the case of NH_3, developing assessment approaches in hot-spot areas is similarly important: there are the areas of acutest environmental impact and regulatory focus, for example in protecting designated nature conservation areas near farm sources. In addition, assessments in hot-spot areas need to consider the fraction that derives from local sources versus that which is of more distant national or transboundary origin.

The inclusion of ammonia hot-spots for Working Group 3 is thus the first time that this issue has been treated specifically in a UNECE Expert Workshop. As a result, a large part of the work was to compare the different approaches that are being implemented in different countries. This considered both: whether similar framework approaches are being taken, as well as the comparability and reliability of the detailed models and methods being implemented. The focus on atmospheric dispersion modelling is in the rural environment rather than the urban environment, so the uncertainty in agricultural NH_3 emissions estimates for specific sites, was considered alongside modelling issues which are relevant in the rural context. The tools considered ranged from detailed models for assessment of the effects of landscape structure, such as the effect of tree belts round farms, to general dispersion models at the landscape scale, as well as the development and application of screening tools.

Most focus at these scales to date has been on the fluxes and impacts of NH_3 dry deposition: guidelines were also considered regarding the contribution of locally enhanced wet and aerosol NH_4^+ deposition. Overall, Working Group 3 sought to address whether common lessons can be learned from the experience in different countries. This included the broad principles of modelling approaches and the interaction with other policy issues (e.g., Habitats Directive, other forms of N pollution).

While modelling was a major focus, the advances in measurement approaches in hot-spot areas were also discussed. This discussion included developing principles for practical monitoring of NH_3 concentrations in hot-spots, as well as the bioindication of nitrogen responses near NH_3 sources. A major review of the issues of ammonia in hot-spot areas was prepared as background to the discussions, and revised following review at the Expert Workshop (Loubet et al. 2009, this volume).

1.5   What Are the Differences Between Mesoscale (Regional) Atmospheric Transport and Chemistry Models in Relation to Their Formulation and Results for Ammonia? (Working Group 4)

This question included the following objectives:

(a) To review emission parameterizations used in the models, establishing comparability, spatial and temporal resolution and uncertainties

(b) To review dispersion, air chemistry and deposition formulations identifying key differences and uncertainties
(c) To assess the overall performance of the models against measurements and against a common reference, leading to recommendations for improving mesoscale models of ammonia transport and deposition

Atmospheric transport and chemistry models are key tools for the assessment of air pollution fluxes and their impacts. Such models typically include NH_3 as one of a suite of pollutants being simulated and as such the degree of attention given to NH_3 is very variable. In many cases, models have been originally developed for other pollutants and thus may not be well suited or developed for ammonia. For example, models originally developed to simulate the dispersion of high stack emissions, may include significant uncertainty or simplification in the estimation of ground level NH_3 concentrations (1–2 m), as measured in monitoring networks. The focus of this Working Group was on the formulation, input and performance of mesoscale (regional or country scale) atmospheric transport models and how these relate to European scale assessments.

The primary input to all atmospheric dispersion models are the emission data. The first question was therefore: what is the quality and comparability of the NH_3 emission data used between different models? The unit NH_3 emission estimates used for mapping may derive from emission factors or a mass flow approach and these differ between countries. But are these differences fully justified by differences in agricultural management practices?

There are a wide range of models from Lagrangian to Eulerian, as well as com-bination approaches. The Working Group asked to what extent to these differences affect the outcomes for modelling of NH_3 dispersion and deposition. Similarly, it was asked what the main differences are in chemical formulations between NH_3 and other pollutants in the models and how does these affect the modelled atmos-pheric transport distance of NH_3 and NH_4^+. Uncertainties in the parameterization of wet deposition were noted, and the Working Group reviewed how well orographic enhancement of wet deposition is currently treated in different models. Finally, it was recognized that modelling dry deposition (or bidirectional exchange) of NH_3 remains a major uncertainty at the regional scale. The group examined the current evidence regarding the effect of SO_2 interactions on NH_3 uptake resistances and deposition velocities and assessed to what extent NH_3 compensation point formulations have been incorporated in the regional models.

This Working Group focused first on the validation of the modelled NH_3 emis-sion data and then summarized the existing differences between models and how these may relate in principle to differences in their performance for NH_3 and NH_4^+. A complete model inter-comparison was not feasible at this stage (due in part to the fact that different regional models run over different domains with varying input data requirements). However, as far as possible, the performance of each of the regional models used was compared with results of the European scale EMEP model, which is used in the calculations under the Convention. The background analysis was prepared by van Pul et al., which is reported here following review at the Expert Workshop (van Pul et al. 2009, this volume).

1.6   Integration and Synergies in the Workshop

The review was conducted first by plenary presentation and discussion of the main background documents at the Expert Workshop. The four Working Groups were then conducted in parallel to maximize discussion of the key issues. This allowed the Expert Workshop to discuss additional scientific contributions from experts. Submitted contributions are included in the present book as short chapters which support the each of the four major reviews.

It should be noted that all of the Working Groups were to some extent interrelated. The major points of interaction between groups are illustrated in Fig. 1.2. To deal with these interactions, the Working Groups were firstly encouraged to foster the exchange of relevant experts between groups during the review process. In addition, Cross-Cutting Groups were established to address two areas of key linkage between topics: the reliability of ammonia emissions data and the policy context within which to address the challenges raised by the environmental effects of ammonia. The key questions and issues faced by the cross-cutting analysis are outlined below.

1.6.1   What Are the Reliability of Ammonia Emissions Data and Abatement Efficiencies? (Cross Cutting Group A)

In this analysis the information considered by each of Working Groups 1, 3 and 4 was synthesized regarding the reliability of ammonia emissions data, including both national totals and spatially distributed estimates for input into atmospheric

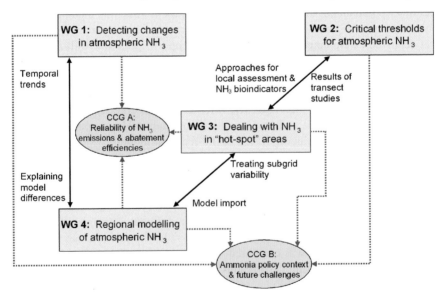

Fig. 1.2 Summary of major interactions and expert exchange between the different Working Groups (wgs) and Cross Cutting Groups (ccgs) at the UNECE Edinburgh Workshop on ammonia

models at farm, landscape and regional scales. In addition, the group addressed the known uncertainties in emission inventories regarding abatement efficiencies and the implications of the atmospheric verification results (WGs 1, 3 and 4). Specifically, the implications were addressed regarding the need to improve the inventory calculations of emissions and the assumed mitigation effectiveness.

1.6.2 What Is the Agricultural and Environmental Policy Context and How Can Scientific Understanding Help Address the Future Challenges to Reduce the Negative Effects of Ammonia? (Cross Cutting Group B)

The second cross-cutting analysis considered the policy context of the ammonia problem, including socio-economic, environmental, institutional and technological aspects. In particular, it addressed the links between the different Working Group themes to summarize the current challenges faced in relation to the different environment and health effects of ammonia emissions. Based on the scientific findings from the different groups, the group addressed the potential role of different policy options to help mitigate these impacts.

1.7 Workshop Conclusions and Recommendations

The UNECE Edinburgh Expert Workshop thus provided a series of major background reviews, supported by short chapters on key recent results. Based on this review, the Working Groups prepared reports of their key conclusions and recommendations, which were reviewed and adopted by the workshop plenary. The conclusions and recommendations from each of the groups are thus presented in Chapter 7. Finally, these conclusions and recommendations provided the basis to draft the Synthesis and Summary for Policy Makers (Chapter 8), which represents the formal product of the workshop provided to the Convention (UNECE 2007).

The summary conclusions and recommendations have already fed into the ongoing reviews of the Convention and the Gothenburg Protocol. Two key examples may be given. Firstly, the new critical levels values presented here have now been adopted into the current revision of the 'Mapping Manual' used as the common basis for calculation of critical thresholds (cf. ICP Modelling and Mapping 2004). Secondly, the recommendations to better link the multiple emissions and effects of the different forms reactive nitrogen have been recognized. In this context, during 2007, the Executive Body of the Convention established a Task Force on Reactive Nitrogen. This new group will examine the nitrogen linkages between emissions, dispersion, impacts and abatement methodologies, including ammonia and nitrogen oxides, and their interactions with nitrous oxide and nitrate leaching etc. The agenda is thus set to address the challenge of developing more holistic analyses within the frame of the wider nitrogen cycle.

References

Aas W. (2005) Workshop on the implementation of the EMEP monitoring strategy (Ed.) (Oslo, 22–24 November 2004). EMEP Co-operative Programme for Monitoring and Evaluation of the Long-Range Transmission of Air Pollutants in Europe, EMEP/CCC 2-2005. NILU, Kjeller, Norway. (http://www.nilu.no/projects/ccc/reports/cccr2-2005.pdf)

Achermann B. and Bobbink R. (2003) Empirical Critical Loads for Nitrogen (UNECE Expert Work-shop, Berne 11–13 November 2002) (Eds.) pp 327. SAEFL, Berne, Switzerland.

Ashmore M.R. and Wilson R.B. (eds.) (1994) Critical levels of air pollutants for Europe. Report of the UNECE Expert Workshop at Egham, UK (23–26 March 1992). Department of the Environment, London.

Bobbink R., Ashmore M., Braun S., Flückiger W., and Van den Wyngaert I.J.J. (2003) Empirical nitrogen critical loads for natural and semi-natural ecosystems: 2002 update. In: Achermann B. and Bobbink R. (eds.) Empirical Critical Loads for Nitrogen (UNECE Expert Workshop, Berne 11–13 November 2002) pp 43–170. SAEFL, Berne, Switzerland. (http://www.iap.ch/publikationen/nworkshop-background.pdf)

Erisman J.W., Bleeker A., and van Jaarsveld J.A. (1998) Evaluation of the effectiveness of the ammonia policy using measurements and model results. Environmental Pollution 102, 269–274.

Horvath L. and Sutton M.A. (1998) Long term record of ammonia and ammonium concentrations at K-puszta, Hungary. Atmospheric Environment (Ammonia Special Issue) 32, 339–344.

ICP Modelling and Mapping (2004) Manual on Methodologies and Criteria for Mapping Critical Levels/Loads and Geographical Areas Where They Are Exceeded. UBA-Texte 52/04. (http://www.oekodata.com/icpmapping/index.html)

Lövblad G. and Sutton M.A. (1997) The requirements for monitoring data for the assessment of transboundary air pollution effects to man and ecosystems. In: J. Schaug and K. Uhse (eds.) EMEP/WMO Workshop on Strategies for Monitoring of Regional Air Pollution in Relation to the Need Within EMEP, GAW and Other International Bodies (Aspenaas Herrgaard, Sweden 2–4 June 1997) pp 83–119. EMEP/CCC Report 10/97. NILU, Kjeller, Norway.

Menzi H. and Achermann B. (2001) UNECE Ammonia Expert Group (Berne 18–20 September 2000) Proceedings (Eds.) 157 Swiss Agency for Environment, Forest and Landscape (SAEFL), Bern (Copy of Working Group 2 report). (http://www.nitroeurope.eu/ammonia_ws/documents/AEG_bern_wg2_report.pdf)

Posthumus A.C. (1988) Critical levels for effects of NH_3 and NH_4+. In UNECE (1988) Proceedings of the Bad Harzburg Workshop, pp 117–127. UBA, Berlin.

Sutton M.A., Asman W.A.H., and Schjørring J.K. (1993) Dry deposition of reduced nitrogen. In: Lövblad G., Erisman J.W. and Fowler D. (eds.) Models and Methods for the Quantification of Atmospheric Input to Ecosystems, pp 125–143. Nordiske Seminarog Arbejdsrapporter 1993:573. Nordic Council of Ministers, Copenhagen (revised version published in: Tellus (1994) 46B, 255–273).

Sutton M.A., Asman W.A.H., Ellerman T., van Jaarsveld J.A., Acker K., Aneja V., Duyzer J.H., Horvath L., Paramonov S., Mitosinkova M., Tang Y.S. Achermann B., Gauger T., Bartnicki J., Neftel A., and Erisman J.W. (2001) Establishing the link between ammonia emission control and measurements of reduced nitrogen concentrations and deposition. In: Menzi H. and Achermann B.(eds.) UNECE Ammonia Expert Group (Berne 18–20 September 2000) Proceedings, pp 57–84. Swiss Agency for Environment, Forest and Landscape (SAEFL), Bern (revised version published in: Environmental Monitoring and Assessment (2003) 82: 149–185).

Sutton M.A., Cape J.N., Rihm B., Sheppard L.J., Smith R.I., Spranger T., and Fowler D. (2003) The importance of accurate background atmospheric deposition estimates in setting critical loads for nitrogen. In: B. Achermann and R. Bobbink (eds.) Empirical Critical Loads for Nitrogen (UNECE Expert Workshop, Berne 11–13 November 2002) pp 231–257. SAEFL, Berne, Switzerland.

UNECE (2007) Review of the 1999 Gothenburg Protocol. Report on the Workshop on Atmospheric Ammonia: Detecting Emission Changes and Environmental Impacts. Report to the 39th Session of the Working Group on Strategies and Review. ECE/EB.AIR/WG.5/2007/3. (http://www.unece.org/env/lrtap/WorkingGroups/wgs/docs39th%20session.htm)

van der Eerden L.J., Dueck T.A., Berdowski J.J.M, Greven H., and van Dobben H.F. (1991) Influence of NH_3 and $(NH_4)_2SO_4$ on heathland vegetation. Acta Botanica Neerlandica 40 (4), 281–296.

van der Eerden L.J., Dueck T.A., Posthumus A.C., and Tonneijck A.E.G. (1994) Assessment of critical levels for air pollution effects on vegetation: some considerations and a case study on NH_3. In: Ashmore M.R. and Wilsons R.B. (eds.) Critical Levels of Air Pollutants for Europe, pp 55–63. Department of Environment, London.

van Jaarsveld J.A., Bleeker A., and Hoogervorst N.J.P. (2000) Evaluatie ammoniak emissieredukties met behulp van metigen en modelberekeningen. RIVM report 722108025. RIVM, Bilthoven, The Netherlands (in Dutch).

Part I
Ammonia Critical Thresholds

Chapter 2
Reassessment of Critical Levels for Atmospheric Ammonia

J. Neil Cape, Ludger J. van der Eerden, Lucy J. Sheppard, Ian D. Leith, and Mark A. Sutton

2.1 Summary

- The existing Critical Levels (CLEs) for NH_3 require revision in the light of new experimental evidence from field-based experiments and surveys.
- The existing annual CLE ($8\,\mu g\,NH_3\,m^{-3}$), when expressed as an equivalent deposition of N to an ecosystem, is less protective than the current Critical Load for most, if not all, European ecosystems and habitats.
- Field-based evidence relating effects on vegetation to NH_3 concentrations measured over one year or longer show that the current annual CLE is set too high.
- A new long-term CLE for the most sensitive vegetation type (lichens and bryophytes) is proposed, based on observed changes to species composition in the field.
- Most of the evidence comes from studies in the UK, but there is corroborative evidence from Switzerland, Italy and Portugal.
- The proposed long-term CLE for NH_3 for ecosystems in which lichens and bryophytes are important is $1\,\mu g\,NH_3\,m^{-3}$.
- There is less evidence for long-term effects of NH_3 on species changes in communities of higher plants; on the basis of expert judgement we propose a long-term CLE for higher plants of $3\,\mu g\,NH_3\,m^{-3}$.

J. Neil Cape,
Centre for Ecology & Hydrology Bush Estate, Penicuik, Midlothian, EH26 0QB,
United Kingdom

Ludger J. van der Eerden,
Foundation OBRAS, Centre for Art and Science, Evoramonte, Portugal

M. Sutton, S. Reis and S.M.H. Baker (eds), *Atmospheric Ammonia,*
© Springer Science+Business Media B.V. 2009

- No assumptions have been made on the mechanism by which NH_3 exposure leads to changes in species composition.
- Several recommendations are made, to address uncertainties relating to the lack of observational data and long-term NH_3 concentration measurement, particularly in southern and eastern Europe.
- There is also need for better understanding of the mechanisms whereby NH_3 affects plants, so that predictive models can be constructed for extrapolation to other types of vegetation and land use in different climatic zones.

2.2 Introduction

The process of arriving at generally accepted, scientifically reliable quality standards for pollutants follows a pathway that covers several decades. The derivation of a Critical Level (CLE) for NH_3 is no exception. A Critical Level is defined as *"the concentration in the atmosphere above which direct adverse effects on receptors, such as plants, ecosystems or materials, may occur according to present knowledge"* (Posthumus 1988).

In the first phase of the process, quality standards are largely based on the "No observable effect concentration" (NOEC) of the most sensitive species tested. This NOEC will tend to decrease over time as the result of discovering receptors with greater sensitivities. In the second phase of the process, with continued investigation, a set of effects data becomes available that allows evaluation of inter-species variability in sensitivity. Quantification of the protection level (e.g. confidence limits or percentage of species protected) becomes possible. In the third phase, models at the ecosystem level based on causal analytical relations become available. The focus on protecting species is then broadened to protecting the functioning of the full system. Up to now, the CLE for NH_3 has been in its first phase, but recent developments mean that it can begin to move into the second phase.

Critical Loads (CLO) are defined as *"a quantitative estimate of deposition of one or more pollutants below which significant harmful effects on specified elements of the environment do not occur according to present knowledge"* (Posthumus 1988). There is a subtle, but important distinction in the phraseology – the CLE marks a lower threshold (above which effects are known to occur), while the CLO marks an upper threshold (below which no effects are known). There are other differences: CLO are usually considered as representing a long-term deposition (10–100 years) whereas CLE are usually defined for periods of up to a year. This distinction is related to the pathways by which deposition and concentration may have effects; deposition (CLO) is usually considered as operating through changing soil or water chemistry, whereas concentrations (CLE) are usually seen as having direct effects on aboveground vegetation, in addition to any effect mediated through dry deposition and subsequent effects through soils and waters. Mathematically, multiplying a CLE with deposition velocity results in a deposition rate which can be compared directly with a CLO. Thus, it might seem superfluous to set both CLEs and CLOs. Current practice is different, however (Table 2.1).

Table 2.1 Current differences in practice between Critical levels (CLE) and Critical Loads (CLO) for N-containing air pollutants

	CLE	CLO
Summarized definition	Concentration above which effects do occur	Deposition below which effects do not occur
Exposure duration:	Short term (1 year or less)	Long term (∓10 years)
Effect of peak exposures:	Included	Neglected
Agent:	Separate CLE for each N-compound	All N-compounds added
Object of interest:	Individual plant species	Natural vegetation or forests; soils and freshwaters
No effect concentration:	Generally: the lowest statistically significant response observed in experiments	Generally: estimate of a "safe" deposition level derived from empirical evidence or modelling.
Goal:	Protection of sensitive plant species	Protecting proper functioning of ecosystems
Combination effects:	Possibility of synergism is considered	Additivity is presumed

This is at least partly due to differences in scientific sources: assessing CLEs has generally been triggered by air quality specialists, while CLOs generally have their basis in ecology. Currently, CLEs are especially useful in emission abatement, while CLOs are useful in nature conservation and in evaluation of risks to soil and water quality. The direct comparison of CLE with CLO relies on knowledge of the appropriate deposition velocity for each land use and vegetation type, and this may not be available, or may be rather uncertain. In some circumstances, therefore, the ability to use a more easily measurable CLE, rather than a CLO, may be an advantage.

CLEs for NH_3 were first defined at the UNECE Bad Harzburg workshop (Posthumus 1988) and then revised at the UNECE Workshop in Egham, UK, in 1993 (Ashmore and Wilson 1994). For a list of earlier discussions see (Fangmeier et al. 1994).

The 1993 Workshop agreed CLEs for different averaging times as follows:

1 h:	3,300 µg NH_3 m^{-3}
1 day:	270 µg NH_3 m^{-3}
1 month:	23 µg NH_3 m^{-3}
1 year:	8 µg NH_3 m^{-3}

2.3 Issues in Assessing CLEs for NH_3

There have been several reviews on NH_3 effects (Fangmeier et al. 1994; Krupa 2003; van der Eerden et al. 1994; WHO 1997). Most of the evidence to be presented below in support of a change in CLEs derives from measurements on vegetation made close to large point sources of NH_3, either in field release experiments or downwind of intensive agriculture. In this section we present some of the key issues involved in the use of such data. Some of these issues have already been raised in earlier reviews – see Box 2.1.

Box 2.1 Points already made in previous documentation of NH$_3$ critical levels

"NH$_3$ primarily acts as a fertilizer, usually increasing shoot growth while reducing or not affecting root growth."

"Internal consequences of exposure to NH$_3$ are increased nitrogen and often altered concentrations of nutrients, amino acids and carbohydrates."

"[L]ong term exposures may ultimately affect the plant's ability to endure other environmental stresses, reducing the chances for survival."

"A future improvement might be the choice of a standard set of effect parameters which are ecologically relevant for the survival of each species within the ecosystem."

(van der Eerden et al. 1994)

"[T]he amount of available data must still be regarded as too small, i.e. the number of observations was too small in many cases, to calculate critical levels for NH$_3$ for certain plant groups."

"[O]nly a very limited number of experimental data to calculate critical levels are available."

"[T]he range of susceptibility to NH$_3$ is suggested to be as follows: natural vegetation > forests > crops."

"The concept of critical levels and critical loads is based on the assumption that the system does not respond to exposures below a certain threshold. However, for nitrogenous air pollution there are good reasons to assume that this threshold is equal to the natural background deposition, because with a low nitrogen input the system will use additional nitrogen."

(Fangmeier et al. 1994)

"[O]f the plants threatened by increased nitrogen deposition, 75–80% are indicator species for low-nitrogen habitats."

"No-observed-effect concentrations (NOECs) are usually lower than critical levels."

"[C]ritical loads focus on functioning of the ecosystem, while critical levels focus on protection of the relatively sensitive plant species."

"Observation of NH$_3$ injury to plants also indicates that this is greatest in winter."

"[G]rowth stimulation is often considered an adverse effect in most types of natural vegetation."

"[N]early all of the information (*used to calculate critical levels*) originating from one Dutch research group. Only a few pollution climates were considered."

"More experiments with lower concentrations are required."

"The assumption that all deposited nitrogen-compounds … act additionally in their impact on vegetation is poorly based on experimental results and is probably not valid for the short term."

"The critical levels for NH$_3$ … are probably only valid for temperate oceanic climatic zones."

"In the Netherlands, for example, all cyanobacterial lichens that were present at the end of the 19th century are now absent. In Denmark, 96% of the lichens with cyanobacteria are extinct or threatened." (WHO, 1997)

2.3.1 Is NH₃ the Dominant Causal Agent of the Effects Attributed to Emissions from Intensive Agriculture?

For the field measurements around point sources (intensive agriculture) it is most likely that NH_3 gas is the causal agent, although the emissions from these sources contain a cocktail of compounds including carbon dioxide, gaseous amines and particles containing nutrients, including nitrogen, phosphorus and potassium.

It is possible that nutrient-containing particles (dust) could play a part, but their rate of deposition does not match the observed patterns of effect, nor of NH_3 concentrations. Large particles ($>10^{-6}$ m diameter) deposit rapidly, and their effects would not be seen beyond a few tens of metres from sources. Smaller particles have much slower deposition rates, and most will be transported long distances before being deposited to the ground or incorporated into clouds and precipitation. Other compounds very probably play minor roles as well, although a final proof is not yet available. CO_2 is emitted in large quantities in co-occurrence with NH_3, but again its impact on the NH_3 effect is probably negligible. Wet N deposition is unlikely to change markedly over short distances from the source; any increases in N deposition measured in throughfall close to point sources are most likely attributable to the removal of soluble dry deposition (mostly NH_3) from plant surfaces rather than direct incorporation into falling rain.

2.3.2 Types of Ammonia Effects and Interfering Factors

NH_3 acts as a macro-nutrient for all biological systems. At low N-status, most plants respond to exposure to small concentrations of NH_3 by increasing their biomass production. With higher levels of NH_3 exposure, some species may develop injury (direct toxicity) or become more susceptible to pests and pathogens, some may disappear from the community because of changes in competition, the N circulation in the system accelerates and the system eventually starts leaching N.

CLEs for NH_3 can serve several applications (e.g. impact assessment of individual sources, protecting a specific ecosystem, environmental policy, or biodiversity protection at the national or EU scale). The types of effects that are considered to be adverse are dependent on this context. For instance, to monitor the productivity of a forestry plantation requires different assessments from those that dictate the decision on whether to permit the extension of a farm and the protection of biodiversity of an adjacent moorland.

Several types of effect could be considered for inclusion in the data base for assessing CLEs. In Section 2.4 a selection is made; the discussion here provides the basis for selecting appropriate effects for assessment, and follows the pathway from plant uptake to effects at the ecosystem level.

2.3.2.1   Uptake

Leaves contain ammonium (NH_4^+) ions in the apoplast, and the apoplast pH determines the NH_3 concentration that is in thermodynamic equilibrium with the apoplast, known as the "compensation point" of the foliage. At low atmospheric NH_3 concentrations the uptake of NH_3 from the atmosphere will be determined by whether the air concentration is higher or lower than the compensation point. In theory, the compensation point may be regarded as the lowest relevant "NOEC". A CLE below this NOEC does not make sense. A plant emits NH_3 if the atmospheric NH_3 concentration is below its compensation point.

The compensation point of plant communities increases with higher and prolonged ammonia exposure (e.g. Sutton et al. 1995). This behaviour is well established for higher plant species, but less certain for lower plant species. Although such an increase in compensation point is an effective adaptation by the plant itself, from the environmental point of view this phenomenon could be considered as an indicator that the habitat is adversely affected. Moreover, the cycle of uptake and re-emission means that NH_3 can be transported over tens of kilometres until it finds its ultimate sink in plants with a lower compensation point. However, as many uncertainties still exist in measuring and interpreting the compensation point this parameter cannot yet be considered for inclusion in the evaluation of a new CLE.

2.3.2.2   Increased N Content

Uptake of NH_3 by plants leads to increased NH_4^+, and subsequent conversion to increased amino acids. The conversion is indicated by increased activities of glutamine synthetase (GS) and other enzymes. This paper presents new results in the scope of NOECs (Section 2.4).

In forestry, a N content of 2% is considered to be a threshold for increased stress sensitivity in trees, although it has not been well validated. The debate continues as to whether it should be 1.5% or 1.8%, or species dependent. The same applies to the 1.2% limit for *Sphagnum* mosses, which is assumed by many to be an effect threshold (Bragazza et al. 2004; Lamers et al. 2000), although it still needs to be rationalised and verified appropriately. Other indicators like foliar N/K, NH_4^+/N, and GS activity or content also lack consensus. Hence, while overall relationships are agreed, quantitative limits remain a matter of discussion rather than a basis for establishing an adverse effect of NH_3.

2.3.2.3   Direct and Indirect Effects

Attempts to distinguish between direct and indirect effects of nitrogen uptake, sometimes defined as effects caused via leaf uptake from the atmosphere or via root uptake from the soil, have proved to be confusing under field conditions and to be largely meaningless. Clear indications exist that when considering the effects of NH_3 gas, direct uptake through the foliage is the dominant pathway. The fact that both *Calluna* (an ericoid shrub with a dense root system) and *Hypnum* (a pleuro-carpous moss with no roots) show similar responses to NH_3 strongly argues for the importance of foliar uptake.

In this respect the terms "primary" and "secondary" effects, referring to "first response" and "change in stress tolerance" are sometimes confused with direct and indirect effects. Exposure to NH_3 leads to a mixture of direct, indirect, primary and secondary effects. Therefore, we propose that all types of eco-physiologically relevant responses could be used as the basis for establishing a CLE. This does not mean that we propose to include all measurable responses in the CLE assessment. We assume that some responses (for instance, increased foliar N) provide useful background information, but not information that is directly applicable to calculating a CLE. The causal relation of the exposure with an ecologically relevant end-point should be clear and quantified, and the ecologically relevant end-point should be used to establish the CLE.

2.3.2.4   Types of Effect at the Plant Level

Field studies show that NH_3 may be implicated in the erosion of the epicuticular wax layer at relatively low concentrations (Kupcinskiene 2001). The relevance of wax erosion for the vitality and function of the plant is, however, poorly understood (Thijse and Baas 1990).

Foliar injury is generally considered to be an adverse effect, even if the 'injury' is seen as enhanced or accelerated senescence. Although it does say something about aesthetics, it may have no direct relationship with the vitality of the plant (i.e. several studies have shown that visible pollutant injury is not necessarily cor-related with long-term performance), and establishing a direct causal relationship to exposure may be difficult. Enhanced rates of needle fall from coniferous trees are a common phenomenon in the neighbourhood of N sources. Although this effect was a major indicator of effects in the era of concern about forest decline, it can also be considered to be an effective adaptation to avoid too much N uptake.

With higher nitrogen input (including NH_3), woody species like trees and heathland shrubs speed up their growth and biomass production, but also reach their maxi-mum life span in fewer years. For forestry plantations that will generally not be a problem, because it probably has no negative impact on wood production per hectare. However, for *Calluna vulgaris* the life span might go down from 40 to 10 years (Berendse, F. 2006 personal communication), possibly resulting in faster rejuvenation and higher costs of maintenance of such habitats. With prolonged elevated N input many tree species lose their apical dominance earlier (e.g. at an age of 25 instead of 50 years).

Plants with a low N status will initially respond to increased N exposure with an increase in growth, but with higher exposures a decrease in reproduction will occur. For example, Leith et al. (2001) found NH_3 exposure to reduce flowering of *Eriophorum vaginatum*, although overall, growth of this species benefited from NH_3 treatment compared with other moorland species. Similar results were found in *Antenaria dioica* and *Arnica montana* (van der Eerden et al. 1991).

The effects of NH_3 on vegetation may be enhanced by interaction with drought. NH_3, being a fertiliser generally causes thinner leaves (increased Specific Leaf Area (SLA), resulting in more evaporation and more uptake capacity for air pollutants), less effective closure of stomata and increased shoot/root ratio (van der Eerden et al. 1991).

Low temperatures increase the solubility of NH_3 in water, and NH_x concentrations will be enhanced; the equilibrium concentration of undissociated NH_3 is twice as great at 5°C as at 20°C. Moreover, at lower temperatures, the detoxification of NH_3 is less effective than at higher temperatures. Apart from evidence from laboratory fumigation experiments, there is some field evidence for interactions of low concentrations of NH_3 with both low temperature (frost) and drought. A reduction in the cover of green shoots of *Calluna vulgaris* at the Whim experimental site in south-east Scotland has been observed after each winter, when the shoots had a bleached appearance. The lowest NH_3 concentration at which this type of damage occurred has decreased with each year of exposure (Sheppard et al., see Chapter 4). The bleaching observed in *C. vulgaris* is most likely due to an impact of NH_3 on winter desiccation (Sheppard and Leith 2002). Frost hardiness experiments indicated that ammonia reduced shoot hardiness. However, the effect was not sufficient to explain the damage observed in the field.

These secondary effects should also be considered when comparing NH_3 concentrations with field responses. Exceeding a NOEC does not guarantee development of injury. To cause injury, it is possible that a slight exceedance must be accompanied by adverse conditions like low light intensity, an early night frost, drought stress, a storm (having more impact on trees with too high a shoot/root ratio), an infestation by pests or disease etc. (van der Eerden et al. 1991).

Interactions with other gaseous pollutants are also poorly understood. There is experimental evidence of increased deposition rates of SO_2 in response to NH_3 (Cape et al. 1995), and of increased NH_3 deposition in response to SO_2 (Shaw and McLeod 1995). However, interaction with SO_2 might be expected to lead to increased deposition to external leaf surfaces, and localised depletion of gas-phase NH_3 near stomata, thereby reducing internal uptake and NH_3 effects.

Interactions with biotic stresses (pathogens, insects) are known to occur at high NH_3 concentrations; see reviews (Fangmeier et al. 1994; Krupa 2003) but little is known about effects at low concentrations.

2.3.2.5   Impacts at the System Level

Increased N input in the soil generally leads to increased microbiological activity, increased nitrate and ammonium concentration in edaphic ground water, increased N immobilisation in litter and increased output of N_2, N_2O and NO etc. All these features are signs of adaptation of the system itself, partly at a cost to the surrounding environment. Obvious adverse effects are exhaustion of the buffer capacity, acidification, and leaching of nitrate and base cations to deeper ground water or surrounding surface water.

There are indications that vegetation can adapt to higher inputs of N without apparent major ecological damage. In field experiments, ombrotrophic bogs in Scotland responded at much lower N exposure levels than heathland in the Netherlands and Germany. This might be explained by adaptation over decades, for instance by the selection of NH_x tolerant ecotypes in central Europe. However, it may also be an artefact due to an inappropriate or polluted control regime in the experiments.

Even if some tolerant species and habitats can apparently adapt, long-term exposure has led to total loss of many species and major changes in plant communities in the Netherlands and elsewhere. The NH_3 triggered transition of heathland into grassland and the degradation of many species rich habitats have been well described. Increasing evidence is developing that epiphytic lichens are particularly sensitive to NH_3, and that shifts in species composition of lichens can be quantitatively related to NH_3 exposure. Section 2.6 presents new information on this matter. We propose to give specific attention to biodiversity, as this proves to be a relatively sensitive criterion that is clearly linked to a measurable adverse effect, i.e. loss of particular species from an ecosystem.

2.3.3 Measuring Height of NH_3 Concentrations

In setting up a CLE based on the air concentration of NH_3 reference must be made to the concentrations to which the vegetation is exposed. There is no standard height used in experimental protocols for measuring NH_3 concentrations, although 1.5 m above ground is usual for short vegetation. If a surface is absorbing NH_3 from the atmosphere, then a marked vertical gradient occurs, with concentrations decreasing towards the surface. The problems caused by the vertical gradient, and the correct methods for assessing the reference height at which concentrations should be measured (Sutton et al. 1997), has been well described for the case of ozone (Pleijel 1998). An example of a vertical gradient in NH_3 is illustrated in Fig. 2.1, which shows the long-term monthly average concentrations of NH_3 in ambient air at the Whim experimental site (Leith et al. 2004a) at several heights above the canopy.

These effects indicate that care must be taken in making NH_3 measurements at an appropriate height above the canopy of the vegetation of interest. This may be of particular concern in complex layered canopies, for example if assessing the concentrations to which forest understorey vegetation is exposed. Consequently, the CLE has to be defined for air concentrations measured sufficiently far above the plant canopy that the uncertainty caused by any vertical gradient is either well-defined (and reported) or negligible.

2.3.4 Relevance of Temporal Variation and Peak Concentrations in NH_3

Generally NH_3 is emitted by local sources at a height of 5 m or lower. The effective emission height from barns and sheds used in intensive animal or poultry production may even be at surface level due to eddies and turbulent "downwash" of air downwind. Grazed and manured land represent surface sources. The ratio between peak and mean concentrations measured close to the surface is relatively high near

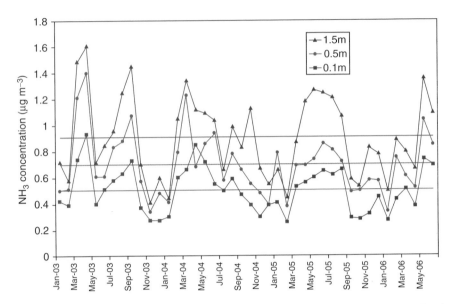

Fig. 2.1 Ambient background NH$_3$ concentrations at the Whim experimental site, at 0.1, 0.5 and 1.5 m above the canopy, with the average vertical gradient in NH$_3$ concentration shown as horizontal lines. The concentrations represent background ambient conditions at a point unaffected by the experimental NH$_3$ release

to NH$_3$ sources. The more isolated the source and the lower the regional background concentration, the higher this ratio is.

A relevant question in the scope of CLE setting is if it matters whether an average exposure level consists of high peaks combined with low levels between the peaks, or of a more constant moderately elevated level. Answers to this question may indicate the effectiveness of emission abatement strategies like:

- Use of a chimney (as this reduces local peaks but increases regional background levels)
- Distribution of emissions over more sources (this has the same effect)
- Limiting emissions during sensitive periods, or when weather conditions inhibit rapid mixing and dispersion

The quantitative impact of peak concentrations is still uncertain (van der Eerden et al. 1998), while current and future agricultural practice requires appropriate protection against short-term peak exposures, especially in regions where the ratio of peak concentrations to mean concentrations is relatively high. Changes in agricultural practice (e.g. manure spreading in spring) may lead to higher peak/mean ratios and short term concentrations than have been measured to date.

The statistical distribution of NH$_3$ concentrations over time from several monitoring sites in northern Europe shows that the annual CLE will be a stricter constraint on exposure than the existing hourly or daily CLEs. However, there are too few monitoring data from southern and eastern Europe to generalise on the relationship between peak concentrations and annual averages.

If NH_3 is not directly metabolised, exposure to a peak concentration results in temporary uptake until the compensation point of the canopy is reached, followed by subsequent release when the peak exposure is followed by a considerably lower air concentration. The compensation point for uptake of atmospheric NH_3 is much smaller for natural vegetation than for agricultural vegetation. Hence, the same phenomenon could take place at a lower concentration level. For this reason one has to be careful about using the results of experiments with constant exposure levels for derivation of CLEs; they are probably underestimating the effects in the field where concentrations fluctuate about the long-term average concentration.

2.4 Procedures for Deriving NOECs from Exposure-Response Relationships

If one accepts that the existence of a 'measurable difference' from background conditions (a NOEC) is an adequate metric to establish a CLE, it is implicit that the 'background' reference level truly represents the non-disturbed state of the system.

Unfortunately, in countries like the Netherlands, field studies cannot make use of a clean control treatment. For those regions 'background' conditions may have to relate to the 19th century rather than to any currently available region, because any effects of low concentrations have already occurred at some time in the past decades. In those regions even the reference levels for controlled experiments are at several micrograms of NH_3 per cubic metre (because of insufficient filtration of incoming air), which are many times greater than air concentrations in remote rural areas in other parts of Europe.

Given these constraints, we are faced with the task of establishing when a measurement at a certain location is significantly greater than the 'background'. The word 'significant' refers to both ecological and statistical significance. Responses in terms of growth, vitality, reproductive fitness, competitive ability, shifts in species composition etc. are obviously of ecological significance. But for some other responses to exposure to NH_3, like enhanced foliar contents of N, arginine or NH_4^+, GS activity etc., the relationships with ecological end points and their dependence on environmental conditions are still insufficiently quantified for them to be used directly as a basis for setting CLEs. They can play a role, however, as corroborative evidence.

In statistical terms, 'significant' means that the measurement exceeds the 'background' value, and has only a small probability (e.g. <5%) of falling within the range of possible values regarded as 'background' – this depends inter alia on the inherent uncertainty of the measurement method and the spatial (and temporal) variability of the measured vegetation. In the past, many of the NH_3 experiments consisted of just a treatment and a control. An analysis of variance was used to test the statistical significance of the effects. The more recent experiments, using multi-level treatments as well as field surveys along a concentration transect, allow the evaluation by regression analysis.

In some cases the exposure/response relationship shows a clear break point – the point above which the slope of the curve is significantly greater than zero. This is an obvious NOEC. However, for most of the new information presented in this paper we have had to use a statistical procedure as indicated in Fig. 2.2. In general, for the data available (see Tables 2.3 and 2.4), a linear or log-linear response curve was most appropriate, although other forms (e.g. sigmoidal) could exist and be used in a similar fashion. The equation of the line that best fits the data was established by means of a least-squares analysis. The 95% confidence limits for the relationship can than be calculated. The upper 95% curve at the lowest exposure concentration estimates the largest value of the response variable that falls within the local 'background' range (the lowest concentration measured; point A in Fig. 2.2). If this response value is extended to higher concentrations, the point where it intersects the fitted curve (point B) indicates the lowest concentration that yields a measurement value above the local 'background' (read from the x-axis at point C). This limiting concentration (C) is then an indication of the NOEC obtained from that data set. This procedure utilises all the information available (in fitting the relationship) while focussing on the lower end of the exposure scale. A measure of the appropriateness of the sampling regime (number of samples at any location) can be ascertained from the relationship between the spread of measurement data about the mean and the range of the fitted curve.

The procedure presented here requires attention to some issues:

- If the true background concentration is not represented (i.e. the lowest measured concentration is above the background concentration) then this technique will tend to overestimate the NOEC.

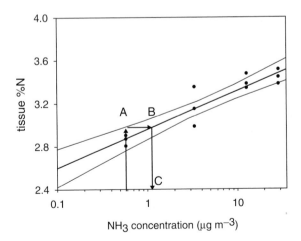

Fig. 2.2 Illustrative example of estimation of the NOEC from measurements at several locations differing in NH$_3$ concentration. The lowest measured concentration is taken as representative of the local 'background' concentration

- This approach relies on the form of the relationship between the measured response variable and the NH_3 concentration. In most of the examples given in Table 2.4 there are relatively few data points, making it difficult to be certain of the appropriate relationship. In general, the best fit is with a linear response to the logarithm of NH_3 concentration, although for some situations a linear:linear response may be better.
- It should be noted that the degree of correlation in the experimental datasets affects the NOEC values derived by this method. Basically, NOEC are higher, not only in the case where the receptor is less sensitive, but also if the quality of the relationship is not sufficient to imply significant effects at lower NH_3 concentrations.
- It should also be noted that the NOEC estimate does not include a safety or "assessment" factor (see Section 2.8.2).

2.5 Procedures for Deriving CLEs from a Set of NOECs

The CLE for NH_3 set in 1992 was assessed using the "envelope" approach (van der Eerden et al. 1994). All experimental results from the exposure of plants to NH_3 (both effective and non effective exposures) are plotted in a time-concentration graph. A curve just under the lowest effective exposure levels indicates the CLE. In this approach the lowest NOEC is the CLE for each exposure period. A disadvantage of this approach is that no information is used from the exposure/response relationships and there is no dependence on inter-species variation in sensitivity.

The approach of assuming the CLE to be equal to the NOEC of the most sensitive species is not very satisfying.

To include variation in sensitivity a statistical evaluation technique could be applied, as is relatively often used in soil and water pollution. It is based on an estimate of the level at which a certain percentage (e.g. 95%) of the species are protected with a certain probability level (e.g. 50% or 95%) (Aldenberg and Slob 1993; Kooijman 1987). In the scope of standard setting it is currently widely in use, also in EU environmental policy, although criticism exists as well.

Because the current and proposed CLEs are based on empirical evidence from measured exposure to NH_3, no assumptions have been made on the pathways or mechanisms of action. This is a similar approach to the estimation of empirical CLOs.

A causal-analytical model that simulates ecosystem responses is to be preferred over statistical approaches. Existing mechanistic models for the performance of forests, heath land and crops have an entry for N. The response of those systems to NH_3 can be simulated if uptake and detoxification capacity can be quantified. Although the first attempts (van der Eerden et al. 1998) were not very promising for several reasons, further research is certainly useful and may result in information that can be used in the next update of CLEs.

We have considered all evaluation types and concluded that for the available data set, and with the applications envisaged, a simple evaluation suffices – assessing the range of NOECs for each vegetation type and setting the CLE at the lower limit of that range.

2.6   Limitations of the CLEs of 1992

In 1992, most of the information on effects of NH_3 on vegetation came from laboratory fumigation experiments in the Netherlands. Those experiments had their drawbacks and limitations. The control treatments were polluted with NH_3 at concentration levels probably around $1–5\,\mu g\ NH_3\ m^{-3}$, which is higher than the ambient levels in most of Europe. The temperature at these experiments were generally around $15–25°C$, the exposure duration was rarely longer than 6 months and the concentration was kept constant. Since then, the available evidence indicates that lower temperatures and longer exposure durations promote adverse effects at lower concentrations. Moreover, fluctuating concentration levels appear to be more influential than constant levels (possibly as a result of decreasing deposition velocity with prolonged higher concentrations, which can influence the compensation point). In addition to laboratory fumigations, the results of some field surveys were used for the basis of the 1992 CLE. But these surveys were executed in regions with background concentrations probably in the range of $3–6\,\mu g\ NH_3\ m^{-3}$. It has become clear that effects can be measured at very much lower long-term average NH_3 air concentrations. Many of the ecological changes resulting from enhanced NH_3 exposure in those regions may have already occurred, long before experiments were conducted.

In later reviews (WHO 1997) the CLE of 1992 was regarded to have serious limitations. It was concluded that, in contrast to the data base used to construct CLOs, the amount of available data was too small to differentiate between plant groups. Few data are available on very sensitive species, for example cyanobacterial lichens, although strong indications exist that NH_3 was a major cause of the extinction of those species in countries with high NH_3 concentrations like Denmark and the Netherlands. It was stated that too little is known about risk conditions like early and late winter frost, drought and unbalanced nutrient supply. The concept of CLE is based on the assumption that the system does not respond to exposures below a certain threshold within a certain time (up to 1 year). However, for NH_3 this threshold may be equal to the natural background deposition in some places. A general conclusion was that the CLE for NH_3 is probably only valid for temperate oceanic climatic zones.

2.7   Evidence for Adverse Effects Below the CLE of 1992

2.7.1   Evidence from Comparing CLOs and CLEs

Comparison of the annual CLE from 1992 with empirical CLOs shows that the CLE is much less stringent than the CLO (Fig. 2.3 and Table 2.2).

The solid diagonal line in Fig. 2.3 shows the predicted deposition for an air concentration of $8\,\mu g\ NH_3\ m^{-3}$ as a function of deposition velocity. For tall vegetation, this equates to $80\,kg\ N\ ha^{-1}\ year^{-1}$, well above most CLOs. The CLE is therefore not protecting vegetation from adverse impacts of N deposition. Even for short vegetation, where deposition is less efficient, the predicted dry deposition

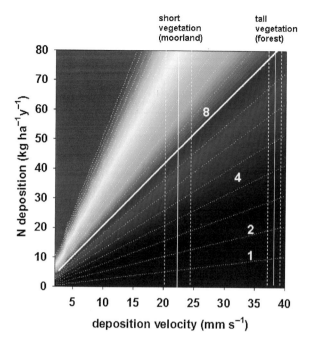

Fig. 2.3 Relation between annual N deposition and deposition velocity for NH_3 for a range of annual average NH_3 concentrations (diagonal dotted lines). Vertical lines show mean and dashed lines show interquartile range for UK conditions for short vegetation (like moorland; V_d 16–32 mm s^{-1}) and tall vegetation (like forest; 33–48 mm s^{-1})

for 8 μg NH_3 m^{-3} is 45 kg N ha^{-1} year^{-1}, which is higher than the empirical CLOs for semi-natural ecosystems. Again the CLE does not protect the ecoystem.

In other words, the CLO for N deposition is exceeded well before the CLE of 8 μg NH_3 m^{-3} is exceeded.

The translation between empirical CLOs (UNECE 2003) and equivalent NH_3 concentrations in air for different vegetation types is given in Table 2.2. This shows that even in the absence of other N components, NH_3 concentrations much lower than the CLE are expected to have significant adverse effects on a wide range of habitats.

The comparison between CLO and NH_3 concentration is based on UK conditions. In general, wind speeds and surface wetness are greater over the UK than mainland Europe. Consequently, deposition velocities are higher in the UK. So, the NH_3 concentrations required to exceed the CLO are higher in other regions of Europe, particularly for sites with low rainfall and low frequency of mist or fog.

With the exception of marine habitats (Table 2.2), the CLE of 8 μg m^{-3} is redundant, i.e. the empirical CLO for N deposition will be exceeded on the basis of NH_3 concentration alone before the CLE is reached, often by a very large margin.

The foregoing discussion and calculations confound two different time-scales: although it is not explicitly set in the definition, the CLEs are meant to be valid for a 1-year exposure or shorter, while CLOs focus on longer term exposures. In the Grange-

Table 2.2 NH$_3$ concentration at which empirical critical load (UNECE 2003) would be exceeded. These are maximum values, representing the case where other components of nitrogen deposition are set to zero. Deposition velocities are based on UK conditions (short vegetation: 16–32 mm s^{-1}, tall vegetation: 33–48 mm s^{-1})

Ecosystem type	Empirical CLO (kg N ha^{-1} year^{-1})	NH$_3$ µg m^{-3}	Comments
Forest			
Forest trees	15–20	1.3–2.3	Forest canopy and exposed surfaces
Ground vegetation	10–15	1.2–1.8	Probably higher; low deposition velocity
Lichens and algae	10–15	0.9–1.8	For exposed surfaces, high deposition velocity
Heathland, scrub and tundra			
Tundra	5–10	0.6–2.4	
Arctic, alpine and subalpine scrub	5–15	0.6–3.6	Lower concentrations for rougher surfaces
Northern wet heath Calluna dominated	10–20	1.2–4.8	
Northern wet heath Erica dominated	10–25	1.2–6.0	
Dry heaths	10–20	1.2–4.8	
Grasslands and tall forb habitats			
Sub-atlantic semi-dry calcareous grassland	15–25	1.8–6.0	
Non-mediterranean dry acid and neutral closed grassland	10–20	1.2–4.8	
Inland dune grasslands	10–20	1.2–4.8	
Low and medium altitude hay meadows	20–30	2.4–7.2	
Mountain hay meadows	10–20	1.2–4.8	
Moist and wet oligotrophic grasslands Molinia	15–25	1.8–6.0	
Moist and wet oligotrophic grasslands Juncus	10–20	1.2–4.8	
Alpine and subalpine grasslands	10–15	1.2–3.6	
Moss and lichen dominated mountain summits	5–10	0.6–2.4	Possibly lower for exposed locations
Mire, bog and fen habitats			
Raised and blanket bogs	5–10	0.6–2.4	
Poor fens	10–20	1.2–4.8	
Rich fens	15–35	1.8–8.4	
Mountain rich fens	15–25	1.8–6.0	
Coastal habitats			
Shifting coastal dunes	10–20	1.2–4.8	
Coastal stable dune grasslands	10–20	1.2–4.8	
Coastal dune heaths	10–20	1.2–4.8	
Moist to wet dune slacks	10–25	1.2–6.0	
Marine habitats			
Pioneer and low-mid salt marshes	30–40	3.6–9.6	

over-Sands Critical Loads Workshop (UNECE 1995), it was noted that the CLO for N *"cannot be assumed to provide a protection period of longer than 20–30 years"*.

The focus of the CLE definition on short term exposures originates from its use in protecting agricultural (annual) crops, rather than on natural vegetation. However, information on the impact of NH_3 on perennial natural vegetation has been available since the early 1990s, indicating that a CLE for NH_3 should cover multi-year exposures. If one extrapolates the relationship between averaging time and CLE derived by van der Eerden et al. (1994) to longer averaging times, the 25-year CLE (analogous to the empirical Critical Load time-scale) would be around 2.5 μg NH_3 m^{-3}.

2.7.2 Experimental Evidence for Effects of Long Term Exposure Below 8 μg NH_3 m^{-3}

In Section 2.6.1 only mathematical extrapolations were presented, but experimental data are now appearing that show a progressive effect with time of exposure to small NH_3 concentrations (Sheppard et al. 2009, Chapter 4).

This section reviews recent experimental and observational data that demonstrate measurable changes in vegetation, compared to 'background' conditions, which are directly attributable to (measured) exposure to NH_3. Results from measurements on vegetation where the NH_3 gas concentration has not been measured are not included (for example, studies where NH_3 concentration was only indicated by quoting data relative to distance from a point source), although they may have a bearing on the spatial range over which such effects can be observed.

One of the most comprehensive datasets is from Sheppard et al. (see Chapter 4 Fig. 2.4), where the tissue %N of the moss *Hypnum jutlandicum* is plotted in response to long-term average NH_3 concentrations after 4.5 years exposure in the field-fumigation experiment at Whim, in south-east Scotland (Leith et al. 2004a). In this case, the large number of data points clearly shows the linear response to a logarithmic increase in NH_3 concentration, and a calculated CLE, as defined above, of 0.8 μg NH_3 m^{-3}.

Experiments in which the response to low concentrations of NH_3 have been recorded are summarised in Table 2.3, with an indication of the lowest NH_3 concentration measured (the 'background' level) and the calculated NOEC using the method described above. These data come from a variety of sources, including measurements around point sources, experimental fumigations and regional gradients.

Some of the NOECs presented in Table 2.3 seem surprisingly high, but this is mainly due to the constraint in the method that with a poorer fit of the regression the estimated NOEC gets higher. Although methods are available to properly cope with this phenomenon, we decided not to use them because for the derivation of a CLE out of a set of NOECs the focus is strongly on the lowest NOECs. For the last entry in Table 2.3, the quantitative role of NH_3 was difficult to assess. Strong correlations were also observed with wet N deposition, and with stemflow and throughfall N content (Mitchell et al. 2005). Given these correlations, it may be considered unsafe

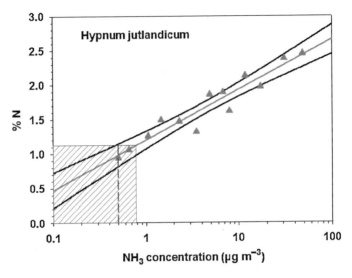

Fig. 2.4 Increase in tissue N concentration of the moss *Hypnum jutlandicum* in response to experimental field-fumigation with NH_3 after 4.5 years of treatment (data from Sheppard et al., Chapter 4)

to set a very low CLE (0.1 μg m^{-3}) based on this dataset. Nevertheless, the study is a useful reminder of other factors influencing the relation of measured NH_3 concentrations with observed effects.

Field surveys on community structures in response to gradients in NH_3 concentrations may also be used to estimate NOECs. Pitcairn et al. (2009, this volume) and Leith et al. (2005) showed parallel gradients in Ellenberg N Index. The more sensitive index derived from the presence/absence of nitrophobe and nitrophile species (Pitcairn et al. 2009, this volume; Wolseley et al. 2009, this volume) suggested that significant changes in species composition occurred at concentrations between 1 and 3 μg NH_3 m^{-3}. More recently published data or data presented in this conference, derived from experiments in the UK, Switzerland, Portugal and Italy, all indicate ecologically significant effects in the range of 0.5–4 μg NH_3 m^{-3} (Table 2.5, Sutton et al. 2009, see Chapter 6, this volume; Rihm et al. 2009, see Chapter 7, this volume; Pinho et al. 2009, see Chapter 10, this volume; Frati et al. 2007).

2.7.3 Experimental Evidence for Impacts of Short Term Exposures

The responses presented in Tables 2.3 and 2.4 are the result of long-term exposure, generally longer than 1 year. Although most measurements suggest that the existing annual CLE will be a stricter constraint on exposure than the hourly or daily CLEs (Sutton et al. 2009, see Chapter 8, this volume), this may not be true in future,

Table 2.3 NOECs for NH$_3$-triggered changes in chemical composition of plant tissue. NOECs were calculated with regression analysis, data from recent experimental studies on the impact of NH$_3$ on vegetation

Measurement	lin/log	Lowest NH$_3$ (µg m^{-3})	NOEC (µg m^{-3})	Source/location	Comment	Reference
Field measurements close to point sources						
Arginine in *Rhytidiadelphus triquetrus*	log	1.6	2.2	Poultry farm/Scottish Borders	Feb 1995–Feb 1996	Pitcairn et al. (2003)
Threonine in *Rhytidiadelphus triquetrus*	lin	1.6	4.0	Poultry farm/Scottish Borders	Feb 1995–Feb 1996	Pitcairn et al. (2003)
Histidine in *Rhytidiadelphus triquetrus*	log	1.6	2.2	Poultry farm/Scottish Borders	Feb 1995–Feb 1996	Pitcairn et al. (2003)
Serine in *Rhytidiadelphus triquetrus*	log	1.6	2.2	Poultry farm/Scottish Borders	Feb 1995–Feb 1996	Pitcairn et al. (2003)
Serine in *Rhytidiadelphus triquetrus*	lin	1.6	4.5	Poultry farm/Scottish Borders	Feb 1995–Feb 1996	Pitcairn et al. (2003)
Glutamic acid in *Rhytidiadelphus triquetrus*	log	1.6	2.8	Poultry farm/Scottish Borders	Feb 1995–Feb 1996	Pitcairn et al. (2003)
Glutamic acid in *Rhytidiadelphus triquetrus*	lin	1.6	5.0	Poultry farm/Scottish Borders	Feb 1995–Feb 1996	Pitcairn et al. (2003)
Aspartic acid in *Rhytidiadelphus triquetrus*	log	1.6	3.5	Poultry farm/Scottish Borders	Feb 1995–Feb 1996	Pitcairn et al. (2003)
Aspartic acid in *Rhytidiadelphus triquetrus*	lin	1.6	5.6	Poultry farm/Scottish Borders	Feb 1995–Feb 1996	Pitcairn et al. (2003)
NH$_4^+$ in *Hypnum cupressiforme*	log	0.6	1.6	Poultry farm/Scottish Borders	Oct–Nov 2002	Pitcairn et al. (2006)
NH$_4^+$ in *Hypnum cupressiforme*	lin	0.6	5.5	Poultry farm/Scottish Borders	Oct–Nov 2002	Pitcairn et al. (2006)
NH$_4^+$ in *Rhytidiadelphus triquetrus*	log	0.6	1.4	Poultry farm/Scottish Borders	Oct–Nov 2002	Pitcairn et al. (2006)
NH$_4^+$ in *Rhytidiadelphus triquetrus*	lin	0.6	4.7	Poultry farm/Scottish Borders	Oct–Nov 2002	Pitcairn et al. (2006)
%N in *Eurynchium striatum*	log	2	2.7	Poultry farm/SW England	86 days NH$_3$ data	Leith et al. (2005)
%N in *Eurynchium praelongum*	log	2	2.6	Poultry farm/SW England	86 days NH$_3$ data	Leith et al. (2005); Pitcairn et al. (2006)
NH$_4^+$ in *Eurynchium praelongum*	log	2	2.2	Poultry farm/SW England	86 days NH$_3$ data	Leith et al. (2005); Pitcairn et al. (2006)
NH$_4^+$ in *Eurynchium striatum*	log	2	10.0	Poultry farm/SW England	86 days NH$_3$ data	Leith et al. (2005)
NH$_4^+$ in *Eurynchium striatum*	log	2	2.8	Poultry farm/SW England	log:log plot	Leith et al. (2005)
%N in *Dryopteris dilatata*	log	3	7.5	Poultry farm 'L'/Central Scotland	July 1995–July 1996	Pitcairn et al. (2002)
%N in ectohydric mosses	log	3	5.0	Poultry farm 'L'/Central Scotland	July 1995–July 1996	Pitcairn et al. (2002)

(continued)

Table 2.3 (continued)

%N in ectohydric mosses	lin	3	9.0	Poultry farm 'L'/Central Scotland	July 1995–July 1996	Pitcairn et al. (2002)
%N in Elder (*Sambucus nigra*)	log	3	9.5	Poultry farm 'L'/Central Scotland	July 1995–July 1996	Pitcairn et al. (2002)
%N in *Flavoparmelia caperata*	log	0.7	1.7	Pig farm/ Italy	2 × 2 weeks	Frati et al. (2007)
Measurements on biomonitors close to a point source						
NH_4^+ in *Lolium perenne*	log	0.6	1.0	Poultry farm/Scottish Borders	biomonitor 38 days	Leith et al. (2009, Chapter 16, this volume)
Total above-ground N per pot	log	0.6	2.0	Poultry farm/Scottish Borders	biomonitor 38 days	Leith et al. (2009, Chapter 16, this volume)
NH_4^+ in *Deschampsia flexuosa*	log	2	2.5	Poultry farm/SW England	biomonitor 86 days	Leith et al. (2009, Chapter 16, this volume)
%N in *Deschampsia flexuosa*	log	2	9.0	Poultry farm/SW England	biomonitor 86 days	Leith et al. (2009, Chapter 16, this volume)
Measurements from controlled field fumigation						
%N in *Hypnum jutlandicum*	log	0.5	0.8	Whim experiment/SE Scotland	4 year (NH_3) at 0.1 m	Sheppard et al. (2006)
%N in *Calluna vulgaris*	log	0.5	1.0	Whim experiment/SE Scotland	4 year (NH_3) at 0.1 m	Sheppard et al. (2006)
%Ca in *Calluna vulgaris*	log	0.5	1.0	Whim experiment/SE Scotland	4 year (NH_3) at 0.1 m	Sheppard et al. (2006)
%Mg in *Calluna vulgaris*	log	0.5	1.3	Whim experiment/SE Scotland	4 year (NH_3) at 0.1 m	Sheppard et al. (2006)
Measurements across regional gradients						
N in epiphytic mosses	log/lin	0.02	<0.1	Atlantic oakwoods, NW UK	Variation in other sources of N	Mitchell et al. (2005)

because of changes in agricultural practice, such as manure spreading. Thus, a short-term (less than 1 year) CLE could be useful as well.

The hourly- and daily-CLEs of 1992 were based on experiments with annual crops, rather then with perennials or species from natural vegetation. The monthly-CLE of 1992 ($23\,\mu g\ NH_3\ m^{-3}$) has more relevance, both in relation to measured concentrations and to the species selected for its definition. This CLE was estimated partly by extrapolating between yearly- and daily-CLEs and partly by specific experimental evidence: one bryophyte species (*Racomitrium lanuginosum*) out of four showed leaf injury at $30\,\mu g\ m^{-3}$ after 23 days. Regression analysis suggested a NOEC well below $30\,\mu g\ m^{-3}$. In another experiment five heathland species were fumigated for 90 days with six concentration levels and regression analysis was used to estimate the NOEC for increased drought sensitivity (indicated by 50% increase in shoot/root ratio). The five NOECs ranged from 13 to $145\,\mu g\ m^{-3}$ (van der Eerden et al. 1991).

2.8 Proposal for New CLEs

Based on the evidence presented here there are good reasons for defining new CLEs, making a distinction between lichens and bryophytes on the one hand and other plant species on the other. The best evidence relates to studies on lichens and bryophytes. A CLE based on these data is appropriate for protecting epiphytic lichens and bryophytes, and ecosystems with significant abundance of ground dwelling lichens and/or bryophytes, such as bogs, fens, heaths and moor land. Exceeding this CLE would result in shifts in species composition, and increased potential for species extinctions. In this context, the CLE is defined as long-term. 'Long-term' refers to a period of several years, sufficient for such changes to occur, but does not guarantee protection beyond 20–30 years.

Table 2.4 (based on changes in species composition) shows that NOECs are in the range of $0.7–4\,\mu g\ NH_3\ m^{-3}$. Table 2.3 only shows NOECs for chemical alterations in response to exposure to NH_3, but also these NOECs generally are in the same range.

Therefore, a long-term CLE for these systems should be set at the lower end of this range. One microgram per cubic metre of NH_3 seems most logical.

For higher plants, there is much less new information, but two studies indicate that the current CLE of $8\,\mu g\ m^{-3}$ is too high as a long-term threshold for higher plants in natural vegetation.

1. Changes in woodland ground flora downwind of an intensive animal unit in SW England suggest a threshold not larger than $4\,\mu g\ m^{-3}$ (Pitcairn et al. 2006).
2. Comparison of the increasing rate of death of *Calluna vulgaris* at a field fumigation experiment in Scotland (Whim bog) with the death rate of the lichen *Cladonia* spp. indicates consistently that *Calluna* death occurs at a concentration 2.2 times that at which *Cladonia* is killed (Sheppard et al. 2009, Chapter 4, this volume); this implies a 'no effect' concentration for *Calluna* of about $2\,\mu g\ m^{-3}$.

Table 2.4 Summary of NOECs of the impact of long-term exposure to NH$_3$ on species composition of lichen communities. NOECs were directly estimated from exposure/response curves or calculated with regression analysis. The data are from recent experimental studies, both field surveys and controlled field experiments on the impact of NH$_3$ on vegetation

Location	Vegetation type	Lowest measured NH$_3$ concentration (µg m^{-3})	Estimated NOEC (µg m^{-3})	Reference
SE Scotland, poultry farm	Epiphytic lichens	0.6	0.7 (on twigs) 1.8 (on trunks)	Leith et al. (2004b) Sutton et al. (2009, Chapter 6, this volume)
Devon, SW England	Epiphytic lichens diversity (twig)	0.8 (modelled)	1.6	Wolseley et al. (2009) Chapter 6, this volume
United Kingdom, national NH$_3$ network	Epiphytic lichens	0.1	1.0	Leith et al. (2005) Sutton et al. (2009, Chapter 9, this volume)
Switzerland	Lichen population index	1.9 (modelled)	2.4	Rihm et al. (2009, this Chapter 7, volume)
SE Scotland, field NH$_3$ experiment, Whim bog	Lichens and bryophytes – damage and death	0.5	<4	Sheppard et al. (2009, Chapter 4, this volume)
Corroborative evidence[a]				
SW England	Epiphytic lichens	1.5	ca. 2	Leith et al. (2005)
South Portugal	Epiphytic lichens	0.5	1	Pinho et al. (2009, Chapter 10, this volume)
Italy, pig farm	Epiphytic lichens	0.7	2.5	Frati et al. (2007)

[a]In these cases NH$_3$ concentration data were available for less than 1 year. This is why these results are categorised as "corroborative evidence".

Based on this evidence, the long-term CLE for higher plants is probably in the range 2–4 µg m^{-3}, so on the basis of expert judgement we propose the long-term CLE for higher plants as 3 µg NH$_3$ m^{-3}.

We assume that this long-term CLE (defined as above) will protect bogs, heathland, woodland ground flora and probably also oligotrophic grassland, from NH$_3$-driven shifts in species composition, and the potential for species extinctions.

There is no additional information on short-term exposures, but we would provisionally retain the existing monthly CLE of 23 µg m^{-3}, to ensure appropriate protection against short-term exposures caused by changes in agricultural practice that give large emissions at certain times of year.

The proposed long-term CLEs do not apply to all ecosystems or habitats, because of a lack of relevant information. However, the proposed long-term CLE for higher plants is likely to be no more restrictive than the existing CLO for most habitats, based on estimating the contribution of 3 µg NH$_3$ m^{-3} to dry deposition of N in most ecosystems. For example, typical values for the UK would be 15–20 kg

N ha^{-1} year^{-1} for short vegetation, and up to 30 kg N ha^{-1} year^{-1} for tall vegetation, in addition to the deposition of other N species (wet and dry).

2.9 Discussion

2.9.1 Species Taken into Consideration

The newly proposed CLE is mainly based on available data for lichens, bryophytes, and species from bogs, heathland and species-rich grasslands. This covers a broader range of species than the previous CLE, in which crops dominated. In an EU workshop on risk assessment, a panel of experts proposed, as a general rule, a minimum of 15 species within a minimum of three taxonomic groups. This has been achieved with the new CLE. To include more species from natural vegetation certainly improves the data base. Such a change can also be motivated by studies showing that species of climax vegetation on N-poor soils are always relatively sensitive to NH_x (and NO_x) (Pearson and Stewart 1993). On the other hand, to select only species that are assumed to be sensitive reduces the information on variation in sensitivity.

An issue that is often raised in setting environmental standards is whether species can respond differently in different experimental settings (and in different EU regions). This could be caused by different conditions or by different ecotypes within a species. The fact that results on lichen responses from Italy, Scotland, The Netherlands, Switzerland and Portugal are similar and all lead to NOECs in the same range supports the assumption, that regional difference in sensitivity is relatively small, at least for epiphytic lichens. Such a comparison cannot yet be made with higher plant species because of a lack of data.

2.9.2 Is a Safety Factor Needed?

In ecotoxicological studies, uncertainty in the final result is created by experimental procedures (especially the translation of laboratory tests to the field situation), the limited number of species and of conditions tested, etc. This uncertainty is generally covered by an "assessment" factor. A factor 10 is most common, which means that the CLE would be 1/10 of the general NOEC.

We assume such an assessment factor is not needed for a number of reasons:

- The NOECs represent sensitive species, rather than a random or average selection.
- The majority of data are from field experiments or field observations.
- Some of the NOECs may contain a slight over estimation of species sensitivity if they are related to NH_3 concentrations measured at a commonly used height of 1.5 m, instead of at canopy level.

The new CLE is based on protection of biodiversity in sensitive ecosystems. Other targets may result in other CLEs. If, for instance the wood production in a forest plantation is the item of interest, the proposed CLE probably is very stringent. If, on the other hand, each measurable sign of NH_3 input is a criterion (even if there is no long-term ecologically relevant effect), the CLE may need to be lower.

2.9.3   Biomonitoring the Potential for an Effect

In any study of the potential effects of air pollutants on vegetation, and in setting the CLE, of concern is the most sensitive species or organism present. In general, there is no way of deciding a priori which of the components of an ecosystem is likely to be the most sensitive, and it may be sufficient to show the potential for an effect, by using a biomonitor, rather than an actual effect on one of the components of the ecosystem. This begs the question as to what is an appropriate biomonitor plant to use, and whether it is surprising if a species able to respond to additional N, from whatever source, gives any indication of the likelihood of harmful effects to the natural ecosystem. However, the ability to exploit additional N is not confined to bio-monitor species, and differential utilisation of additional N may well lead to changes in competition within communities. The data in Table 2.3 clearly show that NH_3 can influence the N content and growth of biomonitors, even at very low concentrations, and over periods as short as a month, with implications for other species.

2.10   Conclusions and Recommendations

The annual CLE of 8 µg NH_3 m^{-3} set in 1992 is of little practical use because it was not defined in terms of sensitive ecosystems, and because it is not as precautionary as empirical Critical Loads for most of the semi-natural habitat types of Europe. Clear evidence has emerged, especially from field studies in the UK, but also from several other EU countries, of effects of NH_3 on vegetation at concentrations well below the current annual CLE. Based on that evidence, a long-term CLE of 1 µg NH_3 m^{-3} can be assumed to be protective for biodiversity of most sensitive ecosystems. A long-term CLE of 3 µg NH_3 m^{-3} is probably protective for biodiversity of systems if bryophytes and lichens are not included.

Despite the progress made in recent years, several uncertainties still remain. We need to develop better methods that allow conversion of CLEs and CLO into each other. In particular, uncertainty exists as to the appropriate deposition velocities (linking CLEs and CLOs) for climatic zones outside the western maritime conditions of Europe, especially for colder and drier climates. Little is known of the quantitative interaction with cold and drought stress, and with other pollutants, particularly at low concentrations of NH_3. Several major gaps in knowledge in this respect are the deposition behaviour and compensation points of NH_3, and distinction within the CLOs for different N compounds.

We recommend:

1. That the measurement height for NH_3 measurements be standardised because of the pronounced vertical gradients in NH_3 concentrations close to vegetation surfaces.

2. Standardisation of lichen biodiversity assessment methods and of classification of acidic/nitrogen preferences of individual species.
3. The use of biomonitors to evaluate CLEs for NH_3 should be investigated.

The current and proposed CLEs are based on empirical evidence for responses of the sensitive species within a community. Directly applicable quantitative information on causal relations and on inter-species variation in sensitivity is still scarce. Deriving methods for linking physiological and biochemical measurements on plants to observed shifts in species composition would greatly assist in detailed analysis of existing information. Ecosystem models that simulate N cycles and biomass production could be useful, but are still in a preliminary state of development for the purpose of assessing CLEs.

References

Aldenberg T., Slob W. (1993) Confidence limits for hazardous concentrations based on logistically distributed NOEC toxicity data, Ecotoxicology and Environmental Safety 25, 48–63.

Ashmore M.R., Wilson R.B. (1994) Critical Levels of Air Pollutants for Europe. Department of the Environment, London, 209.

Bragazza L., Tahvanainen T., Kutnar L., Rydin H., Limpens J., Hajek M., Grosvernier P., Hajek T., Hajkova P., Hansen I., Iacumin P., Gerdol R. (2004) Nutritional constraints in ombrotrophic Sphagnum plants under increasing atmospheric nitrogen deposition in Europe, New Phytol, 163(3), 609–616.

Cape J.N., Sheppard L.J., Binnie J., Arkle P., Woods C. (1995) Throughfall deposition of ammonium and sulphate during ammonia fumigation of a Scots pine forest, Water Air and Soil Pollution, 85(4), 2247–2252.

Fangmeier A., Hadwigerfangmeier A., Vandereerden L., Jager H.J. (1994) Effects of atmospheric ammonia on vegetation - a review, Environmental Pollution, 86(1), 43–82.

Frati L., Santoni S., Nicolardi V., Gaggi C., Brunialti G., Guttova A., Gaudino S., Pati A., Pirintsos S.A., Loppi S. (2007) Lichen biomonitoring of ammonia emission and nitrogen deposition around a pig stockfarm, Environmental Pollution, 146(2), 311–316.

Kooijman S.A. (1987) A safety factor for LC_{50} values allowing for differences in sensitivity among species, Water Research 22, 269–276.

Krupa S.V. (2003) Effects of atmospheric ammonia (NH_3) on terrestrial vegetation: a review, Environmental Pollution, 124(2), 179–221.

Kupcinskiene E. (2001) Annual variations of needle surface characteristics of Pinus sylvestris growing near the emission source, Water Air and Soil Pollution, 130(1–4), 923–928.

Lamers L.P.M., Bobbink R., Roelofs J.G.M. (2000) Natural nitrogen filter fails in polluted raised bogs, Global Change Biology 6, 583–586.

Leith I.D., Sheppard L.J., Pitcairn C.E.R., Cape J.N., Hill P.W., Kennedy V.H., Tang Y.S., Smith R.I., Fowler D. (2001) Comparison of the effects of wet N deposition (NH_4Cl) and dry N deposition (NH_3) on UK moorland species, Water Air and Soil Pollution, 130, 1043–1048.

Leith I.D., Sheppard L.J., Fowler D., Cape J.N., Jones M., Crossley A., Hargreaves K.J., Tang Y.S., Theobald M., Sutton M.A. (2004a) Quantifying dry NH_3 deposition to an ombrotrophic bog from an automated NH_3 field release system, Water Air and Soil Pollution: Focus, 4(6), 207–218.

Leith I.D., van Dijk N., Pitcairn C.E.R., Sheppard L.J., Sutton M.A. (2004b) Bioindicator methods for nitrogen based on transplantation: standardised model plants, in Bioindicator and biomonitoring methods for assessing the effects of atmospheric nitorgen on statutory nature conservation sites, edited by M.A. Sutton, C.E.R. Pitcairn, and C.P. Whitfield, 100–104, JNCC Report 356.

Leith I.D., van Dijk N., Pitcairn C.E.R., Wolseley P.A., Whitfield C.P., Sutton M.A. (2005) Biomonitoring methods for assessing the impacts of nitrogen pollution: refinement and testing, in JNCC Report No. 386, 290, JNCC, Peterborough.

Mitchell R.J., Truscot A.M., Leith I.D., Cape J.N., Van Dijk N., Tang Y.S., Fowler D., Sutton M.A. (2005) A study of the epiphytic communities of Atlantic oak woods along an atmospheric nitrogen deposition gradient, Journal of Ecology, 93(3), 482–492.

Pearson J., Stewart G.R. (1993) The deposition of atmospheric ammonia and its effects on plants, New Phytologist, 125(2), 283–305.

Pitcairn C.E.R., Skiba U.M., Sutton M.A., Fowler D., Munro R., Kennedy V. (2002) Defining the spatial impacts of poultry farm ammonia emissions on species composition of adjacent woodland ground flora using Ellenberg Nitrogen Index, nitrous oxide and nitric oxide emissions and foliar nitrogen as marker variables, Environmental Pollution, 119(1), 9–21.

Pitcairn C.E.R., Fowler D., Leith I.D., Sheppard L.J., Sutton M.A., Kennedy V., Okello E. (2003) Bioindicators of enhanced nitrogen deposition, Environmental Pollution, 126(3), 353–361.

Pitcairn C.E.R., Leith I.D., Sheppard L.J., Sutton M.A. (2006) Development of a nitrophobe/ nitrophile classification for woodlands, grasslands and upland vegetation in Scotland, 21, Centre for Ecology and Hydrology, Penicuik.

Pleijel H. (1998) A suggestion of a simple transfer function for the use of ozone monitoring data in dose-response relationships obtained using open-top chambers, Water Air and Soil Pollution, 102(1–2), 61–74.

Posthumus A.C. (1988) Critical levels for effects of ammonia and ammonium. In Proceedings of the Bad Harzburg Workshop, 117–127, UBA, Berlin.

Shaw P.J.A., McLeod A.R. (1995) The effects of SO_2 and O_3 on the foliar nutrition of Scots pine, Norway spruce and Sitka spruce in the Liphook open-air fumigation experiment, Plant Cell and Environment, 18(3), 237–245.

Sheppard L.J., Leith I.D. (2002) Effects of NH_3 fumigation on the frost hardiness of Calluna - does N deposition increase winter damage by frost?, Phyton-Annales Rei Botanicae, 42(3), 183–190.

Sutton M.A., Fowler D., Burkhardt J.K., Milford C. (1995) Vegetation atmosphere exchange of ammonia: canopy cycling and the impacts of elevated nitrogen inputs, Water Air and Soil Pollution, 85(4), 2057–2063.

Sutton M.A., Perthue E., Fowler D., Storeton-West R.L., Cape J.N., Arends B.G., Möls J.J. (1997) Vertical distribution and fluxes of ammonia at Great Dun Fell, Atmospheric Environment, 31(16), 2615–2624.

Thijse G., Baas P. (1990) 'Natural' and NH_3-induced variation in epicuticular needle wax morphology of *Pseudotsuga menziesii* (Mirb.) Franco, Trees-Structure and Function, 4, 111–119.

UNECE (1995) Mapping and modelling of critical loads for nitrogen - a workshop report Report of the UN-ECE workshop, Grange-over-Sands, 24–26 October 1994, edited by M. Hornung, M.A. Sutton, and R.B. Wilson, 207, Institute of Terrestrial Ecology, Edinburgh.

UNECE (2003) Empirical critical loads for nitrogen, edited by Achermann B. and Bobbink R., 327, SAEFL, Berne.

van der Eerden L.J., Dueck T.A., Berdowski J.J.M., Greven H., van Dobben H.F. (1991) Influence of NH_3 and $(NH_4)_2SO_4$ on heathland vegetation, Acta Botanica Neerlandica, 40 (4), 281–296.

van der Eerden L.J., Dueck T.A., Posthumus A.C., Tonneijck A.E.G. (1994) Critical levels of air pollutants for Europe, edited by M.R. Ashmore, and R.B. Wilson, 55–63, Department of the Environment, Air Quality Division, London,.

van der Eerden L.J.M., de Visser P.H.B., van Dijk C.J. (1998) Risk of damage to crops in the direct neighbourhood of ammonia sources, Environmental Pollution, 102 (1, Supplement 1), 49–53.

WHO (1997) Effects of atmospheric nitrogen compounds (particularly nitrogen oxides) on plants, in Nitrogen Oxides (second edition), Environmental Health Criteria 188, 115–192, World Health Organisation, Geneva.

Wolseley P.A., James P.W., Theobald M.R., Sutton M.A. (2006) Detecting changes in epiphytic lichen communities at sites affected by atmospheric ammonia from agricultural sources, Lichenologist, 38, 161–176.

Chapter 3
Potential for the Further Development and Application of Critical Levels to Assess the Environmental Impacts of Ammonia

Mark A. Sutton, Lucy J. Sheppard, and David Fowler

3.1 Introduction

This paper provides the background that originally argued the case to review the ammonia critical level. These arguments were originally developed in 2003, although the case for reviewing the ammonia critical level was already clearly shown by Burkhardt et al. (1998). They demonstrated that the UNECE ammonia critical level established at Egham in 1992 was much less precautionary than parallel values of critical loads for nitrogen deposition. A differing 'protection timescale' for critical loads and critical levels could be seen partly to explain this. Simple extrapolation of the critical level relationship with averaging time showed that a long-term ammonia critical level (20–30 years) would be of the order of 2.5 μg m^{-3}, a factor of 3 smaller than the Egham annual critical level. Similarly, new evidence from studies of epiphytic lichen communities was emerging, suggesting that the long-term critical level for ammonia could be in the range 0.6–3 μg m^{-3}.

In addition to these arguments, there was a practical case emerging, which highlighted the need to set realistic values of the critical level. The dependence on critical loads requires estimates of nitrogen deposition, which are typically uncertain and difficult to measure in a local or regulatory context. By comparison, it is much easier to measure ammonia concentrations. Therefore, the ammonia critical level has an important role to play as a practical indicator threshold, for example, that can be used to assess local air pollution impacts (e.g. "Natura 2000") and to inform the setting of air quality targets for effects of ammonia on ecosystems.

It was concluded that there was an urgent need to review and update the ammonia critical level. Such a review should ensure that any new critical levels would be kept as simple as possible, as the need is for an operational tool that complements the more detailed critical loads approach. Building on this pre-assessment, the critical level was reviewed in detail at the UNECE Edinburgh workshop. The options to establish air quality targets for ammonia concentrations remain a topic of discussion for future policy development.

Mark A. Sutton,
Centre for Ecology & Hydrology, Bush Estate, Penicuik, Midlothian,
EH26 0QB, United Kingdom

M. Sutton, S. Reis and S.M.H. Baker (eds), *Atmospheric Ammonia,*
© Springer Science + Business Media B.V. 2009

3.2   Definitions and Background

The critical level for ammonia set in 1992 by the UNECE is $8\,\mu g\ m^{-3}$ for an annual mean, $23\,\mu g\ m^{-3}$ for a monthly mean, $270\,\mu g\ m^{-3}$ for a daily mean and $3,300\,\mu g\ m^{-3}$ for an hourly mean (Ashmore and Wilson 1994). This range of values as for other pollutant gases reflects the fact that at increasing NH_3 concentrations environmental effects may be seen over shorter exposure periods (WHO 2000).

Much of the ecological response to ammonia occurs through it contributing additional nitrogen or potential acidity to ecosystems. As a result, the impacts of ammonia are also assessed through the use of critical loads for nitrogen or acidifying deposition. A critical load is the total deposition of an air pollutant (e.g. nitrogen or acidity) below which environmental effects do not occur according to current knowledge.

Nitrogen deposition consists of both oxidized and reduced nitrogen components. Reduced nitrogen (NH_x) includes both ammonia (NH_3) and ammonium (NH_4^+), which are deposited through wet and dry deposition. Oxidized nitrogen (NO_y) includes nitrogen oxides ($NO_x = NO + NO_2$), nitric acid (HNO_3) and nitrate (NO_3^-).

Analysis of the components of nitrogen deposition and comparison with empirical critical loads for terrestrial ecosystems shows that exceedance of critical loads occurs when NH_3 concentrations are much smaller than the NH_3 critical level (Burkhardt et al. 1998). For example, applying a typical deposition velocity (rate of uptake by the ground) for semi-natural vegetation of $15\,mm\ s^{-1}$ and an annual average NH_3 concentration of $2.5\,\mu g\ m^{-3}$ would contribute around $12\,kg\ ha^{-1}\ year^{-1}$ nitrogen deposition. Critical loads for many UK ecosystems are currently estimated to lie in the range 10–$15\,kg\ N\ ha^{-1}\ year^{-1}$ (Achermann and Bobbink 2003). Hence, even if the additional contributions to N deposition from NH_4^+ and NO_y are not counted, the critical load can easily be exceeded by NH_3 concentrations much smaller than the annual NH_3 critical level (Burkhardt et al. 1998).

Given that NH_3 makes a major contribution to critical load exceedance, and that this exceedance occurs at NH_3 concentrations less than the critical level, much more attention has been focused on the development of N critical loads rather than the NH_3 critical level. Using the current UNECE values noted above, the NH_3 critical levels are generally only exceeded in the UK in the immediate vicinity of large livestock farms. This applies both to the annual critical level and the short-term critical levels (e.g. hourly, daily, monthly).

On the basis of the above, it might be thought that there was little potential in pursuing further development of the critical levels approach for ammonia. By contrast there are several reasons why the critical levels approach has significant potential to help the development and implementation of air quality policies.

3.3   Effects of Ammonia on Sensitive Plant Communities

Exceedance of a critical level is often taken to represent a direct toxic effect on a specified biological component. This contrasts with exceedance of a critical load, which usually represents an indirect response, where there may be no direct effects. For example, in adding nitrogen to an ecosystem from the atmosphere, the additional nitrogen may simply modify the competitive ability of different plant groups, leading to a change in species composition. From this perspective, it might be concluded that the critical load remains the correct way to treat the N and acidity effects of low levels of ammonia deposition.

By contrast, a critical level may be seen as simply a measure of the impact of elevated concentrations, as contrasted to a critical load being a measure of the impact of elevated deposition. In this case, it could be argued that expressing indirect environmental effects directly in response to NH_3 concentrations would also be appropriate. In the case of nitrogen deposition, a criticism is that the NH_3 concentration at which indirect effects occur depends on the magnitude of the other components of nitrogen deposition. Hence applying a critical load of 15 kg N ha^{-1} year^{-1} an implied critical level for indirect effects of NH_3 could be calculated, but would depend on local conditions.

Using the estimated NH_3 dry deposition of 12 kg N ha^{-1} year^{-1} from above: with a background $NH_4^+ + NO_y$ deposition of 3 kg N ha^{-1} year^{-1}, the NH_3 critical level (indirect effects) would be around 2.5 μg m^{-3}; with a $NH_4^+ + NO_y$ background deposition of 9 kg N ha^{-1} year^{-1}, the NH_3 critical level would be around 1.25 μg m^{-3}.

In assessing the direct effects of NH_3, however, it may be noted that the UNECE critical level has stood for nearly a decade without reassessment. Recent data suggest that direct effects of NH_3 may occur at smaller concentrations. While the UNECE critical level was based on a toxicological assessment for bryophytes (mosses and liverworts), the recent experience suggests direct effects of NH_3 on lichens at small atmospheric concentrations. An extensive analysis of tree living lichens for the Netherlands suggests that lichens may be classified into species favouring N-poor, naturally acidic bark (Acidophytes) and those species favouring N-rich, more basic bark (Nitrophytes). The occurrence of nitrophytes and the disappearance of the acidophyte species has been shown to be directly proportional to atmospheric NH_3 concentrations (van Herk 1999). The effects appear to relate to direct toxicity in addition to interspecies competition. In particular, NH_3 (being a base) is shown to reduce the acidity of tree bark, leading to the species changes. Among the different deposited nitrogen species, this effect is unique to ammonia.

The effects of ammonia concentrations on nitrophyte lichens in the Netherlands have been detected in the range 5–35 μg m^{-3}, with most of the data being above the currently agreed critical level. Recent data from the UK showing the increase in nitrophyte species are consistent with changes above 8 μg m^{-3}. By contrast, the UK data suggest that the most sensitive acidophyte lichen species are lost at concentrations in the range 0.6–3 μg m^{-3} (Wolseley et al. 2004, 2006; Sutton et al. 2004).

These data indicate that the current critical level for NH_3 is set much too large to allow protection of the most sensitive lichen species. Further data are clearly needed to refine the critical level for ammonia, which could even be as small as $1\,\mu g\ m^{-3}$.

3.4 Timescales of Critical Load and Critical Level Responses

Although both critical loads and critical levels may be expressed with annual values, it is important to consider that the timescales of each are not strictly comparable. The annual critical level refers specifically to a single yearly value, which should not be exceeded. By contrast, the empirical critical loads for nitrogen refer to long term deposition expressed as kg N ha^{-1} year^{-1}. Indicatively, this long term period is considered to represent a period of 20–30 years in the definition of empirical critical loads (Hornung et al. 1995). Because both the critical level and the critical load are expressed on a "yearly" basis, confusion is easy.

The relationship between the UNECE critical levels and the time constant is nearly linear on a scale of log_{10}(critical level) vs log_{10}(time). If the curve is fitted with a polynomial

$$(log_{10}(\text{Critical Level}) = 0.0623\ [log_{10}(\text{time})]2 - 0.9184\ [log_{10}(\text{time})] + 3.5341)$$

then the critical level for a period of 20–30 years would be 2.7–2.4 $\mu g\ m^{-3}$, respectively. Accounting for uncertainty in the extrapolation, the long-term critical level based on the same dataset would be around 2.5 $\mu g\ m^{-3}$ ($\pm 0.5\,\mu g\ m^{-3}$).

It is relevant to consider the issue of timescale in relation to monitoring the exceedance of critical limits. Although it is considered to take an indicative 20–30 years for the effects of critical load exceedance to become apparent, the values reflect an appropriate limit for the assessment of annual or 3-year estimates of deposition and exceedance. In the same way, although exceedance of the 1-year critical level can be tested annually, the long-term (20–30 year) critical level reflects an appropriate level for assessment of future pollution abatement policies.

3.5 Uncertainties in Applying the Critical Loads Approach for NH_3

While the critical loads approach for nitrogen necessarily needs to include NH_3, there are also a number of uncertainties that need to be considered when making the comparison with critical levels for NH_3.

The first and most important uncertainty in the critical loads approach for nitrogen is that different forms of nitrogen may not have the same level of environmental impact per kg of N deposited. Recent evidence from CEH suggests that there are substantial differences. For a given amount of N input (expressed as kg N ha^{-1} year^{-1}) there is mounting evidence that dry deposition has larger impacts on sensitive plants that wet deposition and that NH_x has larger impacts than NO_y (Leith et al. 2002). Although further data are required, the evidence is sufficient to doubt the basic assumption of the critical loads approach, that all forms of N have the same magnitude of impact.

An important additional uncertainty in applying the critical loads approach is the requirement for accurate estimates of atmospheric deposition. Firstly, the experimental and survey studies used to establish empirical critical loads estimates are highly sensitive to varying quality in the atmospheric deposition estimates (Sutton et al. 2003). Secondly, the quality of mapped critical loads exceedance estimates depends centrally on the accuracy of the deposition maps. While substantial effort has been placed to develop robust deposition models for mapping across the UK, at a landscape level (sub 5 km) there are large uncertainties due to variability in NH_3 concentrations. Quantifying deposition rates to complex elements of the landscape (e.g. park trees, hedgerows) is also very uncertain (Loubet et al. 2001; Milford et al. 2001).

3.6   Complementary Benefits of the Critical Levels Approach for NH$_3$

The critical loads approach for nitrogen is well established, and despite the uncertainties it provides an important tool to combine the risk analysis of environmental impacts for different forms of nitrogen and acidifying pollutants.

Part of any response to the uncertainties in the critical loads approach must be further refinement of the methods: effort is needed in quantifying landscape level variability of N deposition, as well as rates of deposition to specific landscape elements. At the same time, information needs to be collected that quantifies the relative dose response relationships of different forms of nitrogen.

Critical levels for NH_3 have received little attention recently, due to the high limits that have been set by the UNECE. Conventional wisdom, based on the existing limits, is that the main issue for NH_3 is exceedance of critical loads for nitrogen.

In contrast to conventional wisdom, several strands of new information combine to suggest that there is merit in revisiting the critical levels approach for NH_3:

(a) Indirect effects of NH_3 to specific vegetation receptors can be expressed in terms of the NH_3 concentration. Based on typical critical loads values, a critical level (expressed as long-term average concentration) for indirect effects would be of the order 1.25–2.5 µg m^{-3}.

(b) There is growing evidence of direct effects of NH_3 on sensitive lichen species, which are related to changes in bark pH independent of N availability (e.g. at 0.6–3 µg m^{-3}).

(c) If the current NH$_3$ critical level is expressed on the same timescale as empirical critical loads, the critical level for NH$_3$ becomes much smaller. Based on the UNECE data, a long-term average (20–30 year) NH$_3$ critical level would be around 2.5 (±0.5) μg m^{-3}.

(d) The increased sensitivity of vegetation to NH$_3$ compared with other forms of N deposition suggests that further attention be given to quantifying the specific impacts of NH$_3$.

(e) With appropriate measurement techniques, concentrations of NH$_3$ can be monitored with much greater accuracy than can rates of N deposition. This means that monitoring of the exceedance of a critical level for NH$_3$ becomes operationally a much easier target than monitoring atmospheric deposition inputs on a site basis. This is especially the case when N deposition inputs are uncertain in the case of proximity to local NH$_3$ sources or when assessing deposition to complex landscape elements (e.g. single trees or hedgerows).

3.7   Potential Operational Application of a Revised Ammonia Critical Level

Further work is necessary to provide the basis to formally revise the UNECE critical level for NH$_3$. However, based on the information summarized here, it is expected the critical level would be around 1–3 μg m^{-3} (long term site average concentration).

The critical level might be expected to differ broadly between major vegetation types, reflecting differences in deposition rates. It is, however, emphasized that the prime purpose of the critical level for NH$_3$ should be as a complementary tool to the critical loads approach. Given that the detail linked to deposition differences is best treated in the critical loads approach, the critical levels approach for ammonia should be seen in contrast as a simple operational tool.

The benefits of the critical level approach for NH$_3$ are particularly expected for site-level environmental impact assessment and for setting of air quality targets.

In estimating the potential impact of new development on sensitive ecosystems, it needs to be shown that the development would not lead to exceedance of either critical loads or the NH$_3$ critical level. Use of the NH$_3$ critical level (with a realistic value) would provide an additional tool to monitor compliance to emissions/ air quality standards. This would be particularly appropriate for sources where the prime emission is of ammonia (e.g. livestock farms).

The critical level for NH$_3$ is also well suited to inclusion in the revision of air quality policies. For example, a standard could be set that the critical level NH$_3$ concentration is not exceeded within the boundaries of relevant sensitive *Sites of Special Scientific Interest* (SSSI's) and *Special Areas of Conservation* (SAC's).

The key advantage of the critical level in each of these respects is that it is much easier to monitor NH$_3$ concentrations than deposition. This would encourage rapid assessment of the impacts of ammonia, which would complement the approach based on estimating deposition and critical loads.

Current NH_3 monitoring in the UK is conducted with a monthly time frequency. This makes it possible to assess easily each of the monthly, yearly and long-term critical levels. By contrast, daily and hourly monitoring of NH_3 concentrations is best done by continuous sampling, which because of costs, is prohibited to one or two sampling sites in the UK. Available hourly NH_3 monitoring data in the UK monitoring suggest that it is more likely for the monthly or annual critical level to be exceeded than the hourly or daily critical level. This further justifies the focus on monthly, yearly and long-term means.

3.8   Conclusions and Recommendations

Based on this preliminary review, is concluded that:

(a) The critical level for NH_3 set in 1992 is too high. A more realistic value (expressed as long-term (20–30 year) average NH_3 concentration) is probably $1-3 \, \mu g \, m^{-3}$.

(b) Per unit of nitrogen deposited, NH_3 is thought to be more damaging to flora than other forms of nitrogen.

(c) Lichens are particularly sensitive to NH_3, and respond not just to increased nitrogen availability, but to the reduction of bark acidity caused by NH_3.

(d) The critical level approach for NH_3 complements the critical loads approach, and has the particular advantage that it allows simple operational assessment through monitoring of NH_3 concentrations.

It is recommended that:

(e) Further effort be placed in quantifying the dose response relationships for NH_3 compared with other forms of nitrogen and that these data also be interpreted in relation to the critical levels approach.

(f) Any revised definition of the critical level for NH_3 be kept as simple as possible, as the need is for an operational tool, which complements the more detailed critical loads approach.

(g) Efforts are given to formally re-evaluate the UNECE critical levels set at Egham in 1992 for NH_3.

(h) Consideration should be given to apply the NH_3 critical level in assessing the protection of statutory nature reserves and in further developing air quality policies.

The conclusions noted above (a–d) and the recommendations (e–g) were addressed at the UNECE Edinburgh Workshop, as summarized by Cape et al. (2009, this volume) and in Chapter 28 of this volume. The formal application of recommendation (h) was beyond the scope of the Edinburgh Workshop, and should be reviewed as part of future policy development.

References

Achermann B. and Bobbink R. (eds.) (2003) Empirical Critical Loads for Nitrogen (UNECE Expert Workshop, Berne 11–13 November 2002) pp 327. SAEFL, Berne, Switzerland.

Ashmore M.R. and Wilson R.B. (eds.) (1994) Critical Levels of Air Pollutants for Europe. (Report of the UNECE Expert Workshop at Egham, UK 23–26 March 1992). Department of the Environment, London.

Burkhardt J., Sutton M.A., Milford C., Storeton-West R.L. and Fowler D. (1998) Ammonia concentrations at a site in S. Scotland from continuous measurements over 2 years. *Atmospheric Environment* 32(3), 325–331. (Ammonia Special Issue)

Hornung M., Sutton M.A. and Wilson R.B. (1995) Mapping and modelling of critical loads for nitrogen - a workshop report (Eds.) (Report of the UN-ECE workshop, Grange-over-Sands, 24–26 October 1994). Institute of Terrestrial Ecology, Edinburgh, 207 (ISBN 1 870393 24 4).

Leith, I.D., Pitcairn C.E.R., Sheppard L.J., Hill P.W., Cape J.N., Fowler D., Tang Y.S., Smith R.I. and Parrington J.A. (2002). A comparison of impacts of N deposition applied as NH_3 or as NH_4Cl on ombrotrophic mire vegetation. *Phyton-Annales Rei Botanicae* 42(3), 83–88.

Loubet B., Milford C., Sutton M.A. and Cellier P. (2001) Investigation of the interaction between sources and sinks of atmospheric ammonia in an upland landscape using a simplified dispersion-exchange model. *Journal of Geophysical Research (Atmospheres)* 106, 24, 183–224, 195.

Milford C., Hargreaves K.J., Sutton M.A. Loubet B. and Cellier P. (2001) Fluxes of NH3 and CO2 over upland moorland in the vicinity of agricultural land. *Journal of Geophysical Research (Atmospheres)* 106, 24, 169–224, 181.

Sutton M.A., Cape J.N., Rihm B., Sheppard L.J., Smith R.I., Spranger T. and Fowler D. (2003) The importance of accurate background atmospheric deposition estimates in setting critical loads for nitrogen. In: B. Achermann and R. Bobbink (eds.) Empirical Critical Loads for Nitrogen (UNECE Expert Workshop, Berne 11–13 November 2002) pp 231–257. SAEFL, Berne, Switzerland.

Sutton M.A., Leith I.D., Pitcairn C.E.R., van Dijk N., Tang Y.S., Sheppard L.J., Dragosits U., Fowler D., James P.W. and Wolseley P.A. (2004) Exposure of ecosystems to atmospheric ammonia in the UK and the development of practical bioindicator methods. In: Wolseley P.A. and Lambley P.W. (eds.) Lichens in a Changing Pollution Environment, English Nature Workshop, pp 51–62. English Nature Research Reports, No 525 (ISSN 0967-876X).

Wolseley P.A., James P.W., Sutton M.A. and Theobold M.R. (2004) Using lichen communities to assess changes in sites of known ammonia concentrations. In: Wolseley P.A. and Lambley P.W. (eds.) Lichens in a Changing Pollution Environment, English Nature Workshop, pp 89–98. English Nature Research Reports, No 525 (ISSN 0967-876X).

Wolseley P.A., James P. W., Theobald M. R. and Sutton M.A. (2006) Detecting changes in epiphytic lichen communities at sites affected by atmospheric ammonia from agricultural sources. *The Lichenologist* 38(2), 161–176.

van Herk C.M. (1999) Mapping of ammonia pollution with epiphytic lichens in the Netherlands. *Lichenologist* 31, 9–20.

WHO (2000) Air Quality Guidelines. (http://www.euro.who.int/document/e71922.pdf).

Chapter 4
Long-Term Cumulative Exposure Exacerbates the Effects of Atmospheric Ammonia on an Ombrotrophic Bog: Implications for Critical Levels

Lucy J. Sheppard, Ian D. Leith, Alan Crossley, Netty van Dijk, J. Neil Cape, David Fowler, and Mark A. Sutton

4.1 Abstract

A line source of ammonia, simulating NH_3 emissions from an intensive live-stock unit, was established in 2002 on an ombrotrophic bog, Whim bog in the Scottish Borders. The site is at 55° 46'N, 3° 16'W and (based on 4.5 years) has a mean monthly temperature of 8.8°C and mean annual rainfall 971 mm, supporting a *Calluna Eriophorum* vegetation, NVC M19. Release of NH_3 is controlled by meteorological conditions *i.e.* wind direction and speed. NH_3 concentrations were measured at >10 distances along the transect of release using passive ALPHA samplers, fixed at two or more heights above the vegetation. Effects were recorded for the ericoids, mosses, *Sphagnum* and *Cladonia* lichens with respect to cover changes, visible damage, nutrients and interactions with abiotic stress. These data, collected over >4 years, have been used to evaluate the Critical Level (CLE) for NH_3 and compared with values derived from short–term (weeks) exposures at high concentrations in controlled conditions. Results suggest the annual Critical Level of $8 \, \mu g \, m^{-3}$ is too high to protect sensitive species experiencing long-term exposures to NH_3. In particular, they show exposure periods longer than 1 year lead to detrimental effects at lower mean NH_3 concentrations through relationships between CLE and exposure period that are still linear after 4.5 years exposure. These results highlight the need to set a CLE representing the mean concentration of NH_3 in air for long term protection of sensitive habitats (e.g. over 20–30 years), that will be substantially smaller than the current annual CLE.

Lucy J. Sheppard
Centre for Ecology & Hydrology, Bush Estate, Penicuik, Midlothian,
EH26 0QB, United Kingdom

4.2   Introduction

Since the potentially damaging effects of enhanced N deposition were recognised in the early 1980s, there has emerged the need to protect vulnerable ecosystems and provide quantifiable tools on which to base legislation. Critical Loads, based on the annual N dose to a unit area (kg N ha^{-1}y^{-1}) an ecosystem could sustain for at least 20–30 years before incurring adverse effects, were set for different ecosystems and periodically updated to take account of new evidence and understanding (e.g. Hornung et al. 1995; UNECE 2003). However, these took no account of the form of the N deposition i.e. reduced versus oxidised, nor did they distinguish between wet and dry deposition. Critical Levels for ammonia targeted effects thresholds for short-term exposures of <1 year, and focus on the protection of relatively sensitive plants at the species level (World Health Organization 1997). Initially, the concept of a gaseous Critical Level (CLE) came from the risk of chemical accidents when the receptor would be exposed to extremely high concentrations for a relatively short time. In reality in the field, the exposure scenario for most plants will be intermittent and probably the exposure concentrations will be less extreme, the exposure frequency higher and the duration for most perennial species will be the lifetime of the species, not withstanding the lifetime of the source.

Most of the experiments that generated the data on which Critical Levels for ammonia (CLE$_{NH3}$) were set were short-term fumigations < 6 months, involving high, uniform concentrations > 25 μg m^{-3}, that excluded the potential for bidirectional exchange processes involving the compensation point, in controlled, relatively well watered warm conditions. Plants studied ranged from N demanding to those that had evolved under conditions of low N availability (Van der Eerden 1982; Van der Eerden et al. 1991). In the field, the main sources of atmospheric NH$_3$ pollution are agricultural enterprises associated with intensive livestock rearing, manure stores and land spreading of manures. Ammonia emissions, leading to dry deposition will affect vegetation in a chronic manner via an elevated background throughout the life cycle of the species and via intermittent acute concentration exposure periods when the wind direction is from the source. The lifetime of most species, other than commercially grown crop plants is generally a minimum of 1 year. The consequences of intermittent, long-term exposure to fluctuating concentrations have not, so far, been addressed in the setting of the CLE$_{NH3}$.

In 2002, CEH Edinburgh established a field release experiment on an ombrotrophic bog to simulate atmospheric NH$_3$ exposures downwind of an intensive livestock unit (Leith et al. 2004). For 4.5 years, effects on a range of receptors, both above and below ground have been quantified. In this paper we report a selection of these responses and focus on their implications for reevaluating the CLE$_{NH3}$. The provision of different reference parameters and plant species has enabled us to examine the robustness of CLE setting (see Cape et al. 2009, see Section 2.2, this volume).

4.3 Methods

4.3.1 Ammonia Fumigation

A detailed description of the site is given in Sheppard et al. (2004b) and the field release system in Leith et al. (2004). Ammonia is released from a line source, 10 m pipe in two 5 m sections each drilled with 100 small holes, fixed 1 m off the ground, across the prevailing wind direction. The NH_3 is piped from a liquid NH_3 cylinder, mixed with ambient air and dispersed in the air downwind of the pipe. Flow is controlled with a mass flow controller and the cylinder release valve from a *Campbell 23X* logger in response to the sonic anemometer registering the correct wind sector (180–215°) and wind speeds >2 m s⁻¹. Controlling the release from the logger ensured the timing and exposure duration were recorded, so that approximate exposure concentrations could be estimated. Passive alpha samplers (Tang et al. 2001) recorded monthly exposure concentrations 0.1 and 0.5 m above the vegetation at >10 distances along a 105 m transect. Additional samplers recorded ambient NH_3 at the UK NH_3 network standard monitoring height of 1.5 m and at 1 m, along the transect in order to calculate the vertical concentration gradient. As a precautionary measure a minimum wind speed > 2 m s⁻¹ was set for release. Meteorological variables measured include rainfall, wind speed and direction, solar and net radiation, air and soil (10, 20 cm) temperature and water table.

This field release system releases annually 120 kg NH_3 over 560 h, when the wind direction is appropriate (9.9% of the year). Ammonia release from broiler hens raised on litter is typically 0.05 kg NH_3 a⁻¹ per bird. Thus, the exposure at Whim bog is equivalent to being downwind of a NH_3 source comprising 23,760 broiler hens (120/0.05 × 9.9).

4.3.2 Vegetation and Assessments

The vegetation is unmanaged, dominated by degenerating *Calluna* and *Eriophorum vaginatum* (NVC M19) growing on an acid (pH 3.3–3.9 H_2O) deep peat with a low (rainfall dependent, 9–11 kg N ha⁻¹ y⁻¹) N deposition history. Permanent quadrats (0.3 by 0.3 m divided into 16 equal squares), 3 per distance were marked and individual species cover assessed annually in each square. Zero cover was recorded where species were absent, and the condition of the plant was also recorded e.g. for *Calluna* (stems, green, dead and brown shoots). Percent cover was averaged for the 16 squares, providing 3 cover estimates per distance. A weighted cover index, taking into account the original cover, was then calculated,

year 1 cover – start cover/((start + year 1)/2),

for each distance. In addition condition was assessed regularly for the whole transect area, recording any species that appeared damaged. Over the 4.5 years since exposure began in May 2002, foliar nutrients and stress resistance to cold and drought have also been recorded using standard protocols (Sheppard and Leith 2002; Sheppard et al. 2004b).

4.4   Results and Discussion

4.4.1   'Real World' Ammonia Exposure Scenarios

Equipment failure has been minimal, therefore the exposure pattern shown in Fig. 4.1. (at two measuring points, 8 and 60 m) provides a realistic indication of mean monthly ammonia concentration ([NH$_3$]), month by month, that would be consistent with [NH$_3$] concentrations downwind of an intensive livestock unit. The obvious message is that [NH$_3$] show considerable variation from month to month and that inter-annually the monthly patterns also differ. Differences in [NH$_3$] mostly reflected the number of potential release hours i.e. wind from appropriate direction, but some variation was also caused by wind speed, which affects mixing and deposition and temperature (Asman et al. 1998 and references therein). These variations in monthly NH$_3$ concentrations have important implications for N assimilation and potential effects. High winter concentrations are often more damaging because assimilation i.e. detoxification through the provision of C skeletons, is reduced at low temperatures (Van der Eerden 1982). Van der Eerden (1982) also suggested the importance of photon flux density is important, inferring plants were more sensitive in the dark. While we could investigate this for this experiment, the observation is at odds with the view that most of the NH$_3$ flux occurs via the stomata, which would be expected to

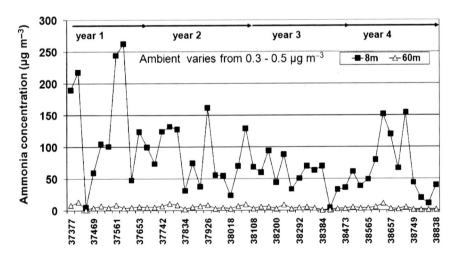

Fig. 4.1 Monthly NH$_3$ concentrations (μg m^{-3}), 8 and 60 m along the ammonia release transect, measured at 0.1 m above the vegetation, at Whim bog from May 2002 through to May 2006

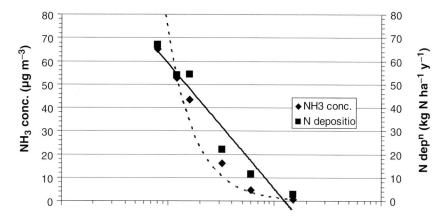

Fig. 4.2 Gradient of NH$_3$ concentrations (μg m^{-3}) and estimated NH$_3$-N deposition (kg N ha^{-1} year^{-1}) along the Whim bog ammonia release transect in 2004 R^2 (concentration) = 0.9571, y = 3401.7 × −1.6677; R^2 (deposition) = 0.9363, y = −23.385 Ln(x) + 112.84

close in the dark (Jones 2006). Concentration-dependent effects may therefore differ according to whether the plants experience the NH$_3$ at times when their detoxification capacity is impaired.

Ammonia concentrations declined exponentially from the NH$_3$ source (Fig. 4.2.). At this site, with this source strength, the NH$_3$ concentrations fall back to the ambient concentration (0.4 μg m^{-3}) measured at 0.1 m ~ 105 m downwind of the source pipe. Given that the conditions for fumigation are only fulfilled for <14% each month, cumulatively 6.7% of the year, the mean [NH$_3$] during exposure periods to which plants are exposed significantly exceeds the monthly concentrations described in Fig. 4.2. However, in the field where most [NH$_3$] monitoring is undertaken with passive samplers, we have to rely on the overall monthly mean values as a working surrogate that reflect these temporal fluctuations.

Figure 4.2 also shows how estimated NH$_3$-N deposition varies with distance for this mixed bog community. N deposition was estimated for the same year 2004 as the measured [NH$_3$] taking into account stomatal opening (solar radiation), [NH$_3$], and wind speed. A concentration-dependent deposition velocity was calculated from flux chamber measurements using a plant community sourced from the Whim bog field site (Jones 2006). Although, there remains uncertainty in these values, they highlight the importance of cuticular saturation effects at high concentrations.

4.4.2 Changes in the Vegetation in Response to Ammonia Concentrations

This site has been unmanaged for nearly a century and the *Calluna*, while predominantly degenerate, has also layered, so that the stand represents plants of wide ranging age and (it appears) sensitivity to [NH$_3$]. A reduction in the cover of

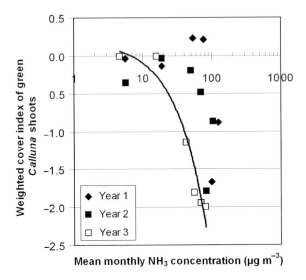

Fig. 4.3 Change in the weighted cover index for green *Calluna* shoots in response to the mean monthly NH₃ concentration exposure prior to assessment. (R² for year 3 = 0.9424, y = −0.0297x + 0.2008)

green *Calluna* shoots was observed each year (Fig. 4.3). Far fewer green shoots were observed after each winter rather, many shoots had a bleached grey/white appearance. The reduction in cover of green *Calluna* shoots was related to the [NH₃], with a greater reduction in green shoot cover at high NH₃ concentrations. This relationship improved as the length of exposure increased (Fig. 4.3) so that in year 3 the R^2 = 0.94. During the first 2 years of fumigation the threshold [NH₃] for loss of green shoots was difficult to define as the relationship was poor but by year 3 had fallen to ~7 μg m⁻³.

Over the whole transect we have identified, at regular intervals, how far along the transect bleaching of >85% of the *Calluna* canopy has occurred, then plotted the mean NH₃ concentration at this distance against the time taken for the damage to extend further from the source (Fig. 4.4). After 4 years of fumigation, this 'critical' [NH₃] still exceeds the annual CLE$_{NH3}$ of 8 μg m⁻³ proposed by Van der Eerden et al. (1991). However, the reduction of log [CLE$_{NH3}$] with log [time], which is clearly demonstrated, shows how effects build up over a period exceeding 1 year, identifying the need for a critical level for long term protection of the habitat. If the fitted line continues to fall at the same rate, the CLE $_{NH3}$ would have only protected the *Calluna* for 6–7 years.

The bleaching observed in *Calluna* is most likely due to an interaction between NH₃ and a secondary stress, possibly winter desiccation (Sheppard and Leith 2002). *Calluna* shoots, especially those on older plants, appear to be rather susceptible to winter desiccation (Watson et al. 1966), which these authors describe as winter browning, even in the absence of known NH₃ exposure. Bannister (1964) found that, under conditions of high photon flux and low relative humidity in winter, *Calluna* showed a much greater reduction in relative turgidity than *Erica*. The

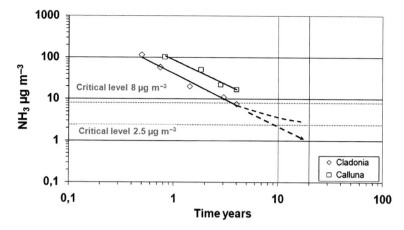

Fig. 4.4 Ammonia concentrations (μg NH_3 m^{-3}) measured at 0.1 m above the vegetation causing 100% death of *Cladonia portentosa* and 85% death of *Calluna vulgaris* in relation to duration of exposure (years). The relationships between the NH_3 concentration causing death each year have R^2 values of 0.98 and 0.97 for *Cladonia* and *Calluna* respectively (y = 40.923 $\times$ −1.2735; y = 86.401 $\times$ −1.1864). The dotted lines show possible extrapolation of the *Cladonia* measurements to a 10–20 year exposure

Table 4.1 Lethal temperature (°C) causing 50% shoot death (LT_{50}) in current year *Calluna* shoots subjected to a range of freezing temperatures in a purpose built cabinet and assessed from a threshold relative conductivity based on the leakage of ions after 24 h.

Preceding mean [NH_3] (μg m^{-3})	Jan 2003	Preceding mean [NH_3] (μg m^{-3})	Mar 2003	Preceding mean [NH_3] (μg m^{-3})	Feb 2004
147.6	−19.1	135.2	−17.8	110.4	−19.1
87.8	−18.5	81.6	−19.4	71.7	−18.5
59.2	−17.9	56.1	−22.0	51.4	−16.2
20.8	−18.2	19.95	−27.4	18.8	−15.6
5.96	−20.2	5.74	−25.2	5.63	−20.6
0.46	−20.8	0.45	−28.2	0.49	−21.8

bleached appearance of the shoots suggests that NH_3 has affected several aspects of the metabolism of *Calluna*, in addition to increasing susceptibility to desiccation.

Frost hardiness experiments were conducted in years' 1 and 2 (Table 4.1). These showed that exposure to NH_3 reduced shoot hardiness, with log [NH_3] explaining up to, a maximum of 70% of the variation in LT_{50}, depending on sampling time. However, the reductions in levels of hardiness caused by NH_3 exposure were not sufficient to explain the damage following the winter temperatures experienced by these plants in the field. Over those two winters minimum temperatures did not fall below −10°C. Detrimental effects of NH_3 exposure on the ability of *Calluna* shoots to conserve water and recover from drought have been reported (Van der Eerden et al. 1991). Drying experiments showed that *Calluna* shoots that had been exposed to NH_3 for two seasons, closed their stomata at a 20% lower water content than ambient shoots,

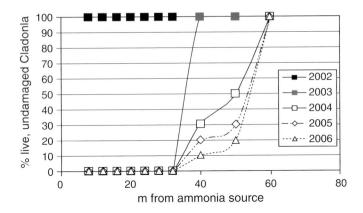

Fig. 4.5 Percentage of live, undamaged *Cladonia portentosa* along the ammonia release transect at Whim bog since 2002, pre NH_3 release, and during subsequent years (2003–2006)

although the relationship with $[NH_3]$ was relatively poor (Sheppard, LJ 2004 unpublished). Interestingly, *Erica tetralix*, *Vaccinium myrtillus*, and *Empetrum nigrum*, which naturally close their stomata at a higher water content than *Calluna*, appear to be thriving along the transect, unaffected by NH_3. *Calluna* therefore appears to be predisposed to NH_3 damage via its restricted ability to regulate its water status.

On this bog the matt forming lichen *Cladonia portentosa* was the species most adversely affected by the elevated $[NH_3]$. A range of apparently toxic effects were recorded, the first, within a few months. For example, pinking of the thallii, which was reversible, happened when a high photon flux and temperature occurred simultaneously with several hours of NH_3 release (Sheppard et al. 2004a). As observed for *Calluna*, damage was observed further and further from the source with critical concentrations decreasing linearly with time (Figs. 4.4 and 4.5). The annual CLE_{NH3} of $8\,\mu g\ m^{-3}$ provided only 3 years protection. Conversely, if the current relationship were to continue, it would imply that a CLE_{NH3} of $1–2\,\mu g\ m^{-3}$ would be needed to protect these lichens for 10–20 years NH_3 exposure (Fig. 4.4). Such a calculation is of course uncertain, but conversely it includes no safety factor (assessment factor). Up to now, the consequences of NH_3 exposure scenarios similar to those operating in the field have not been assessed experimentally. This experiment shows that, when plants (that were hitherto growing in a relatively N clean atmosphere) are exposed to NH_3 for >1 year, the annual CLE_{NH3} must be adjusted downwards to take account of the detrimental effects of longer-term exposures.

Sphagnum capillifolium and *Hypnum jutlandicum* also declined within 20 m of the source, and overall the effects worsened with time. However unlike *Calluna* and *Cladonia*, which have not recovered, complete death of these two quite different mosses was difficult to establish. It appears that *Hypnum*, at least, can produce new green shoots from brown, apparently dead, stems and is thus able to reestablish very quickly under moist conditions. At present, following a wet autumn/winter and in the absence of a summer drought *Hypnum* has now reappeared over the first 20 m. Likewise *Sphagnum* is looking less damaged in 2006 than it did in 2005 when the capitula on the green forms of *S. capillifolium* (possibly unprotected by the carotenoids present in the red forms) were easily dislodged. However, *Sphagnum* has not recovered within 16 m of the source where monthly mean [NH$_3$] are ~44 µg m^{-3}.

Effects on Nutrient concentrations: Apart from the obvious visible damage to *Calluna*, the nutrient concentrations in current year shoots showed significant effects of NH$_3$. In October 2006 concentrations of N (for apparently healthy green shoots) showed a significant exponential relationship with log [NH$_3$], R^2 = 0.78***, while the base cations decreased with log [NH$_3$], for Ca^{2+}, R^2 = 0.86*** and for Mg^{2+}, R^2 = 0.72**. Neither K nor P was affected by the [NH$_3$]. Effects of NH$_3$ on base cation composition were observed by Bobbink et al. (1992), who concluded that base cations were leached from the canopy in equivalent amounts to the NH$_4^+$ taken up. In *Hypnum*, only %N dry wt was significantly related to log [NH$_3$], R^2 = 0.90***, while cation concentrations were not related to log [NH$_3$].

4.5 Conclusions

- The ammonia field release, transect experiment reported here provides a range of NH$_3$ concentrations for comparative purposes, utilizing exposure scenarios that typify those found downwind of agricultural point sources. The experiment is unique in allowing the interactions with environmental factors to be assessed under field conditions and with an appropriate temporal distribution of NH$_3$ concentrations.
- The long-term nature of this experiment, (with, so far, >4 years fumigation) has shown for several sensitive species that the [NH$_3$] thresholds for damage decrease significantly over periods longer than 1 year.
- At present, critical levels for ammonia have been estimated mainly for short term exposures of 1 h, 1 day, 1 month and 1 year. The linear relationship between [NH$_3$] lethal and exposure time plotted on log scales shows that there is a need to establish a critical level that represents the mean NH$_3$ concentration for long-term protection of sensitive species. Agreement needs to be reached on a suitable "protection period", but an indicative period of 20–30 years would be consistent with the empirical critical loads approach (Hornung et al. 1995).
- The present results show that after 1 year exposure 70 µg m^{-3} leads to 85% death of *Calluna* while 25 µg m^{-3} leads to complete death of *Cladonia portentosa*. These values are larger than the current annual critical level, but conversely, it should be noted that the measures of effect are rather extreme (e.g. complete death). These extreme measures were chosen as the simplest to assess in this experiment, but the possibility of more subtle adverse effects with exposure to lower NH$_3$ concentrations cannot be excluded.
- After 2 years 85% death of *Calluna* occurred at 32 µg m^{-3}, while complete death of *Cladonia portentosa* occurred at 14 µg m^{-3}. After 4.5 years exposure these values were 17 and 7 µg m^{-3} respectively.
- It is not possible to say at present how long the linear log [NH$_3$]: log [time] relationship will continue. However, if the relationship is extended, then for 10

years exposure adverse effects would be expected at $6\,\mu g\ m^{-3}$ for *Calluna* and $2.5\,\mu g\ m^{-3}$ for *Cladonia*. After 25 years exposure effects would be expected at mean concentrations of $1.8\,\mu g\ m^{-3}$ for *Calluna* and $0.7\,\mu g\ m^{-3}$ for *Cladonia*. This extrapolation might be considered as precautionary, while conversely it is based on rather extreme adverse effects and does not incorporate a safety factor.
- These results point to the need to establish a long term CLE_{NH3} for protecting the integrity of lichen rich moorland and bog habitats at $\sim1\text{--}2\,\mu g\ m^{-3}$.

Acknowledgements The Natural Environment through the GANE programme supported the site initially, now work is supported through the NERC-DEFRA Terrestrial Umbrella – CPEA 18 and CEH. William Sinclair Horticulture Ltd. is thanked for access to Whim bog, Catherine Ball, Trevor Blackhall and David Coots helped with frost and drying experiments.

References

Asman W.A.H., Sutton M.A., Schoerring J.K. (1998) Ammonia: emission, atmospheric transport and deposition. New Phytol. 139, 27–48.

Bannister P. (1964) The water relations of certain heath plants with reference to their ecological amplitude 11. Field studies. J. Ecol. 52, 481–497.

Bobbink R., Heil G.W., Raessen M.B.A.G. (1992) Atmospheric deposition and canopy exchange processes in heathland ecosystems. Env. Poll. 75, 29–37.

Jones M.R. (2006) Ammonia deposition to semi-natural vegetation. Ph.D thesis. University of Dundee.

Hornung M., Sutton M.A., Wilson R.B. (eds.) (1995) Mapping and Modelling of Critical Loads for Nitrogen – A workshop Report, Report of the UN-ECE Workshop, Grange over sands, (24–26 Oct.) Institute of Terrestrial Ecology, Edinburgh.

Leith I.D., Sheppard L.J., Fowler Cape J.N., Jones M., Crossley A., Hargreaves K.J, Tang Y.S., Theobald M., Sutton M.A. (2004) Quantifying dry NH_3 deposition to an ombrotrophic bog from an automated NH_3 release system. Water Air Soil Poll. Focus 4, 207–218.

Sheppard L.J., Leith I.D. (2002) Effects of NH_3 fumigation on the frost hardiness of Calluna – does N deposition increase winter damage by frost? Phyton 42, 183–190.

Sheppard L.J., Leith I.D., Crossley A. (2004a) Effects of enhanced N deposition on Cladonia portentosa; Results from a field manipulation study. 'Nitrogen in the Environment'. British Lichen Society Workshop, Nettlecombe, Somerset. In: Lambeley P.W. and Wolseley P.A. (eds.) Lichens in a Changing Environment. English Nature Research Report 525, 84–89.

Sheppard L.J., Crossley A., Leith I.D., Hargreaves K.J., Carfrae J.A., van Dijk N., Cape J.N., Sleep D., Fowler D., Raven J.A. (2004b) An automated wet deposition system to compare the effects of reduced and oxidised N on ombrotrophic bog species: practical considerations. Water Air Soil Poll. Focus 4, 197–205.

Tang Y.S., Cape J.N.C., Sutton M.A. (2001) Development and types of passive samplers for NH_3 and NO_2. Sci. World 1, 275–286.

UNECE (2003) In: Achermann B. and Bobbing R. (eds.) Empirical Critical Loads for Nitrogen. Expert workshop Berne 2002. Proceedings SAEFL Berne.

Van der Eerden L.J.M (1982) Toxicity of ammonia to plants. Agr. Environ. 7, 223–235.

Van der Eerden L.J.M, Dueck TH. A., Berdowski J.J.M., Grevan H., Van Dobben H.F. (1991) Influence of NH_3 and $(NH_4)_2SO_4$ on heathland vegetation. Acta Bot. Neerl. 40, 281–296.

Watson A., Miller G.R., Green F.H. (1966) Winter browning of heather (*Calluna vulgaris*) and other moorland plants. Trans. Bot. Soc. Edinburgh. 40, 195–203.

World Health Organization (1997) Nitrogen Oxides (Second Edition) Ch 4 Effects of atmospheric nitrogen compounds (particularly nitrogen oxides) on plants. Environmental Health Criteria Series 188. Geneva 15–191.

Chapter 5
The Application of Transects to Assess the Effects of Ammonia on Woodland Groundflora

Carole E.R. Pitcairn, Ian D. Leith, Netty van Dijk, Lucy J. Sheppard, Mark A. Sutton, and David Fowler

5.1 Introduction

Increased emissions of ammonia from intensive farming have been identified as the major cause of many of the changes in species composition in plant communities in Europe (Bobbink et al. 1993; Sutton et al. 1993). Transect studies at intensive livestock farms in the UK, have provided important information on the impact of NH$_3$ and deposited N on woodland groundflora (Pitcairn et al. 1998, 2002, 2003, 2005). Several parameters have been measured to understand the basis of species composition change in woodland vegetation and to identify possible biomonitors of vegetation change (Pitcairn et al. 2003; Sutton et al. 2004; Leith et al. 2005).

The value of 3 biomonitors of N impacts, are described in this paper, for 2 different woodland sites impacted by poultry farm ammonia emissions, Earlston, a Scots pine open plantation in southern Scotland, and Piddles Wood, a mixed deciduous woodland in south west England.

5.1.1 Biomonitor Methods

Tissue N concentration of a range of plant species, has been shown to be closely related to atmospheric NH$_3$ concentrations and N deposition. The total tissue N and soluble NH$_4$ concentration of mosses are particularly closely linked to atmospheric N inputs (Pitcairn et al. 2003).

Ellenberg devised a comprehensive indicator system for vascular plants of central Europe (Ellenberg 1979; Ellenberg et al. 1991) to describe the response of individual species to a range of ecological conditions (light, temperature, continentality, moisture, pH and nitrogen). To obtain a mean Ellenberg N Index for a sampling location, a N score is allocated to each plant species, so that the overall community has a score on a scale of nutrient poor (1) to nutrient rich (10).

Carole E.R. Pitcairn
Centre for Ecology & Hydrology Bush Estate, Penicuik, Midlothian,
EH26 0QB, United Kingdom

M. Sutton, S. Reis and S.M.H. Baker (eds), *Atmospheric Ammonia,*
© Springer Science+Business Media B.V. 2009

In these transect studies, indices were determined using the modified values for British vascular plants (Hill et al. 1999) and indicator values from Siebel (1993) for bryophytes.

A **Nitrophobe/Nitrophile** classification for Scotland for vascular plant and bryophyte species was developed by Pitcairn et al. (2006), based on the use of indices of nitrophyte and acidophyte lichens in biomonitoring for NH_3 detection (van Herk 1999; Wolseley and James 2002; Leith et al. 2005). The classification, which identifies species at the extremes of the Ellenberg N scale, was developed using the UK National Vegetation Classification (Rodwell 1992), intensive literature searches and expert judgement. In general, the nitrophobe species tended to be vernal species, orchids, woodrushes and several bryophyte species typical of acid woodlands and moorland. Nitrophiles tended to be 'weed species' such as *Urtica dioica*, *Rumex obtusifolia*, *Heracleum* spp., *Epilobium* spp. and grasses typical of arable or disturbed land together with a few bryophytes, which prefer N enriched habitats.

5.2   Methods

5.2.1   Site Description

Earlston woodland comprises a 30 year old *Pinus sylvestris* plantation with some *Betula pubescens* and mature *Fagus sylvatica* bordering the farm track, and a ground flora of ferns, herbs and mosses (Pitcairn et al. 1998). The woodland surrounds a poultry farm (approximately 120,000 broilers farmed on a 44 day cycle, emitting approximately 4,800 kg N year^{-1}) which has been operating for over 20 years, and the woodland edge is 16 m from the farm buildings on the downwind side.

Piddles Wood is a broadleaved, mixed and yew lowland woodland, lying on raised ground south of the River Stour in north Dorset (Leith et al. 2005). The site is an SSSI, notified in 1985 because of its substantial oak woodland with coppiced hazel understorey. A poultry farm (approximately 100,000 birds) abuts on the south west edge of the woodland and has been operating for 10 years. The study area lies mainly on fairly light acid soils, but areas of calcareous soil also occur. The groundflora is therefore very species rich.

5.2.2   Ammonia Monitoring

Ammonia was measured continuously at Earlston for 12 months between February 1995 and April 1996, at 6 sites (16, 30, 46, 76, 126, and 276 m from the farm) through woodland adjacent to the farm using open-ended passive diffusion tube samplers as described in Pitcairn et al. (1998). Further monitoring took place at the same sites in autumn 2002 using ALPHA samplers (Tang et al. 2001). At Piddles Wood, ammonia was measured continuously for 3 months between April and July 2004, at 5 sites (5, 20, 40, 100 and 250 m from farm), using ALPHA samplers.

5.2.3 Plant Chemical Analysis

Tissue N concentration: Samples of *pleurocarpous* mosses were washed in deionised water, dried and ground and then analysed for %N content using a Vario-EL elemental CN Analyzer.

5.2.4 Vegetation Surveys and Species Indicies

Groundflora surveys of the 2 woodlands were made at different distances from the 2 poultry farms, corresponding where possible with ammonia monitoring points, and Ellenberg N Index (Ellenberg 1979; Ellenberg et al. 1991) was determined for each distance. At Earlston, the vegetation was surveyed in strips 50 × 2 m, perpendicular to the ammonia monitoring transect. Seven strips were surveyed at 16, 36, 46, 76, 126 and 276 m downwind of the buildings, corresponding with the downwind ammonia monitoring points and including 1 extra strip at 36 m. All species readily observed were recorded, allocated the appropriate Ellenberg value and a mean unweighted Ellenberg N value based on presence of species, was obtained for each distance. A mean abundance weighted value for each species was obtained for each distance by recording percentage species cover in 2 × 2 m squares along each strip. At Piddles Wood, 5 strips (12 × 2 m) were surveyed at 5, 20, 40, 100 and 250 m from the farm and mean unweighted Ellenberg N values obtained. Abundance weighted values were obtained for 4 strips at 5, 20, 40, and 100 m from the farm.

5.3 Results and Discussion

5.3.1 NH_3 Concentrations

Annual mean ammonia concentrations along the 2 woodland transects, declined exponentially with distance from the farms. Concentrations close to the poultry farms were very high (at Earlston, 16 m from farm, annual mean NH_3 concentration of 29 µg m^{-3}; at Piddles Wood, 5 and 20 m from the farm, 3 month mean NH_3 concentration of 101 and 31 µg m^{-3} respectively), declining to around 1.5 µg m^{-3} at around 250 m from both farms.

5.3.2   Tissue N

At both sites, tissue N concentrations of species including trees, herbs, ferns, mosses and lichens were elevated close to the livestock buildings where NH$_3$ con-

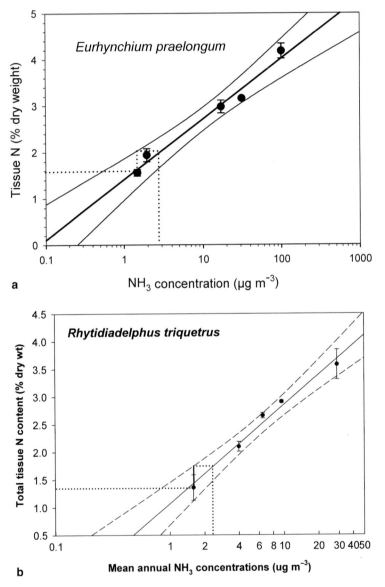

Fig. 5.1 Total tissue N concentrations of pleurocarpous mosses along a defined NH$_3$ gradient transect from the poultry unit at (a) Piddles Wood, Dorset and (b) Earlston, southern Scotland

centrations were large and decreased with distance from the buildings.

Figure 5.1 shows that total tissue N concentration in mosses was elevated at both sites at annual mean NH_3 concentrations above 1.5 µg m^{-3}. Using the NH_3 threshold method (Cape et al. 2009, this volume) the thresholds for the effects at Piddles Wood (from 95% confidence limits) were detected down to 2.5 µg m^{-3} for *Eurhynchium praelongum*. At Earlston, the threshold was slightly lower at around 2.2 µg m^{-3} for *Rhytidiadelphus triquetrus*. The threshold above background is an indication of the lowest threshold concentrations for effects.

5.3.3 Species Change

Enhanced tissue N concentrations may affect plants indirectly through increased susceptibility to damage by frost, drought, pet and pathogen (Krupa 2003). This will also lead to changes in the competitive advantages of certain plant groups and hence to changes in the composition of a plant community.

Species diversity was reduced close to livestock buildings. Species such as *Deschampsia flexuosa, Holcus lanatus, Hedera helix, Chaerophyllum temulentum, Lamiastrum galeobdolon, Urtica dioica* and *Chamaenerion angustifolium* were abundant close to the buildings. More sensitive species such as *Oxalis acetosella, Galium odoratum, Potentilla erecta, Dactylorhiza fuchsii* and *Conopodium majus* and *pleurocarpus* mosses became more abundant as distance from the farm increased. Major changes occurred within 50 m of the livestock buildings.

The Ellenberg N index was able to distinguish differences in composition downwind of both poultry farms (Figs. 5.2 and 5.3). A mean Ellenberg Index of >4.5 suggests a change in species composition of woodland ground flora but error bars were large. Even at concentrations of 100 µg m^{-3} as measured 5 m from Piddles Wood poultry farm, the mean Ellenberg Index was only 5.6. The mean abundance-weighted Ellenberg indicator value for both transects showed only slight trends with distance from the livestock buildings.

The Nitrophobe/Nitrophile classification provided a more sensitive indication of vegetation change. Results for Piddles Wood based on subtracting nitrophile presence from nitrophobe presence (Fig. 5.4), showed that the flora is dominated by nitrophiles at mean concentrations of NH_3 above 3 µg m^{-3}. When subtracting total nitrophile cover from total nitrophobe cover, the flora was dominated by nitrophiles at NH_3 concentrations close to 1 µg m^{-3}. However it must be pointed out that the Piddles Wood survey took place in July 2004 when only remnants of vernal species such as wood anemone were visible. In addition, changes in soil type, canopy cover, leaf litter levels would also affect cover.

The results for Earlston woodland (Fig. 5.5) based on species number, showed groundflora to be dominated by nitrophiles at mean annual NH_3 concentrations above 9.5 µg m^{-3}. When based on species cover, nitrophiles dominated at concentrations above 1.6 µg m^{-3} NH_3.

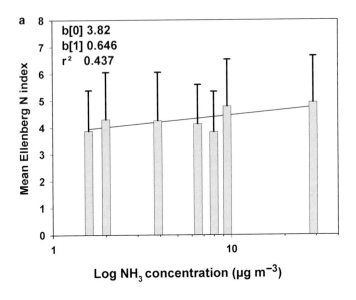

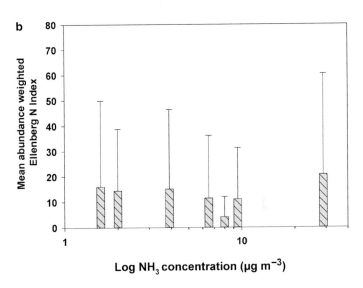

Fig. 5.2 Ellenberg Nitrogen values for woodland ground flora in the vicinity of Earlston poultry farm. (a) Mean unweighted N index, (b) Mean abundance weighted N index. Indicator scales from Siebel (1992) were used for bryophytes. For higher plants and ferns, indicator scales are from Hill et al. (1999). Error bars are standard deviations. (Pitcairn et al. 2002)

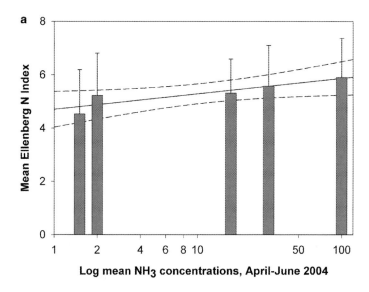

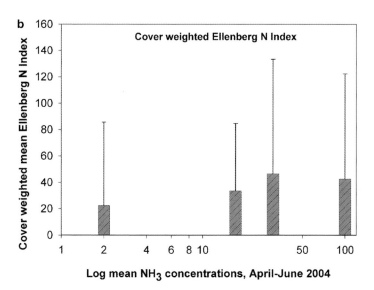

Fig. 5.3 Ellenberg Nitrogen indices of woodland ground flora downwind of Piddles wood poultry farm. (a) Mean unweighted nitrogen index, (b) Mean abundance weighted nitrogen index. Indicator scales from Siebel (1992) were used for bryophytes throughout. For higher plants and ferns, indicator scales are from Hill et al. (1999). Error bars are standard deviations (Pitcairn et al. 2002)

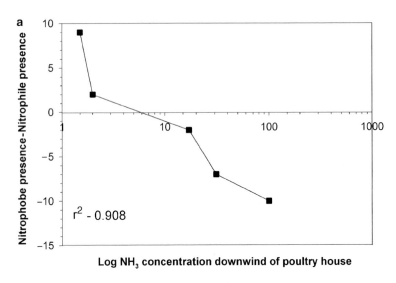

Log NH₃ concentration downwind of poultry house

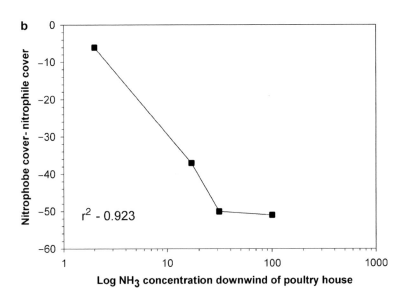

Fig. 5.4 Eutrophication estimate for Piddles Wood, (a) nitrophobe-nitrophile presence, (b) nitro-phobe-nitrophile cover. Ammonia measurements available for 5 sites, cover estimates available for 4 sites

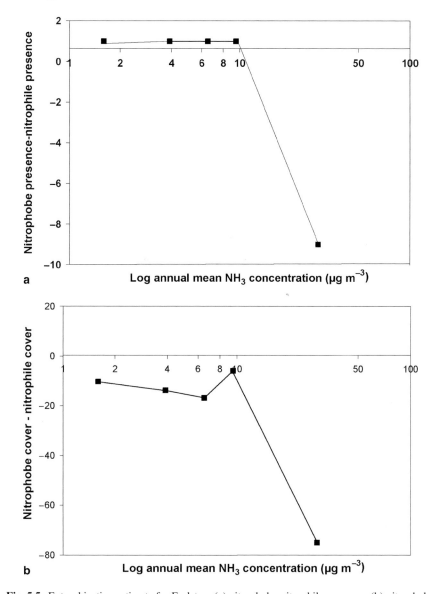

Fig. 5.5 Eutrophication estimate for Earlston, (a) nitrophobe-nitrophile presence, (b) nitrophobe-nitrophile cover

5.4 Conclusions

- Tissue N concentrations in moss groundflora species are significantly increased by exposure to ammonia emissions from 2 poultry farms.
- The threshold for effects of NH_3 concentrations on tissue N indicated a critical level of 2–2.5 µg m^{-3} for *Eurynchium praelongum and Rhytidiadelphus triquetrus*.

- Tissue N concentration of mosses may be calibrated to give an indication of N inputs to the ecosystem and the potential for vegetation change.
- The Nitrophile/Nitrophobe Index provides a more definitive indication of the impacts of NH_3 on woodland groundflora than the Ellenberg Index.
- The Nitrophile/Nitrophobe Index based on species cover is a more sensitive indication of the impacts of NH_3 on woodland groundflora than that based on species presence.
- The transect studies through woodland adjacent to 2 poultry farms show that concentrations of NH_3 greater than $2\,\mu g\ m^{-3}$ can adversely affect the species composition of the groundflora.

References

Bobbink, R., Boxman, D., Fremstad, E., Heil, G., Houdijk, A., Roelofs, J. (1993) Nitrogen eutrophication and critical load for nitrogen based upon changes in flora and fauna in (semi)-natural terrestrial ecosystems. In: Critical Loads for Nitrogen, Proceedings of a UN-ECE workshop at Lökeberg, Sweden. 6–10 April 1992, 111–159. Nordic Council of Ministers, Copenhagen, Denmark.

Ellenberg, H. (1979) Indicator values of vascular plants in Central Europe. Scripta Geobotanica 9, 7–122.

Ellenberg, H., Weber, H.E., Dull, R., Wirth, V., Werner, W., Paulissen, D. (1991) Zeigerwerte von Pflanzen in Mitteleuropa. Scripta Geobotanica 18, 1–248.

Hill, M.O., Mountford, J.O., Roy, D.B, Bunce, R.G.H. (1999) Ellenbergs' indicator values for British plants. ECOFACT Volume 2, Technical Annex. ITE Monkswood, Huntingdon. Department of the Environment, Transport and the Regions, London.

Leith, I.D., van Dijk, N., Pitcairn, C.E.R., Wolseley, P.A., Sutton, M.A. (2005) Biomonitoring methods for assessing the impacts of nitrogen pollution: refinement and testing. JNCC Report No. 386, Peterborough, UK.

Krupa, S.V. (2003) Effects of atmospheric ammonia (NH3) on terrestrial vegetation: a review. Environmental Pollution 124, 179–221.

Pitcairn, C.E.R., Leith, I.D., Sheppard, L.J., Sutton, M.A., Fowler, D., Munro, R.C., Tang, S., Wilson, D. (1998) The relationship between nitrogen deposition, species composition and foliar nitrogen concentrations in woodland flora in the vicinity of livestock farms. Environmental Pollution 102, 41–48.

Pitcairn, C.E.R., Skiba, U.M., Sutton, M.A., Fowler, D., Munro, R., Kennedy, V.K. (2002) Defining the spatial impacts of poultry farm ammonia emissions on species composition of adjacent woodland groundflora using Ellenberg indicators, nitrous oxide and nitric oxide and foliar nitrogen as marker variables. Environmental Pollution 119, 9–21.

Pitcairn, C.E.R., Fowler, D., Leith, I.D., Sheppard, L.J., Sutton, M.A., Kennedy, V., Okello, E. (2003) Bioindicators of enhanced nitrogen deposition. Environmental Pollution 126 (3), 353–361.

Pitcairn, C.E.R., Leith, I.D., van Dijk, N., Sutton, M.A. (2005) Refining and testing the Ellenberg index biomonitoring method at the intensive sites. In: Leith I.D., van Dijk N., Pitcairn C.E.R., Wolseley P.A., Sutton M.A. (eds.) Biomonitoring Methods for Assessing the Impacts of Nitrogen Pollution: Refinement and Testing. JNCC Report No. 386, Peterborough, UK.

Pitcairn, C.E.R., Leith, I.D., Sheppard, L.J., Sutton, M.A. (2006) Development of a nitrophobe/nitrophiles classification for woodlands, grasslands and upland vegetation in Scotland. CEH Edinburgh, Report to the Scottish Environment Protection Agency.

Rodwell, J.S. (1992) British Plant Communities. Cambridge University Press.

Siebel, H.N. (1993) Indicatiegetallen van blad -en levermossen. IBN-rapport 047, Wageningen.

Sutton, M.A., Pitcairn, C.E.R., Fowler, D. (1993) The exchange of ammonia between the atmos-
    phere and plant communities. Advances in Ecological Research 24, 301–393.
Sutton, M.A., Pitcairn, C.E.R., Leith, I.D., Sheppard, L.J., van Dijk, N., Tang, Y.S., Skiba, U.,
    Smart, S., Mitchell, R., Wolseley, P., James, P., Purvis, W., Fowler, D. (2004) Bioindicator and
    biomonitoring methods for assessing the effects of atmospheric nitrogen on statutory nature
    conservation sites. In: Sutton M.A., Pitcairn C.E.R., Whitfield C.P. (eds) JNCC (JNCC Report
    No: 356), Peterborough, UK.
Tang, Y.S., Cape, J.N., Sutton, M.A. (2001) Development and types of passive samplers for NH_3
    and NO_x. The ScientificWorld 1, 513–529.
van Herk, C.M. (1999) Mapping of ammonia pollution with epiphytic lichens in the Netherlands.
    Lichenologist 31, 9–20.
Wolseley, P.A., James, P.W. (2002) Using lichens as biomonitors of ammonia concentrations in
    Norfolk and Devon. British Lichen Society Bulletin 91, 1–5. http:/www.nhm.ac.uk/botany/
    lichen/twig (accessed March 2006).

Chapter 6
Estimation of the Ammonia Critical Level for Epiphytic Lichens Based on Observations at Farm, Landscape and National Scales

Mark A. Sutton, Pat A. Wolseley, Ian D. Leith, Netty van Dijk, Y. Sim Tang, P.W. James, Mark R. Theobald, and Clare Whitfield

6.1   Introduction

In past decades, a huge amount of information was collected on the sensitivity of lichens to atmospheric sulphur dioxide. As the concentrations of sulphur dioxide have decreased following emission control measures, much more attention is now focusing on the possibility of direct effects of ammonia. In the Netherlands, a change in lichen populations through the 1980s and 1990s was originally attributed to an increase in ammonia emissions, although subsequently it was found difficult to separate these changes from the effects of parallel decreases in sulphur dioxide emissions (van Dobben and Ter Braak 1998). However, several subsequent studies across Europe have shown that ammonia is having substantial and unambiguous effects on epiphytic lichen populations (van Herk 1999; Sutton et al. 2004a, b; Wolseley et al. 2004, 2006; Frati et al. 2007; Pinho et al. 2009, this volume).

Our field measurements in the UK have trialled a number of indices to summarize the response of lichens to excess nitrogen deposition and specifically ammonia (Sutton et al. 2004a, b; Leith et al. 2005; Wolseley et al. 2004, 2006). Of these, we have found the approach of distinguishing two functional species groups, 'nitrophytes' and 'acidophytes', to be most successful. This approach was originally developed by van Herk (1999) and consists of establishing lists of the known species which prefer a high supply of reactive nitrogen, the nitrophytes, and the known species which avoid a high supply of reactive nitrogen, the acidophytes. The name for the latter group reflects the fact that, as gaseous ammonia is typically the driving variable, high nitrogen supply tends to increase bark pH, which is naturally acidic under clean conditions. The fact that ammonia tends to increase bark pH in the field indicates that the effect of NH_3 as a base dominates over any nitrification on bark surfaces (which could potentially acidify the surface, cf. Sutton et al. 1993).

In further developing the acidophyte – nitrophyte approach, we have made several key changes to the methodology (see Wolseley et al. 2009, this volume).

Mark A. Sutton
Centre for Ecology & Hydrology, Bush Estate, Penicuik, Midlothian,
EH26 0QB, United Kingdom

M. Sutton, S. Reis and S.M.H. Baker (eds), *Atmospheric Ammonia,*
© Springer Science+Business Media B.V. 2009

Specifically, we combined the indices for acidophytes (L_A) and nitrophytes (L_N), making an overall nitrogen index (L_{AN}); we simplified the scoring method, allowing it to be implemented by focusing on more-easy-to-identify macro-lichens; and we extended the method from tree trunks to also record lichens on twigs. The last is important, since twigs have a faster turnover time, and may respond more rapidly to pollution changes, while the data indicate that lichen communities on twigs are even more sensitive to ammonia than lichens on trunks.

In this short paper, we summarize the results of several of the field studies made within the UK comparing effects at farm, landscape and national scales. Specifically, we use these data to investigate a suitable ammonia 'critical level' concentration, above which significant adverse effects on epiphytic lichen populations can be observed.

6.2   Field Sites and Methods

6.2.1   Landscape Scale Assessment

In a first assessment (Wolseley and James 2002; Wolseley et al. 2006) we tested the acidiphyte-nitrophyte method at the landscape scale, also providing a first test of the acidophyte-nitrophyte method for twigs, building on the twig sampling approach developed earlier by Wolseley and Pryor (1999). Two contrasting landscapes were investigated to allow for possible climatic differences: Thetford in Norfolk, eastern England ('continental' climate) and North Wyke in Devon, south west England ('oceanic' climate). The Thetford study area was chosen because of the availability of monitoring of atmospheric ammonia concentrations across the landscape, which was conducted as part of a detailed analysis of landscape-level variability in nitrogen fluxes (Theobald et al. 2004). The North Wyke site was chosen because of the availability of ammonia concentrations at one location; for other locations in that landscape a modelling approach had to be used to estimate the ammonia concentrations. Further details of the sites and modelling approach are reported by Wolseley et al. (2006). In both of these studies we recorded the lichen populations growing on oak (*Quercus petraea*).

6.2.2   Farm Scale Assessment

In the second phase we wanted to look on more detail at how ammonia concentrations and lichen populations varied with distance from single farm point sources. In a first analysis (Sutton et al. 2004a, b) we measured ammonia and recorded lichen populations through a woodland downwind of a well-studied chicken (broiler) farm in southern Scotland (see Pitcairn et al. 2002). In this study, we recorded the lichens on trunks of Sitka spruce (*Picea sitchensis*) and on twigs of birch (*Betula pubescens*) from 10 m up to 300 m downwind of the farm, as well as compared this with a clean reference location nearby. As part of the wider study, further testing of the method was made downwind of a poultry farm in Dorset, southwest England and downwind of a motorway, at Happenden, south west Scotland (Leith et al. 2005), though the results of these studies are not included here. The former had extremely high ammonia concentrations, while at the latter site, the maximum ammonia concentrations were only slightly above background.

6.2.3   National Scale Assessment

Having established the basis of the trunk and twig sampling using a relatively simple protocol, we then extended the analysis to consider the effects of ammonia on lichens at the national scale (see Leith et al. 2005). For this study, we selected 32 sites across the UK covering a range of climatic conditions and known ammonia concentrations, with these sites being selected as a subset of the UK National Ammonia Monitoring Network (Sutton et al. 2001; Tang et al. 2009, this volume). To conduct the sampling, basic training in the sampling method for macro-lichens was provided to staff of conservation and pollution regulatory agencies for the devolved regions of the UK. The agency staff then conducted the field survey, with reference specimens provided to lichenologists to confirm uncertain identifications. In the UK survey, preference was given for the selection of naturally acid barked trees, where they were present at sites. Most data were collected for lichens on oak, but other tree species included: birch (*Betula pubescens*), Sikta spruce (*Picea sitchensis*), sycamore (*Acer pseudoplatanus*) and ash (*Fraxinus excelsior*).

6.2.4   Recording Approach

Details of the sampling approach are provided by Wolseley et al. (2009, this volume). In summary the method consisted of recording lichens: (a) on four sides of each tree trunk for a group of typically five trees at each site and (b) on up to 10 twigs on accessible branches at each site, where the branch was divided in to three subsections covering the first 3 m from the tip. The presence of main indicator species were used to derive the L_A and L_N scores as shown in Table 6.1, for the UK-wide assessment. For the landscape and farm scale surveys conducted by expert lichenologists, crustose lichen species were also recorded. The full species list of nitrophytes and ascidophytes according to van Herk (1999) and applied in the landscape assessment are reported by Wolseley et al. (2006).

Based on the L_A and L_N indices, we calculated the overall acidophyte-nitrophyte index L_{AN} for each site simply as: $L_{AN} = L_A - L_N$. A positive value thus indicates a site with lower ammonia concentrations and dominated by acidophyte lichen species, while a negative value indicates a site with higher ammonia concentrations, dominated by nitrophytes. Values of L_{AN} were calculated both for tree trunks and for twigs.

Table 6.1 List of macrophyte (foliose, fruticose) acidophyte and nitro-phytes lichen species as included in the L_{AN} index for the UK and farm-scale survey. Note that not all species were recorded as present at the sites

Acidophyte species	Nitrophyte species
Bryoria spp.	*Hyperphyscia adglutinata*
Cladonia spp.	*Phaeophyscia orbicularis*
Evernia prunastri	*Physcia adscendens*
Flavoparmelia caperata	*Physcia caesia*
Hypogymnia spp.	*Physcia tenella*
Hypotrachyna laevigata	*Xanthoria candelaria*
Parmelia saxatilis	*Xanthoria parientina*
Platismatia glauca	*Xanthoria polycarpa*
Pseudevernia furfuracea	
Usnea spp.	

In parallel to the lichen scoring, we also considered it important to record bark pH as this gives a very useful indicator to interpret the lichen response to ammonia. For the bark of trunks, small samples were collected, stored in paper bags and the pH recorded in the laboratory using a flat-head electrode, according to a method adapted from Kermit and Gauslaa (2001). For the twigs, these were cut into 6 cm lengths, with the ends sealed with paraffin wax, and placed in a tube with 6 ml of 25 mM KCl, with pH being recorded after 1 h, and checked after 2 h to ensure a stable value had been obtained. Further details of this approach are provided by Wolseley et al. (2006).

In summary, the main advances in our method as applied consisted of: (a) simplified recording scheme, usable by non-experts (restriction to macrolichens; simplified scoring system), (b) consideration of both lichens on twigs and trunks, (c) calculation of an overall lichen nitrogen index ($L_{AN} = L_A - L_N$), and (d) parallel recording of L_{AN} with bark pH for both twigs and trunks.

6.2.4.1 Derivation of the Ammonia Critical Level

The results from our field observations can be shown simply as regressions between the lichen nitrogen index (L_{AN}) and the potential driving parameters: site mean atmospheric ammonia concentration or bark pH. Potentially, interactions may also occur between bark pH and NH_3 concentration in affecting L_{AN}. However, the most important response for the determination of the critical level is the L_{AN} response to NH_3 concentration.

As also explained by Cape et al. (2009, this volume), a critical level can be derived as the NH_3 concentration at which the observed L_{AN} value is significantly lower than the L_{AN} value for background NH_3 concentrations. This can be expressed graphically (see Fig. 6.3 and following), as the point at which the least-squares regression best-fit line shows a value of L_{AN} which is smaller than the minimum value of the confidence interval for L_{AN} at the cleanest study location. It is obvious from this approach that it is only possible to derive a realistic critical level where the local background NH_3 concentration is substantially smaller than the 'true' critical level. Ammonia concentrations are highly variable, and regional background values far from sources are of the order of 0.03–0.3 µg m⁻³, according to time of year (see e.g. Sutton et al. 2001). By contrast, local background concentrations in the Netherlands are typically larger than 3 µg m⁻³, due to the ubiquity of different agricultural sources. As a result, field studies to characterize the ammonia critical level, need as far as possible to include clean reference locations. Failure to do so will lead to overestimation of the critical level that is estimated by this approach. The regression approach used here determines a statistically significant effect from field data. Therefore, it should also be noted that, depending on the size and scatter of the dataset available, the true critical level may again be smaller than determined by the method used here.

6.3 Results and Discussion

6.3.1 Landscape Scale

The results of our first analysis at the Thetford landscape demonstrated a clear relationship between mean ammonia concentration and bark pH. This is illustrated in Fig. 6.3, comparing the bark pH of oak trunks and twigs against the measured mean ammonia concentration for each site. Since the measurements were conducted in a flat landscape, (c. 8 × 12 km) systematic climatological differences between sites can be neglected. Overall, the data show that the deposition of ammonia (being alkaline) significantly increases bark pH. It is worth to note that, at low ammonia concentrations, the twigs had a naturally higher pH than the tree trunks at the same locations. This difference may be associated with increased availability and leaching of base cations from the bark of twigs than from trunks.

Wolseley et al. (2006) showed the results of L_A and L_N scores in relation to the measured and modelled NH_3 concentrations for the Thetford and North Wyke landscapes. In general lower L_A values were found for North Wyke above 1–1.5 µg m⁻³ (see Fig. 4A, C in Wolseley et al. 2006). (The cleanest location at this site – the local background – was estimated to be an ammonia concentration of 0.9 µg m⁻³). For the purpose of general interpretation of the lichen responses, the modelled NH_3 concentration estimates for North Wyke may be considered as sufficient. However, since the purpose in this paper is to estimate an ammonia critical level, we considered that that these model estimates were only sufficient to provide supporting evidence, and do not analyze them further.

For the Thetford results, where actual NH_3 monitoring data were available, total lichen diversity for twigs was found to reduce below c. 2 µg m⁻³ (Wolseley et al. 2006, Fig. 3D), although few sites had smaller NH_3 concentrations (see Fig. 6.1, and the local background was 1.4 µg m⁻³). The response of L_A, LN and L_{AN} to NH₃ concentration was less clear at Thetford, particularly due to an outlier site with very high NH₃ concentrations (9.2 µg m⁻³), where very few lichens were present, leading to small values of both L_A and L_N. In fact, acidophyte species on twigs were only recorded at 3 out of the 12 sites surveyed, representing 3 of the 5 cleanest sites (NH₃ < 2.2 µg m⁻³), suggesting that at higher NH₃ concentrations, acidophytes on oak twigs are completely eradicated at this location.

Given the dependence of the North Wyke ammonia data on modeling, in Fig. 6.2 we present the L_{AN} scores from both Thetford and North Wyke in relation to bark pH which was measured at all the sampling locations. This demonstrates the clear relationship between bark pH and the presence of nitrogen sensitive and nitrogen loving epiphytic lichens. The differences in bark pH are both due to natural inter-tree differences and the effect of ammonia on bark pH at each site. Thus high NH₃ concentrations increase bark pH (Fig. 6.1), which adversely affects the lichen populations (Fig. 6.2). However, caution is needed before concluding that all of the ammonia effect on lichens is mediated through changes in bark pH. Ammonia may have two affects: (a) a pH effect (most significant at large NH₃ concentrations) and (b) a eutrophication effect (most significant at low NH₃ concentrations), as indicated by further analysis of the UK scale data.

The results from North Wyke and Thetford, are thus supportive of a critical level in the region of 1–2 µg m⁻³, but they can only be considered as supporting data because (a) the North Wyke data were dependent on model estimates of NH₃ concentration and (b) the Thetford data are rather scattered, and subject to a high background NH₃ concentration. Despite these limitations, these studies are useful in demonstrating clear ammonia interactions at the landscape scale, with broadly similar results for contrasting oceanic and continental climates.

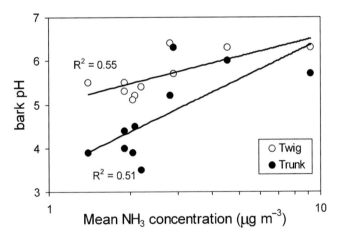

Fig. 6.1 Relationship between the pH of oak bark and mean ammonia concentration from a local survey at the Thetford landscape in Norfolk, UK (cf. Wolseley et al. 2006, Fig. 3B)

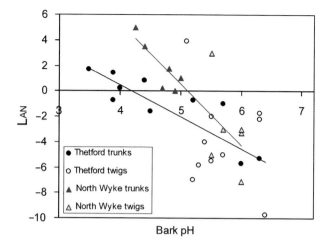

Fig. 6.2 Relationship between recorded Lichen index L$_{AN}$ and bark pH from local surveys of oak at the landscape scale. Lichen observations at Thetford, eastern England were made at locations across an intensively studied landscape of approximately 8 × 12 km accompanied by NH$_3$ concentration measurements (see Fig. 6.1). At North Wyke, south west England, observations were made at different distances (up to 1 km) from a livestock farm containing mainly cattle. The fitted lines show the trunk and twig data combined for Thetford (R^2 = 0.42) and for North Wyke (R^2 = 0.67)

6.3.2 Farm Scale

Even clearer responses between lichens and ammonia concentrations were found at the farm scale study in southern Scotland. Although, measurements were made at fewer sites, lower critical levels than implied by the initial landscape studies were estimated both due to: (a) the existence of lower background concentrations occurring in Scotland, and (b) the lower degree of scatter in this dataset. The results for the L$_{AN}$ scores are shown in Figs. 6.3 and 6.4 for trunks and twigs respectively.

It is worth considering carefully the approach to derive the critical level from these examples. In Fig. 6.3, the cleanest background measurement was from measurements on site at 0.58 µg m^{-3}. The tie-line 'a' to the lower 95% confidence limit, shows the L$_{AN}$ score which would be significantly different. Thus tie-line 'b' shows the point at which this is met on the regression best-fit line, while tie-line 'c' shows that this occurs at 1.7 µg m^{-3}, which is therefore the concentration above which (with 95% confidence) the L$_{AN}$ score is shown to have decreased compared with the local background conditions.

The results for twigs show a smaller significant ammonia concentration at 0.7 µg m^{-3}. This is both due to the higher sensitivity of twigs to NH$_3$, as seen by the overall lower values of L$_{AN}$ for similar NH$_3$ concentrations, and due to the availability of twig measurements at a cleaner background location (0.3 µg m^{-3}), which was not recorded for trunk lichens. Much of the NH$_3$ effect at this site may have been mediated by changes in bark pH: the responses of trunk and twig pH to NH$_3$ concentration

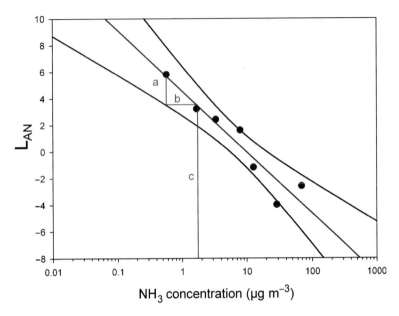

Fig. 6.3 Response of lichens on trunks, expressed as the index L_{AN}, to ammonia concentrations from a farm scale assessment in southern Scotland. The tie lines indicate the estimation of a critical level from these data. The 95% confidence interval of the regression line is shown

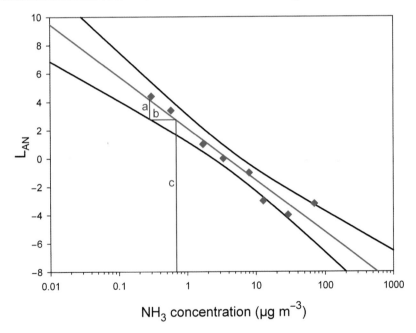

Fig. 6.4 Response of lichens on twigs, expressed as the index L_{AN}, to ammonia concentrations from a farm scale assessment in southern Scotland. The tie lines indicate the estimation of a critical level from these data

were not significantly different, with the overall relationship highly significant (R^2= 0.86, n=12).

6.3.3 National Scale

The national scale assessment analyzed the effects of rather lower ammonia concentrations than the preceding example. The range of mean ammonia concentrations at the UK sites was 0.04 µg m^{-3} (at Ariundle, north west Scotland) to 7.5 µg m^{-3} (at Bedlington, Suffolk, England). The UK-wide measurements included both areas of agricultural fields, as well as forests and other semi-natural land. However, the lichen recording, as well as the underpinning ammonia monitoring, avoided sampling in the immediate vicinity of farm sources, in order to be more representative of wider conditions across the country.

The L_{AN} values for each site were found to be correlated both to bark pH and mean ammonia concentration, each explaining around half the variance. Further analysis showed that a multiple regression of L_{AN} against both NH_3 concentration and bark pH explained substantially more of the variance (data not shown). This would not have been expected if all of the effect of the ammonia were mediated through bark pH. It therefore implies that there is an additional effect, for example, which might be attributed to a parallel eutrophication effect of the deposited nitrogen. It should be noted that there was little spatial correlation with climatic parameters, while detailed analysis of these interactions, also with total, wet and dry S and N deposition, will be reported elsewhere.

For the present purpose, of estimating an ammonia critical level, it is sufficient to examine the single-factor relationships between L_{AN} and mean NH_3 concentrations (Figs. 6.5–6.7). As with the landscape and farm scale studies, higher NH_3 concentration was associated with lower values of L_{AN}, as acidophyte lichen communities disappeared to be replaced by nitrophytes, such as *Xanthoria* and *Physcia* spp. (Table 6.1).

The most precise response was found by limiting the dataset to just the bark of oak trees ($R^2 = 0.62$). However, this is more than offset by the benefits of a larger dataset when including the results for all the tree species surveyed for lichen populations. For this reason we consider here only the results of the full dataset. Figure 6.8, shows that for lichens growing on trunks, the UK survey was able to detect a significant response at a mean NH_3 concentration of 1.15 µg m^{-3}. For lichens on twigs a significant response was detected at 1.7 µg m^{-3}. It is important to comment on the difference between these values. Overall, the L_{AN} values were lower for twigs than for trunks (Figs. 6.5, 6.6). Thus L_{AN} value of zero is predicted at 1.6 µg m^{-3} for twigs and at 2.8 µg m^{-3} for trunks. It is therefore clear, that this dataset also supports lichens on twigs as being more sensitive to ammonia than lichens on trunks. The reason for the larger estimate of a significant effect of ammonia for twigs, is simply due to the increased scatter of the twig dataset ($R^2 = 0.36$) compared with the trunk dataset ($R^2 = 0.51$). Combining this information, it is therefore clear that

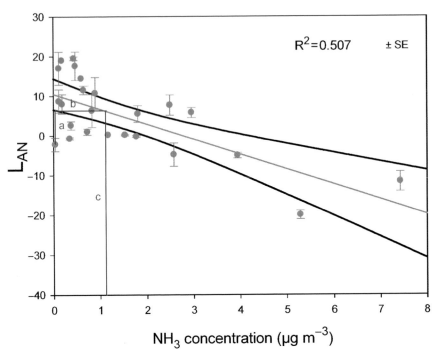

Fig. 6.5 Relationship between the abundance of sensitive lichens (L_{AN} index) on tree trunks and NH_3 concentrations across the UK. The 95% confidence limits of the regression line are shown

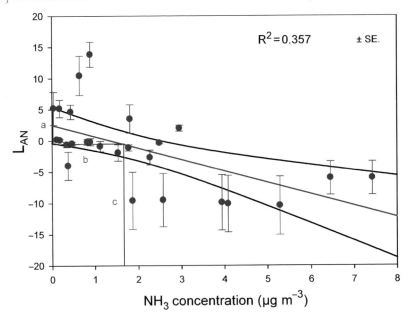

Fig. 6.6 Relationship between the abundance of sensitive lichens (L_{AN} index) on twigs and NH_3 concentrations across the UK. The 95% confidence limits of the regression line are shown

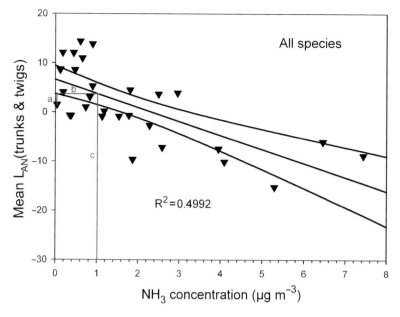

Fig. 6.7 Relationship between the abundance of sensitive lichens (L_{AN} index) and NH_3 concentrations across the UK: The mean of twig and trunk scores. The 95% confidence limits of the regression line are shown

the value of $1.7\,\mu g\ m^{-3}$ represents an overestimate of the actual critical level for lichens on twigs. One option to analyze this further would be to investigate further ways of normalizing the results for natural differences in bark pH, thereby reducing the scatter in the L_{AN} response to NH_3 concentration.

Given the dependence of the minimum significant ammonia concentration on the scatter in the dataset, another simple approach was tested of averaging the twig and trunk L_{AN} results for each site, and comparing this with the mean ammonia concentration (Fig. 6.7). The averaging reduces the scatter in the data, with the result that a lower significant concentration is estimated of $1.02\,\mu g\ m^{-3}$. It should also be noted that this represents a substantial effect on the L_{AN} value, being equivalent to a 40% reduction in the positive L_{AN} value compared with clean conditions.

6.4 Conclusions and Application

Each of the landscape, farm and UK scale studies showed substantial effects of ammonia on the composition of epiphytic lichen communities. Sensitive lichen species, such as *Usnea* (Old Man's Beard), *Bryoria* (Troll's Beard) and *Cladonia* were found to be lost at even modest ammonia concentrations, while nitrogen living species, particularly *Xanthoria* (Yellow Rosette Lichen) increased at their expense.

In all cases, these field data demonstrate that adverse effects on the sensitive lichen communities are occurring at concentrations much smaller than the previously-set UNECE annual critical level for ammonia of $8 \mu g \, m^{-3}$ (van der Eerden et al. 1994). Although, the landscape-scale study has limited value for setting a new critical level (because of uncertainties in the reference ammonia concentration estimates, and a high background concentration), these data indicate effects in the region of $1-2 \mu g \, m^{-3}$. Given the uncertainties, particularly the high background values, it is likely that these are overestimates of the true critical level.

More weight can be assigned to the results of the farm scale and UK scale studies. These are, in fact, quite consistent with the indicative results of the landscape scale study. For the farm scale study, significant effects on the twig lichen communities were detected at $0.7 \mu g \, m^{-3}$, while significant effects on trunk lichens wee detected at $1.7 \mu g \, m^{-3}$. The lower threshold for lichens on twigs is broadly consistent with the increased sensitivity of twig acidophyte lichens to those on trunks. However, it should be noted that a lower value than $1.7 \mu g \, m^{-3}$ would probably have been estimated for the trunks at that site, had measurements at the clean reference site been available for inclusion in the analysis.

The values of significant effect thresholds derived from the UK survey are very similar. Here the results were apparently counter-intuitive in that a lower significant threshold was estimated for lichens on trunks ($1.2 \mu g \, m^{-3}$) compared with lichens on twigs ($1.7 \mu g \, m^{-3}$). However, the overall lower value of L_{AN} for lichens on twigs compared with trunks, and the higher scatter in the twig dataset show that this is simply a result of the statistical method used. The results can therefore be summarized as: a significant response of lichens on trunks to ammonia at $1.15 \mu g \, m^{-3}$, while the real threshold for effects on twigs will be less than this. A simple approach to reduce scatter was applied by taking the mean L_{AN} response for twigs and trunks. As a result of the averaging and more precise response, a significant reduction in L_{AN} was estimated at a smaller threshold of $1.0 \mu g \, m^{-3}$.

Combining the results of the farm scale and UK scale studies, and bearing in mind the supporting data from the landscape scale studies, it is clear that the critical level for epiphytic lichens is much less $2 \mu g \, m^{-3} \, NH_3$. Indeed each of the component studies (Scottish farm study, Thetford, North Wyke, UK survey) showed acidophyte lichens on twigs to be largely eradicated with values above this concentration.

Overall, the farm scale study showed effects on twig lichens above $0.7 \mu g \, m^{-3}$, while the results for the average of twig and trunk lichens in the UK study showed significant effects above $1.0 \mu g \, m^{-3}$. Setting a new ammonia critical level requires the comparison of these results with other datasets (see Cape et al. 2009, this volume). However, on the basis of the datasets presented here, a critical level of $1 \mu g \, m^{-3}$ would be an appropriately conservative estimate.

If $1 \mu g \, m^{-3}$ were taken as a critical level for NH_3, based on the UK results, it can be seen that this equates to a 40% loss of the positive L_{AN} compared with clean conditions (Fig. 6.7). The results can also be expressed simply as a loss of the NH_3 sensitive species. Using the mean L_A score for twigs and trunks from the UK analysis (Fig. 6.9), indicates a reduction of the sensitive acidophyte lichens of 46% (compared with $0 \mu g \, m^{-3}$), or 43%, when compared with a realistic regional background of $0.1 \mu g \, m^{-3}$. For comparison, the mean loss of acidophyte lichens (twigs and trunks) at $2\,\mu g\ m^{-3}$ is 79% or 77% using a reference of 0 or $0.1\,\mu g\ m^{-3}$, respectively.

It is relevant to compare these results with the UK distribution of atmospheric ammonia. In the UK the National Ammonia Monitoring Network includes measurements

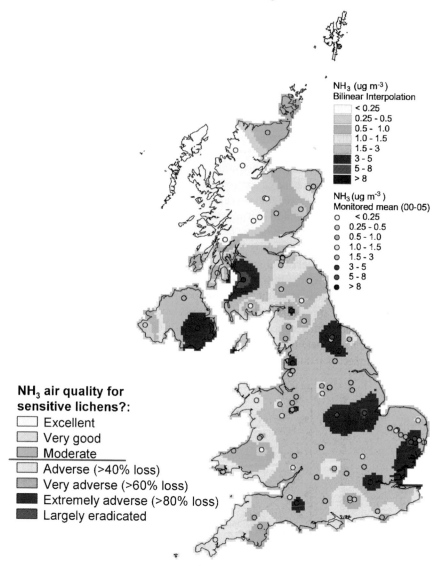

Fig. 6.8 Map of ammonia concentrations across the UK interpolated from the UK National Ammonia Monitoring Network (2000–2005). The potential Air Quality Scale for lichens illustrated is based on a critical level of $1\,\mu g\ m^{-3}$ long term mean concentration (—), with the loss figures based on the reduction of sensitive acidophyte lichens expressed through the L_A index values from the UK wide survey (mean of data from twig and trunk sampling). It should be noted that acidophyte lichen communities on twigs are more sensitive than on trunks and were generally eradicated where NH_3 exceeded $2\,\mu g\ m^{-3}$

at ca. 100 sites (Sutton et al. 2001). These can be used to calibrate atmospheric dispersion models (NEGTAP 2001; Fournier et al. 2005), or simply interpolated to show the typical concentrations. In principle, the former approach is considered to provide the best estimates for fine scale studies, but it is accepted that for the present purpose it is desirable to show the measurements directly. Figure 6.8 illustrates the point measurements with a bilinear interpolation of the UK ammonia monitoring results. Since there is substantial scatter at the local scale, the map should not be used to infer ammonia concentrations at a specific location. However, it provides valuable overview of the regional differences.

Overall, Fig. 6.8 shows that around 65% of the UK would exceed an ammonia critical level of 1 μg m⁻³, although there are substantial regional differences, with the highest percentage exceedance for Northern Ireland and England, and the lowest percentage exceedance for Scotland. The very large exceedance should rightly be taken as a significant concern, if habitats with sensitive lichens are to be maintained in favourable condition. However, it should also be remembered, that such large exceedance values are not new. For example, similar large percentage exceedance values are found across the UK when considering critical loads for nutrient nitrogen deposition (NEGTAP 2001), with much of the critical load exceedance being due to the dry deposition of gaseous ammonia.

Based on a critical level of 1 μg m⁻³, a potential air quality scale for lichens in the UK could be considered, as shown in the caption to the side of Fig. 6.8. The initial banding noted here is based on the following points: (a) a critical level of 1 μg m⁻³, above which significant adverse effects are shown according to the current datasets, (b) the consideration that concentrations much lower than the critical level, give lichen communities a better degree of protection against exceeding the critical level, (c) the percentage of reductions in acidophyte lichens (mean for twigs and trunks,

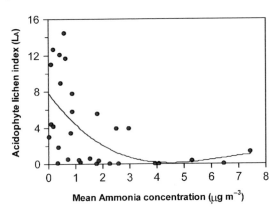

Fig. 6.9 Relationship between the mean value for twigs and trunks of the acidophyte lichen index (L_A) and mean atmospheric ammonia concentration, as derived from the UK-wide survey. The line is fitted with as $L_A = -0.0368x3 + 0.714x2 - 4.196x + 7.89$, $R^2 = 0.32$, where x is ammonia concentration in micrograms per cubic metre. The increase in fitted L_A at high ammonia concentrations is not representative, being influenced by a single outlier point

as expressed by the L_A index) occurring above 1, 1.5, and $3 \mu g \, m^{-3}$ as derived from Fig. 6.9, and (d) the observation that at above $5 \mu g \, m^{-3}$ there are generally few if any acidophyte lichens remaining ($L_A = 0-1.5$ in the UK-wide dataset).

Acknowledgements We are grateful for research funding from the Joint Nature Conservation Committee (JNCC), the Scotland and Northern Ireland Forum For Environmental Research (SNIFFER), Scottish Natural Heritage (SNH), Natural England and the Centre for Ecology & Hydrology (CEH). The measurements in the landscape study areas were supported by the UK NERC GANE research project LANAS and the UK Terrestrial Umbrella research project, jointly funded by the Air and Environment Quality Division of the UK Department for Environment, Food and Rural Affairs (AEQ Division, Defra) and CEH. For the UK scale observations, we thank the field officers of the respective agencies for their work and enthusiasm in testing the lichen recording method, as well as Defra (AEQ Division) for underpinning funding of the UK National Ammonia Monitoring Network. The paper and its presentation was made possible with support from the NinE program of the European Science Foundation and COST 729.

References

Fournier N., Weston K.J., Dore A.J., Sutton M.A. (2005) Modelling the wet deposition of reduced nitrogen over the British Isles using a Lagrangian multi-layer atmospheric transport model. Q. J. Roy Meteor. Soc. 131, 703–722.

Frati L., Santoni S., Nicolardi V., Gaggi C., Brunialti G., Guttova A., Gaudino S., Pati A., Pirintsos S.A., Loppi S. (2007) Lichen biomonitoring of ammonia emission and nitrogen deposition around a pig stockfarm. Environ. Pollut. 146(2), 311–316.

Kermit T., Gauslaa Y. (2001) The vertical gradient of bark pH of twigs and macrolichens in a Picea abies canopy not affected by acid rain. Lichenologist 33, 353–359.

Leith I.D., van Dijk N., Pitcairn C.E.R., Wolseley P.A., Whitfield C.P., Sutton M.A. (2005) Biomonitoring methods for assessing the impacts of nitrogen pollution: refinement and testing, JNCC Report 386. www.jncc.gov.uk/page-3886

NEGTAP (2001) National Expert Group on Transboundary Air Pollution. Department of Environment, Food and Rural Affairs, London.

Pitcairn C.E.R., Skiba U.M., Sutton M.A., Fowler D., Munro R., Kennedy V.K. (2002) Defining the spatial impacts of poultry farm ammonia emissions on species composition of adjacent woodland groundflora using Ellenberg indicators, nitrous oxide and nitric oxide and foliar nitrogen as marker variables. Environ. Pollut. 119, 9–21.

Sutton M.A., Pitcairn C.E.R., Fowler D. (1993) The exchange of ammonia between the atmosphere and plant communities. Adv. Ecol. Res. 24, 301–393.

Sutton M.A., Tang Y.S., Dragosits U., Fournier N., Dore T., Smith R.I., Weston K.J., Fowler D. (2001) A spatial analysis of atmospheric ammonia and ammonium in the UK. Sci. World 1 (S2), 275–286.

Sutton M.A., Leith I.D., Pitcairn C.E.R., van Dijk N., Tang Y.S., Sheppard L.J., Dragosits U., Fowler D., James P.W., Wolseley P.A. (2004a) Exposure of ecosystems to atmospheric ammonia in the UK and the development of practical bioindicator methods. In: Wolseley P.A., Lambley P.W. (eds.) Lichens in a changing pollution environment. English Nature workshop, pp 51–62. English Nature Research Reports, No 525 [ISSN 0967–876X].

Sutton M.A., Pitcairn C.E.R., Whitifield C.P. (2004b) Bioindicator and biomonitoring methods for assessing the effects of atmospheric nitrogen on statutory nature conservation sites (Eds.) JNCC Report 356, 247 www.jncc.gov.uk/page-3236

Theobald M.R., Dragosits U., Place C.J., Smith J.U., Sozanska M., Brown L., Scholefield D., Del prado A., Webb J., Whitehead P.G., Angus A., Hodge I.D., Fowler D., Sutton M.A. (2004) Modelling nitrogen fluxes at the landscape scale Water Air Soil Poll.: Focus 4(6), 135–142.

van der Eerden L.J.M, Dueck T.A. Posthumus A.C., Tonneijck A.E.G. (1994) Assessment of critical levels for air pollutant effects on vegetation: some considerations and a case study on NH$_3$.
In: Ashmore M.R., Wilson R.B. (eds.) Critical Levels of Air Pollutants for Europe. Proceedings
of the UNECE Workshop on Critical Levels, Egham, pp 55–63. Air Quality Division,
Department of the Environment, London.

van Dobben H.F., Ter Braak C.J.F. (1998) Effects of atmospheric NH3 on epiphytic lichens in the
Netherlands: the pitfalls of biological monitoring. Atmos. Environ. 32 (3), 551–557.

van Herk C.M. (1999) Mapping of ammonia pollution with epiphytic lichens in the Netherlands.
Lichenologist 31, 9–20.

Wolseley P.A., James P.W. (2002) Using lichens as biomonitors of ammonia concentrations in
Norfolk and Devon. Br. Lichen Soc. Bull. 91, 1–5.

Wolseley P.A., Pryor K.V. (1999) The potential of epiphytic twig communities on Quercus petraea
in a Welsh woodland site (Tycanol) for evaluating environmental changes. Lichenologist 31,
41–61.

Wolseley P.A., James P.W., Sutton M.A., Theobold M.R. (2004) Using lichen communities to
assess changes in sites of known ammonia concentrations. In: Wolseley P.A., Lambley P.W.
(eds.) Lichens in a Changing Pollution Environment. English Nature workshop. pp 89–98.
English Nature Research Reports, No 525 [ISSN 0967–876X].

Wolseley P.A., James P.W., Theobald M.R., Sutton M.A. (2006) Detecting changes in epiphytic
lichen communities at sites affected by atmospheric ammonia from agricultural sources.
Lichenologist 38(2), 161–176.

Chapter 7
Mapping Ammonia Emissions and Concentrations for Switzerland – Effects on Lichen Vegetation

Beat Rihm, Martin Urech, and Kathrin Peter

7.1 Introduction

This summary presents new maps of ammonia emissions and concentrations produced for Switzerland, as well as selected results of a study investigating the effect of ammonia on the vegetation of lichens.

The mapping of ammonia was initiated in the early 1990s with the aim to calculate the deposition of nitrogen and subsequently the exceedance of critical loads of acidity and nutrient nitrogen in Switzerland with high spatial resolution (i.e. 1 × 1 km^2). In the first approach, ammonia emissions were mapped using livestock statistics spatially related to the municipalities and land-use maps (FOEFL 1994, 1996). Ecosystem-dependent depositions were directly calculated from the amount of ammonia emitted within a radius of 10 km. In this 'budget' model, the overall ratio of emission to deposition was constrained to a fixed value (45%). As systematic measurements of ammonia concentrations did not exist at that time, the results were checked mainly with deposition measurements (e.g. BUWAL 1994) and EMEP results.

Later, the ammonia emission inventory was improved by using livestock statistics spatially related to the locations of farms, and the atmospheric dispersion of ammonia was modelled by an empirical function of concentration vs. distance with a spatial resolution of 1 ha. A good correlation was obtained between modelled and measured concentration values for 17 sites (Rihm and Kurz 2001). In 1999, a continuous ammonia monitoring program was started with more than 30 sites. It allowed to refine the method by adjusting the function (in the case of the emitting cell) so that an optimum correlation was obtained (Thöni et al. 2004).

Since then the emission inventory was further improved using detailed livestock statistics (SFSO 2003), stratified emission factors for agricultural emissions (Reidy et al. 2007) and modelled ammonia emissions from traffic. In addition, the ammonia emissions in neighbouring countries were included (adjacent EMEP-grids).

Beat Rihm
Meteotest, Fabrikstrasse 14, CH-3012 Berne, Switzerland

M. Sutton, S. Reis and S.M.H. Baker (eds), *Atmospheric Ammonia,*
© Springer Science+Business Media B.V. 2009

7.2 Mapping of Emissions

Agricultural emissions were calculated with livestock statistics of the year 2000 (SFSO 2003). These include the approximate location of the farm building, the numbers of animals for 24 livestock categories and the farm class, which is a combination of region, altitude and type of production (cattle/crops/mixed/special). The emission factors from Reidy et al. (2007) are stratified in the 24 livestock categories, 36 farm classes and 5 emission stage (housing, hardstandings, grazing, manure storage and manure application).

It was assumed that the emissions from housing, hardstandings and manure storage originate from the farm buildings (point source). The emissions from grazing and manure application were assumed to be distributed evenly over the agricultural areas in each municipality (area source) since fields cannot be allocated to single farms on the national scale. There are around 3,000 Swiss municipalities with an average area of 14 km^2. The agricultural areas are identified by a land-use map with a resolution of one hectare and 24 categories (SFSO 1997). Figure 7.1 shows a detail of the resulting map of emissions modelled on a hectare-grid, which also includes non-agricultural sources.

In 2000, the total ammonia emissions estimated by Reidy et al. (2007) were 44.6 kt NH$_3$-N year^{-1} (including 3.4 kt NH$_3$-N from non-agricultural sources). On the average these are 10.8 kg N ha^{-1}.

7.3 Modelling of Concentrations

The model used to calculate the dispersion of NH$_3$ is a typical 'simpler model' (SM) implemented in a geographical information system (GIS). It is based on an analytical relationships describing the decrease of ammonia concentration with distance from the source: $C = 15.197 \, D^{-1.9332}$, where C is the annual mean concentration (μg m^{-3}) induced by 1 kg NH$_3$ emitted per year from a point source at a distance D (in meters). The function describes a distance-profile developed by Asman and Jaarsveld (1990).

They applied the model for different countries in Europe (The Netherlands, United Kingdom, Belgium, Denmark, Sweden). Although the Swiss climate differs from the climate in those countries, the results correspond quite well to monitoring data (Fig. 7.2). Probably there are some contrary effects compensating each other: On one hand, the low wind speed in Switzerland would lead to higher concentrations on the ground. On the other hand, the well mixed land-use pattern (cropland/grassland, forests, and settlements) has a high surface roughness on the landscape level. Furthermore, the mixed pattern of sources and sinks in large parts of the country leads to a relatively high local deposition and with low wind speeds and complex topography, the influence of large-scale wind directions is weakened by local effects such as thermal and cold air flow.

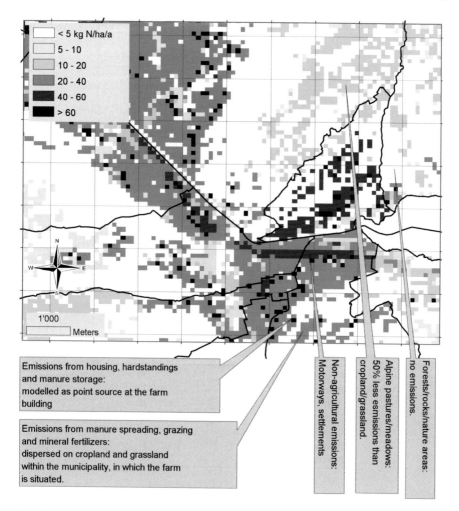

Fig. 7.1 Ammonia emission (NH_3-N) in a typical pre-alpine situation with a valley (dark gray), forested slopes (light gray) and alpine pastures/rocks (white)

The modelled annual mean concentration of ammonia in Switzerland is $1.30\,\mu g\;m^{-3}$ (minimum $0.1\,\mu g\;m^{-3}$, 95 percentile $3.5\,\mu g\;m^{-3}$, maximum $114\,\mu g\;m^{-3}$).

7.4 Calculation of Deposition

Deposition of ammonia was calculated from the concentration map by applying land-use specific deposition velocities (Table 7.1). The area-weighted mean deposition velocity for Switzerland is $17\,mm\;s^{-1}$ (including the large high-alpine areas without vegetation).

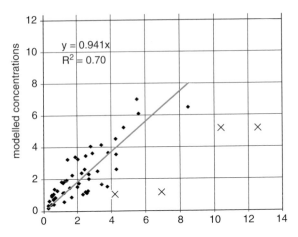

Fig. 7.2 Modelled ammonia concentrations vs. concentrations observed at 53 monitoring sites between 2000 and 2004 (Thöni et al. 2004, modified). Sites situated very near to farms or other sources are excluded from the regression function (cruxes) as they can hardly be modelled based on the national emission map. Units: annual mean, μg m^{-3}

Table 7.1 Deposition velocities of ammonia used for different land-use types in Switzerland

Land-use type	mm s^{-1}
Coniferous forest	30
Deciduous forest	22
Agricultural land	12
Surface water	20
Unproductive vegetation	20
Settlement	8
No vegetation	5

The total ammonia deposition calculated by this procedure amounts to 23.9 kt NH$_3$-N year^{-1} (or 5.8 kg NH$_3$-N ha^{-1} year^{-1}). The wet deposition and aerosol deposition modelled on a 1 × 1 km^2 raster (FOEFL 1996; Rihm and Kurz 2001) are 22.1 and 2.9 kt NH$_3$-N year^{-1}, respectively. Thus, the total Swiss deposition of reduced nitrogen amounts to 48.9 N year^{-1} which can be compared with the result of EMEP for the year 1998 (48.3 kt N).

7.5 Effects on the Vegetation of Lichens

The modelled estimates of ammonia concentrations and deposition have been used in several studies focusing on ecological effects or risks.

As an example, Fig. 7.3 shows the relationship between the vegetation of epiphytic lichens and modelled ammonia concentrations (annual mean in 2000)

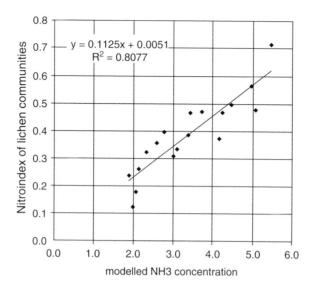

Fig. 7.3 Vegetation of lichens (Nitroindex) vs. modelled ammonia concentrations ($\mu g\ m^{-3}$)

in three regions of the Swiss Plateau (Urech M. and Peter K. 2006, unpublished). The frequency of 40 lichen species was determined on 648 deciduous trees. It is measured by a standardised grid with 10 equal squares. The frequency is defined as the number of grid squares populated by a certain species. Thus, the frequency values of a lichen species are between 0 and 10.

In order to evaluate eutrophication, the so-called *Nitroindex* was defined as follows: the frequencies of four anitrophilous species (*Evernia prunastri*, *Pseudevernia furfuracea*, *Hypogymnia physodes* and *Parmelia saxatilis*) are subtracted from the frequencies of five nitrophilous species (*Physcia adscendens*, *Phaeophyscia orbicularis*, *Physconia grisea*, *Xanthoria parietina* and *Xanthoria fallax*) and divided by the sum of all 40 frequency values of the tree. Resulting values for the Nitroindex range from −1 to +1 with higher values indicating higher eutrophication.

The annual mean of ammonia concentration was modelled for the location of each tree. The trees were spatially aggregated in 18 sub-regions, each encompassing approximately 35 trees. The mean values of ammonia and Nitroindices for these sub-regions are shown in Fig. 7.3.

There is a clear correlation between the lichen-based Nitroindex and NH_3 concentrations ($R^2 = 0.81$). Lichens are not only very sensitive to air pollutants in general. They seem to be also an appropriate bio-indicator for eutrophying effects of NH_3.

Acknowledgements This study is financed by the Swiss Federal Office for the Environment (FOEN). We would like to thank Beat Achermann (FOEN, Air Pollution Control and NIR Division) for his support and the fruitful discussions in preparing this study.

References

Asman W. A. H., Van Jaarsveld H. A. (1990) A Variable-Resolution Statistical Transport Model Applied for Ammonia and Ammonium. National Institute of Public Health and Environmental Protection (RIVM), Bilthoven, The Netherlands.

BUWAL (ed.) (1994) Stickstoffeintrag aus der Luft in ein Naturschutzgebiet. Bundesamt für Umwelt, Wald und Landschaft, Umwelt-Materialien Nr. 28. Berne.

FOEFL (ed.) (1994) Critical Loads of Acidity for Forest Soils and Alpine Lakes – Steady State Mass Balance Method. Federal Office of Environment, Forests and Landscape, Bern. Environmental Series Air, 238, 68 S.

FOEFL (ed.) (1996) Critical Loads of Nitrogen and Their Exceedances - Eutrophying Atmospheric Deposition. Federal Office of Environment, Forests and Landscape, Bern. Environmental Series Air , 275, 74p.

Reidy B., Rihm B., Menzi H. (2007) A New Approach for the Calculation of the Swiss Inventory on Ammonia Emissions from Agriculture Based on a Stratified Farm Survey and Farm-Specific Model Calculations. Atmospheric Environment, doi:10.1016/j.atmosenv.2007.04.036 Vol. 42, May 2008, pp. 3266–3276.

Rihm B., Kurz D. (2001) Deposition and Critical Loads of Nitrogen in Switzerland. Water Air and Soil Pollution, 130, 1223–1228.

SFSO (1997) Arealstatistik der Schweiz, Erhebung 1992/97 (BN24). Swiss Federal Statistical Office (SFSO), Neuchâtel, Switzerland.

SFSO (2003) Livestock Statistics 2000. Database extract by D. Bohnenblust, Swiss Federal Statistical Office (SFSO), Neuchâtel, Switzerland.

Thöni L., Brang P., Braun S., Seitler E., Rihm B. (2004) Ammonia Monitoring in Switzerland with Passive Samplers: Patterns, Determinants and Comparison with Modelled Concentrations. Environmental Monitoring and Assessment, 98, 95–107.

Olendrzyński K., Dębski B., Skośkiewicz J., Kargulewicz I., Fudała J., Hławiczka S., Cenowski M. (2004) Inwentaryzacja emisji do powietrza SO2, NO2, NH3, CO, pyłów, metali ciężkich, NMLZO i TZO w Polsce za rok 2002. IOŚ.

Olendrzyński K., Dębski B., Kargulewicz I., Skośkiewicz J., Cieslinska J., Fudała J., Hławiczka S., Cenowski M. (2007) Emission Inventory of SO2, NOx, NH3, CO, PM, NMVOCs, HMs, and POPs in Poland in 2005. UNECE-EMEP/Poland Report, National Emission Centre, IOŚ, 42.

Urech M., Peter K., 2006. Räumlicher und Zeitlicher Zusammenhang zwischen Flechtenvegetation und Stickstoffbelastung. Projektbericht zuhanden des BAFU, Abteilung Luftreinhaltung und NIS, Sektion Luftqualität, Bern (unpublished).

Chapter 8
Over Which Averaging Period Is the Ammonia Critical Level Most Precautionary?

Mark A. Sutton, Addo van Pul, Ferd Sauter, Y. Sim Tang, and Laszlo Horvath

8.1 Introduction

The ammonia critical level set at the UNECE Egham Workshop in 1992 (van der Eerden et al. 1994) was based on mean NH_3 concentrations for different averaging periods. These values were derived from experimental studies and a toxicological model, so that a high value for the critical level was given for short exposure periods (acute exposure), and a low for long exposure periods (chronic exposure). The critical level values set applied to all vegetation types and are summarized in Table 8.1.

It may be noted that in a subsequent review, WHO (2000) recommended the discontinuation of the monthly and hourly critical levels in regulatory assessment, although the arguments for doing so were not clearly explained. For the purpose of the present scientific assessment, all the time periods are used (i.e., hourly, daily, monthly, yearly).

An important question in using critical levels is where to place the emphasis in assessment and development of regulations. For example, should there be a strong focus on the annual or the hourly critical level? If the main focus in on the annual level, then ambient monitoring with a low temporal resolution (e.g. monthly monitoring) is sufficient to assess the extent to which the critical level is met. Monitoring of this kind can be reliably achieved using low-cost methods (e.g., Ferm and Rodhe 1997; Sutton et al. 2001; Tang et al. 2001; Tang and Sutton 2003). Conversely, if there should be a strong focus on using the daily or hourly critical levels, this implies the need for continuous monitoring of ammonia at many sites. Such continuous monitoring is possible, but is very expensive using current techniques (e.g., Wyers et al. 1993; Burkhardt et al. 1998; Erisman et al. 2001).

Considering these points, it is extremely relevant to identify over which averaging period the NH_3 critical level is most precautionary. That is, for any location, which critical level will tend to be exceeded first, the hourly, daily, monthly or the annual value? If the annual critical level is most precautionary, this has the advantage that

Mark A. Sutton
Centre for Ecology & Hydrology, Bush Estate, Penicuik, Midlothian, EH26 0QB, United Kingdom

M. Sutton, S. Reis and S.M.H. Baker (eds), *Atmospheric Ammonia,*   93
© Springer Science + Business Media B.V. 2009

Table 8.1 Ammonia critical level values set for different averaging periods following the review in 1992 at the UNECE Egham Workshop (van der Eerden et al. 1994)

Exposure period	Critical level set at Egham ($\mu g\ m^{-3}$)
1 h	3,300
1 day	270
1 month	23
1 year	8

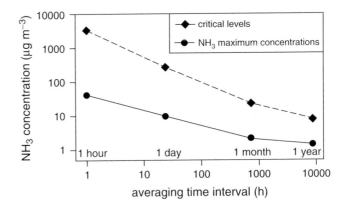

Fig. 8.1 Comparison of the ammonia critical levels set at Egham (van der Eerden et al. 1994) for different averaging periods with measured maximum ammonia concentrations recorded at Bush Estate, near Edinburgh. The critical level was not exceeded over any averaging period, but was closest to exceedance (as a fraction of the critical level) for the annual average

appropriate NH_3 air monitoring strategies can be devised which are implemented at much smaller cost. This is important if ammonia monitoring is to be considered in a regulatory context, as the principles of "better law-making" should encourage the use of cost-effective approaches in environmental regulation.

The present short paper examines examples of existing ammonia air monitoring data, from sites where these are available at a high time resolution, to see whether clear recommendations can be made.

8.2   Background

The background to this study was provided by an earlier analysis of continuous ammonia monitoring at a rural location in southern Scotland (near the Edinburgh Research Station of CEH, Bush Estate, Penicuik), as reported by Burkhardt et al. (1998). Based on 15 min time resolution measurements over 18 months, the authors calculated the maximum measured concentrations for each of the hourly, daily, monthly and annual values. They then compared these values with the ammonia critical level (see Fig. 8.1).

Burkhardt et al. (1998) found that the critical level was not exceeded for any averaging period for the Bush site. However, they also found that the value closest to exceedance (expressed as percentage value/critical level) occurred for the annual critical level, while the hourly maximum was the smallest percentage fraction of the critical level (i.e. was furthest from reaching exceedance). This is illustrated in Fig. 8.1 by the fact that, when plotted on a log-log scale the two lines of maximum reported concentrations and the critical levels increasingly diverge at lower averaging times. The shapes of the two profiles were also extremely similar, being linear up to 1 month, which can presumably be related to the statistics of different averaging periods. Essentially, the conclusion of Burkhardt et al. (1998) was that if the annual critical level is not exceeded, then the critical levels for shorter periods will also not be exceeded, i.e., the annual value is the most precautionary.

Two main caveats exist regarding this conclusion from Burkhardt et al. (1998). The first is that the assessment was derived from only one site. At other sites with a different mix of ammonia sources (and different geometric standard deviation to the air concentrations), the overall relationship might be different. Secondly, the annual average concentration had a value which was much smaller than the critical level, and the temporal distribution of pollutant exposure might be different at higher values.

8.3   Methods

To account for these issues in the present analysis, we examined several datasets on measured ammonia concentrations, including sites in a wide range of climates (comparing the UK, with the Netherlands and Hungary) and including sites (from the Netherlands) with extremely high NH_3 concentrations. We calculated the maximum recorded mean concentrations for each of the averaging periods: annual, monthly, daily, and, where available, hourly. The new datasets were analyzed for the period 2000–2005. In the Netherlands, data were obtained using the AMOR continuous wet denuder system, which is a more automated model of the AMANDA system (Wyers et al. 1993). For Hungary, the measurements were collected using the standard reference method, being manual daily ammonia denuders (Ferm 1979), which have been in operation at the Hungarian monitoring site since 1981 (Horvath and Sutton 1998).

8.4   Results

The results of the analysis are summarized in Fig. 8.2. For each of the eight sites considered, it is clear that the slope of the line of maximum measured ammonia concentration vs averaging period is smaller than that of the difference in the critical level. This means that the critical level becomes more precautionary over longer averaging times, and that, of the different Egham critical levels, the annual value is the most precautionary.

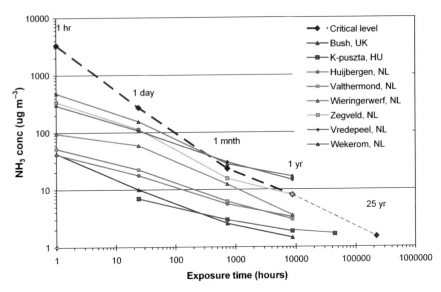

Fig. 8.2 Relationship between the Egham ammonia critical levels (van der Eerden et al. 1994) for different averaging periods (hourly, daily, monthly, annual) and the maximum ammonia concentrations measured at different sites in the UK, Hungary and the Netherlands. The dashed line indicates a hypothetical extrapolation of the critical level to an averaging period of 25 years (giving 1.5 µg m^{-3}), which is comparable with the "protection period" for empirical critical loads for nitrogen deposition (Hornung et al. 1995)

Actual exceedance of the Egham critical levels occurred at two sites (Vredepeel and Wekerom in the Netherlands) for both annual and monthly periods. However, consistent with the overall result, the exceedance (as a percentage) at these sites was smaller for the monthly maximum than the annual, while there was no exceedance of the daily and hourly critical levels at any of the monitoring sites considered.

While the general pattern is clear, there are some minor exceptions, shown by scatter in the lines. For example, at Wieringerwerf, in the Netherlands, the maximum monthly value rather than the maximum annual value was closest to the critical level (as a percentage).

8.5   Discussion

Overall, the annual value is found to be the most precautionary of the Egham ammonia critical levels (van der Eerden et al. 1994). In Fig. 8.2, the slopes of the different lines are extremely similar, which may be attributed to similar geometric standard deviations for atmospheric ammonia concentrations due to dispersion in the vicinity of many ground level sources. This finding supports the original conclusions of Burkhardt et al. (1998) as having been broadly representative.

Nevertheless, the results for Bush actually show one of the highest slopes in Fig. 8.2. This may be due to this being a site which is relatively close to a few large NH_3 sources (livestock farm 300 m distant), but being in an otherwise clean location with low ammonia emissions. The consequence is that the conclusion from Burkhardt et al. (1998) was conservative, and is even more strongly made by the comparison with new data.

By contrast to Bush, one of the smallest slopes in Fig. 8.2 was recorded for K-puszta in Hungary. This may reflect K-puszta as being a background site, several kilometres from sources. As a result, dispersion from farm point sources may make a smaller contribution to ammonia concentrations at this site, while vegetation surface-atmosphere exchange plays a larger role in controlling NH_3 concentrations (Horvath and Sutton 1998; Sutton et al. 2003). As vegetation represents a highly diffuse source of trace emissions, air concentrations dominated by this source are expected to be characterized by a smaller geometric standard deviation.

Overall, the finding of this study is that the environmental assessment can safely be focused on the annual critical level as being the most precautionary of the existing Egham critical levels (as compared with the hourly, daily and monthly values). Hence long-term monitoring using (e.g. monthly sampling and annual averages) is sufficient for ammonia critical level assessment, when using the Egham values reported by van der Eerden et al. (1994).

With this perspective, it is also relevant to consider even-longer averaging times than those considered in the Egham critical levels. It has already been noted that the slopes of the lines in Fig. 8.2 for measured concentration maxima vs averaging period are substantially smaller than the slope of the Egham critical levels vs averaging period. This suggests that multi-year, or "long term" critical levels are likely to be more precautionary than the annual critical level. As a result a long term critical level for ammonia needs to be considered in revising the Egham ammonia critical levels.

In this context, it is important to note that the Egham critical levels provide a maximum "protection period" of only 1 year exposure. This may be compared with the protection period currently assumed for empirical critical loads of nitrogen deposition. In that case, it has been stated that the critical load cannot be assumed to give protection for periods longer than 20–30 years (Hornung et al. 1995). Hence, if ecological effects of gaseous ammonia accumulate over periods longer than 1 year (for which evidence is presented elsewhere in this volume), then the Egham critical levels are less precautionary than critical loads simply due to differences of averaging period. As a result of this difference, it is not surprising that the current empirical critical loads (Achermann and Bobbink 2003) are much more precautionary than the Egham ammonia critical loads (van der Eerden et al. 1994). For example, Burkhardt et al. (1998) showed that, even though the critical level was not exceeded their measurements (Fig. 8.1), the nitrogen critical load would have been significantly exceeded.

To address these issues in a simple way, we show on the log-log plot of Fig. 8.2 a hypothetical linear extension of the Egham critical load to a 25 year protection period. We use this period as it is midway between 20–30 years, so as to be broadly comparable with the empirical critical loads. It should be emphasized that in this paper we make no statement about the ecological/biological justification for such and extrapolation. Such a "long term critical level" must also be justified from the findings from ammonia effects studies. However, it can be seen that a critical level of the order of $1.5 \mu g \ m^{-3}$ is obtained purely on the basis of extrapolation of the Egham critical level to a period comparable with the critical loads. If such a value is combined with typical deposition velocities to calculate NH_3 dry deposition, and is then added to other sources of nitrogen deposition (e.g. wet deposition), it is seen to be much more comparable with the current critical loads, which are typically 10–20 kg N ha^{-1} $year^{-1}$ for many semi-natural habitats (Achermann and Bobbink 2003).

From Fig. 8.2, it can bee seen that substantial exceedance of the long-term critical level can be expected. It is therefore vital that revision of the critical levels should include consideration of a long-term value for ammonia. A target protection period of c. 20–30 years is appropriate (as consistent with the empirical critical loads). Such a long-term critical level can then be compared with on-going NH_3 monitoring data and future projections. In this way the approach more closely matches that taken in impact assessments for empirical critical loads of nitrogen.

While, in general, the annual critical level is the most precautionary of the Egham values (hourly, daily, monthly, annual), one potential exception is for sites that occur in the vicinity of occasional field spreading of manures. Such sites can show very occasional monthly peak ammonia concentrations, as for example discussed by Tang et al. (2009, see Section 3.3, this volume). This effect may explain why at one site in the present analysis (Wieringerwerf, NL), the monthly critical level was closest to the UNECE Egham critical level. For this reason, while the main emphasis for future critical levels should be on annual and long term values, it is advisable to retain a monthly critical level.

In practice, NH_3 monitoring for assessment of critical levels can therefore be made using monthly time-integrated monitoring. This is a great advantage as it means that future ammonia monitoring activities for regulatory impact assessment linked to effects on vegetation can be conducted using low cost monitoring methods (e.g. passive diffusion samplers, or low-cost denuders, Ferm and Rodhe 1997; Kirchner et al. 1999; Sutton et al. 2001; Tang et al. 2001). Of course, high time resolution NH_3 monitoring remains extremely useful for understanding processes, the statistical relationships (e.g. as examined in this paper) and source-receptor relationships (Sutton et al. 2003). However, as such continuous measurements are rather expensive, they can be focused on a few key air monitoring stations where the aim is to develop an integrated understanding of atmospheric concentrations and surface-atmosphere exchange fluxes.

Acknowledgements We are grateful for funding of the monitoring activities from UK Defra, the Dutch Ministry of the Environment and the Hungarian Meteorological Service. This analysis was conducted with support from COST 729, NitroEurope and the ESF NinE programme.

References

Achermann B., Bobbink R. (eds.) (2003) Empirical critical loads for nitrogen. (Proceedings of an Expert Workshop, 11–13 November 2002, Berne). Environmental Documentation No. 164. Swiss Agency for the Environment, Forests and Landscape, Berne.

Burkhardt J., Sutton M.A., Milford C., Storeton-West R.L., Fowler D. (1998) Analysis of ammonia concentrations at a site in S. Scotland from continuous measurements over 2 years. Atmos. Environ. 32 (3), 325–332.

Erisman J.W., Otjes R., Hensen A., Jongejan P., van den Bulk P., Khlystov A., Mols H., Slanina S. (2001) Instrument development and application in studies and monitoring of ambient ammonia. Atmos. Environ. 35, 1913–1922.

Ferm M. (1979) Method for determination of atmospheric ammonia. Atmos. Environ. 13, 1385–1393.

Ferm M., Rodhe H. (1997) Measurements of air concentrations of SO_2, NO2 and NH3 at rural and remote sites in Asia. J. Atmos. Chem. 27, 17–29.

Hornung M., Sutton M.A., Wilson R.B. (1995) Mapping and modelling of critical loads for nitrogen - a workshop report (Eds.) (Report of the UN-ECE workshop, Grange-over-Sands, 24–26 October 1994). Institute of Terrestrial Ecology, Edinburgh, 207

Horvath L., Sutton M.A. (1998) Long term record of ammonia and ammonium concentrations at K-puszta, Hungary. Atmos. Environ. 32, 339–344.

Kirchner M., Braeutigam S., Ferm M., Haas M., Hangartner M., Hofschreuder P., Kasper-Giebl A., Roemmelt H., Striedner J., Terzer W., Thoeni L., Werner H., Zimmerling R. (1999) Field inter-comparison of diffusive samplers for measuring ammonia. J. Environ. Monit. 1, 259–265.

Sutton M.A., Tang Y.S., Miners B., Fowler D. (2001) A new diffusion denuder system for long-term, regional monitoring of atmospheric ammonia and ammonium. Water Air Soil Poll.: Focus 1, 145–156.

Sutton M.A., Asman W.A.H., Ellerman T., van Jaarsveld J.A., Acker K., Aneja V., Duyzer J.H., Horvath L., Paramonov S., Mitosinkova M., Tang Y.S., Achermann B., Gauger T., Bartnicki J., Neftel A., Erisman J.W. (2003) Establishing the link between ammonia emission control and measurements of reduced nitrogen concentrations and deposition. Environ. Monit. Assess. 82 (2), 149–185.

Tang Y.S., Sutton M.A. (2003) Quality management in the UK national ammonia monitoring network. In: Borowiak A., Hafkenscheid T., Saunders A., Woods P. (eds.) (Proceedings of the International Conference: QA/QC in the Field of Emission and Air Quality Measurements: Harmonization, Standardization and Accreditation, held in Prague, 21–23 May 2003) pp 297–307. European Commission, Ispra, Italy [ISBN: 92-894-6523-9].

Tang Y.S., Cape J.N., Sutton M.A. (2001) Development and types of passive samplers for NH_3 and NO_x. Sci. World 1, 513–529.

van der Eerden L.J.M, Dueck T.A., Posthumus A.C., Tonneijck A.E.G. (1994) Assessment of critical levels for air pollutant effects on vegetation: some considerations and a case study on NH_3. In: Ashmore M.R., Wilson R.B. (eds.) Critical Levels of Air Pollutants for Europe. (Proceedings of the UNECE Workshop on Critical Levels, Egham) pp 55–63. Air Quality Division, Department of the Environment, London.

WHO (2000) Air Quality Guidelines for Europe (2nd edition). WHO Regional Publications, European Series No. 91., World Health Organization, Regional Office for Europe, Copenhagen. www.euro.who.int/ document/e71922.pdf

Wyers G.P., Otjes R.P., Slanina J. (1993) A continuous-flow denuder for the measurement of ambient concentrations and surface-exchange fluxes of ammonia. Atmos. Environ. 27, 2085–2090

Chapter 9
Macrolichens on Twigs and Trunks as Indicators of Ammonia Concentrations Across the UK – a Practical Method

Patricia A. Wolseley, Ian D. Leith, Netty van Dijk, and Mark A. Sutton

9.1   Introduction

Lichen community composition on acid-barked trees has been shown to respond to increasing atmospheric ammonia (NH_3) concentrations by loss of acidophyte species and an increase in nitrophyte species. A simple method of sampling selected acidophyte and nitrophyte lichens on trunks and twigs of trees in the vicinity of ammonia monitoring sites across the UK allowed us to test the correlation of lichen communities with ammonia concentrations across the climatic and vegetation zones of the UK.

Sites were selected and field staff from the conservation and regulatory agencies introduced to standard lichen sampling and identification techniques at a workshop co-organised by NHM and CEH Edinburgh. L_{AN} (Lichen Acidophyte Nitrophyte) values were calculated for all sites, based on the frequency of acidophyte and nitrophyte macrolichens on trunks and twigs. Bark samples from trunk and twig were collected and surface bark pH measured in the lab, to test the correlation of acidophyte and nitrophyte communities with bark pH.

L_{AN} values on tree trunks and twigs are correlated with NH_3 concentrations across the UK. However, acid-barked trees, e.g. Oak(*Quercus*), provide a better correlation with NH_3 than basic-barked species. Lichens on twigs respond to increasing ammonia at lower concentrations than lichens on trunks, and considerably lower than the annual critical level ($8 \, \mu g \, NH_3 \, m^{-3}$), confirming that the twig L_{AN} can act as a sensitive early warning system of changes in atmospheric NH_3.

Lichens have long been used as indicators of environmental health and in particular as indicators of air quality. In the 1960s and 1970s lichen communities were observed to be highly correlated with sulphur dioxide deposition, and a number of different methods were developed for their use as bioindicators across Europe, involving indicator species and lichen frequency (Hawksworth and Rose 1970; Nimis et al. 1990; VDI 2005; Asta et al. 2002). The majority of these methods

P. A. Wolseley
Natural History Museum, London SW7 5BD, United Kingdom

M. Sutton, S. Reis and S.M.H. Baker (eds), *Atmospheric Ammonia,*
© Springer Science + Business Media B.V. 2009

were devised to detect a loss of lichen diversity under conditions of increasing SO_2. However, it became apparent that under conditions of falling SO_2 and increasing nitrogen deposition that diversity in urban areas was increasing but that the lichen community was considerably altered (Hawksworth and McManus 1989). This meant that simple diversity or frequency scales could no longer be used to detect changes in atmospheric quality.

In the Netherlands there was concern over the effects of ammonia on lichen communities (van Dobben 1996), and this led to the development of acidophyte and nitrophyte indicator scales correlating loss of acidophyte lichens and increase in nitrophyte species with increasing ammonia (van Herk 1999). Further work by van Herk (2001) demonstrated that an increase in nitrophytes was also correlated with increasing bark pH and that long distance nitrogen deposition was affecting sensitive lichen communities across Europe (van Herk et al. 2003).

The effects of environmental conditions on colonisation of new substrata such as twigs was observed on oak twigs in Wales (Wolseley and Pryor 1999). When frequency of lichens on both twigs and trunks of standard oak trees was tested across known gradients in ammonia concentrations in Thetford and Devon, results showed that at mean NH_3 concentrations $> 2\,\mu g\ m^{-3}$ the twigs were dominated by nitrophytes while acidophytes continued to be present on the trunks as relics of former conditions (Wolseley et al. 2006).

The overall concern that nitrogen was a major factor in causing changes in vegetation led to the wider sampling of epiphytic lichens on trunks and twigs across Britain and to the testing of a simplified method using frequency of macrolichens (Sutton et al. 2004, 2005). This was trialled across Britain using non lichenologists to sample macrolichens on twigs and trunks of trees in the vicinity of sensitive sites where ammonia concentrations were monitored (Leith et al. 2005; Wolseley et al. 2007). The method and results of this survey are summarized here together with implications for the wider use of lichens as indicators in urban and rural Britain.

The objectives were to test a simple method of sampling selected macrolichens on trunks and twigs of trees in order to establish the correlation with ammonia concentrations across the climatic and vegetation zones of the UK.

9.2   Method

Thirty sites were selected in the vicinity of ammonia monitoring stations where adjacent trees were available, and where local staff could undertake the recording of lichens.

A training session was held at CEH Bush, co-organised by NHM and CEH, where participants were introduced to standard lichen sampling and identification techniques and kits for recording issued (Leith et al. 2005).

9.2.1 Sampling Method

- **Trunks**: Tree trunks were sampled on 4 cardinal points using a ladder of 5 consecutive $10\,cm^2$ quadrats placed at a height of 1.50 m on each aspect of the trunk (Fig. 9.1). Macrolichens were recorded as present in each quadrat providing a maximum possible frequency of 20 for each species.
- **Twigs**: Up to 10 exposed and accessible twigs in the same area were sampled, if possible on the same trees as used for trunk sampling. Macrolichen presence was recorded in 3 units along the twig; 0–50, 51–100 cm and 100–150 cm. Where a lichen occurred in all 3 units the score was doubled, the maximum score for any species = 6.

Fig. 9.1 Ladder quadrat placed on oak tree trunk at Bush Estate, Midlothian

Table 9.1 Macrolichens used as indicators

Acidophyte macrolichens	Nitrophyte macrolichens
Cladonia species	Hyperphyscia adglutinata
Evernia prunastri	Phaeophyscia orbicularis
Flavoparmelia caperata	Physcia adscendens
Hypogymnia species	Physcia tenella
Platismatia glauca	Xanthoria candelaria
Pseudevernia furfuracea	Xanthoria parietina
Usnea and/or Bryoria species	Xanthoria polycarpa

Scores for indicator species for acidophytes (L_A) and nitrophytes (L_N) (Table 9.1) were totalled for trunks and twigs and mean values calculated for each site. The index for the site was calculated by subtracting L_N from L_A values, giving an index of Lichen Atmospheric Nitrogen (L_{AN}) where high values indicate good air quality and low or negative values indicate high nitrogen and poor air quality.

9.2.2 Bark pH

Bark samples > 1 cm in diameter were collected from trunks. Bark pH was measured using a Jencons flat-tip electrode on the bark surface moistened with 10% KCl. Unbranched lengths of twigs > 10 cm were cut, and placed in paper bags. Measurement of twig bark pH was adapted from Kermit and Gauslaa (2001), where 6 cm lengths of twigs were cut from the samples, the ends sealed with paraffin wax, and soaked in test tubes in 10 ml 10% KCl for c.1 h and the pH of the solution measured.

9.3 Results

Sampling was undertaken at 30 sites on 9 different tree species including oak (14), birch (5), ash (3) hawthorn (2) willow (2) pine (1), poplar (1), sitka spruce (1), sycamore (1).

L_{AN} values for trunks varied from 19.5 in Scotland to −20 in Norfolk adjacent to intensive farming units, and for twigs from 13.83 in SW England to −10.4 in Norfolk.

L_{AN} values for trunks of oak with a low pH (Fig. 9.2b) showed a better correlation ($R^2 = 0.63$) with NH_3 concentrations than that for all tree species sampled ($R^2 = 0.51$) Fig. 9.2a, due to variation in bark pH between tree species.

A provisional air quality banding for N was used to distinguish 5 grades of lichen condition in all sites for both trunks and twigs from extremely poor to good.

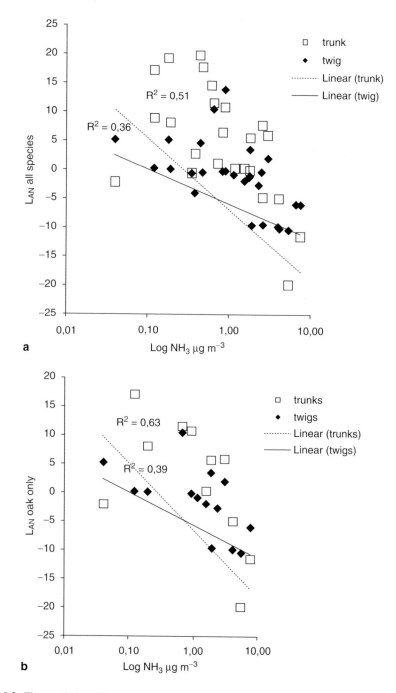

Fig. 9.2 The correlation of L_{AN} with NH_3 is improved for acid-barked tree trunks e.g. oak in (**b**) than for all tree species sampled in the data set (**a**) due to varying bark pH of different tree species

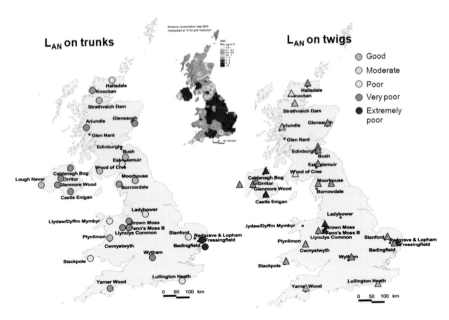

Fig. 9.3 Diagrams of UK ammonia recording sites showing lichen scales for each site on trunks and twigs from good to extremely poor, compared with ammonia concentrations recorded in 2000 interpolated at 10km grid resolution. Scales for lichens on trunks: >14 good, 7 to 14 moderate, 0 to 7 poor, 0 to −7 very poor, >−7 extremely poor; and for lichens on twigs: >5 good, 0 + to 5 moderate, 0 to −5 poor, −5 to −10 very poor, >−10 extremely poor

The scales were adjusted for the variation in frequency counts between trunks and twigs. The results show the correspondence of the lichen index to areas of high ammonia concentrations in Norfolk, Northern Ireland and the midlands, and that the lower values on twigs were associated with an increase in nitrophytes and the decrease in acidophytes on twigs (Fig. 9.3). This is due to the continued presence of acidophytes on trunks that were established in previous environmental conditions while on twigs acidophytes are unable to colonise the new bark of twigs in the present environmental conditions which favour colonisation and establishment of nitrophytes.

9.4 Conclusions

The combination of acidophyte and nitrophyte frequency as L_{AN} on trunks and twigs of trees is correlated with NH_3 concentrations across the UK.

Trees with an 'acid' bark of low pH provide a better correlation of a negative L_{AN} score with increasing NH_3 concentration than basic-barked species such as ash, sycamore and poplar due to the effect of bark pH on lichen communities.

Lichens on twigs respond to increasing ammonia at lower concentrations than lichens on trunks, and considerably lower than the critical level, so that the combination of trunk and twig L_{AN} can act as an early warning system of changes in atmospheric NH_3.

Acknowledgements This work was part of a project commissioned by JNCC and Scotland and Northern Ireland Forum for Environmental Research (SNIFFER) to assess specific biomonitoring methods for wider application for pollution impact assessment by the conservation agencies at designated nature conservation sites (Leith et al. 2005). We thank all participants who took part in this survey.

References

Asta J., Erhardt W., Ferretti M., Fornasier F., Kirschbaum U., Nimis P.L., Purvis O.W., Pirintsos, S. Scheidegger C., van Haluwyn C., Wirth V. (2002) Mapping lichen diversity as an indicator of environmental quality. In: Nimis P.L., Scheidegger C., Wolseley P.A. (eds.) Monitoring with Lichens - Monitoring Lichens. Nato Science Series. IV. Earth and Environmental Sciences, Kluwer, Dordrecht, The Netherlands, 273–279.

Hawksworth D.L., McManus P.M. (1989) Lichen recolonization in London under conditions of rapidly falling sulphur dioxide levels, and the concept of zone skipping. Botanical Journal of the Linnean Society 100(2): 99–109.

Hawksworth D.L., Rose F. (1970) Qualitative scale for estimating sulphur dioxide air pollution in England and Wales using epiph ytic lichens. Nature 227: 145–148.

Kermit T., Gauslaa Y. (2001) The vertical gradient of bark pH of twigs and macrolichens in a Picea abies canopy not affected by acid rain. Lichenologist 33: 353–359.

Leith I.D., van Dijk N., Pitcairn C.E.R., Wolseley P.A., Whitfield C.P., Sutton M.A. (2005) Biomonitoring methods for assessing the impacts of nitrogen pollution: refinement and testing. JNCC Report No. 386, Peterborough, UK.

Nimis P.L., Castello M., Perotti M. (1990) Lichens as biomonitors of sulphur dioxide pollution in La Spezia (Northern Italy). Lichenologist 22: 333–344.

Sutton M.A., Pitcairn C.E.R., Leith I.D., van Dijk N., Tang Y.S., Skiba U., Smart S., Mitchell R., Wolseley P., James P., Purvis W., Fowler D. (2004) Bioindicator and biomonitoring methods for assessing the effects of atmospheric nitrogen on statutory nature conservation sites. In: Sutton M.A., Pitcairn C.E.R., Whitfield C.P., JNCC Report No: 356, Peterborough, UK.

Sutton M.A., Leith I.D., Pitcairn C.E.R., van Dijk N., Tang Y.S., Sheppard L.J., Dragosits U., Fowler D., James P.W., Wolseley, P.A. (2005) Exposure of ecosystems to atmospheric ammonia in the UK and the development of practical indicator methods. In: Lambley P., Wolseley P.A. (eds.) Lichens in a changing pollution environment. English Nature Research Report 525: 51–62.

van Dobben H.F. (1996) Decline and recovery of epiphytic lichens in an agricultural area in The Netherlands (1900–1988). Nova Hedwigia 62(3–4): 477–485.

van Herk C.M. (1999) Mapping of ammonia pollution with epiphytic lichens in the Netherlands. Lichenologist 31: 9–20.

van Herk C.M. (2001) Bark pH and susceptibility to toxic air pollutants as independent causes of changes in epiphytic lichen composition in space and time. Lichenologist 33: 419–441.

van Herk C.M., Mathijssen-Spiekman E.A.M., de Zwart D. (2003) Long distance nitrogen air pollution effects on lichens in Europe. Lichenologist 35: 347–359.

VDI (2005) Biological measurement procedures for determining and evaluating the effects of ambient air pollutants on lichens (bio-indication) - mapping the diversity of epiphytic lichens as indicators of air quality. VDI Verein Deutscher Ingenieure (The Association of Engineers), Kommission Reinhaltung der Luft im VDI und DIN - Normenausschuss KRdL (Commission on Air Pollution Prevention of VDI and DIN - Standards Committee) 3957, Part 13, Berlin. http://www.vdi.de/vdi/presse/mitteilungen_details/index.php?ID=1015947

Wolseley P.A., Pryor K.V. (1999) The potential of epiphytic twig communities on Quercus petraea in a welsh woodland site (Tycanol) for evaluating environmental changes. Lichenologist 31: 41–61.

Wolseley P.A., James P.W., Theobald M.R., Sutton M.A. (2006) Detecting changes in epiphytic lichen communities at sites affected by atmospheric ammonia from agricultural sources. Lichenologist 38: 161–176.

Wolseley P.A., Leith I., van Dijk N., Sutton M. (2007) Macrolichens on twigs and trunks as indicators of ammonia concentrations across the UK – a practical method. In: 20th Task Force Meeting of the ICP Vegetation: Programme, Abstracts (Dubna, March 5–9, 2007). Dubna: JINR: 73 [abstract].

Chapter 10
Assessment of Critical Levels of Atmospheric Ammonia for Lichen Diversity in Cork-Oak Woodland, Portugal

Pedro Pinho, Cristina Branquinho, Cristina Cruz, Y. Sim Tang, Teresa Dias,
Ana Paula Rosa, Cristina Máguas, Maria-Amélia Martins-Loução,
and Mark A. Sutton

10.1 Introduction

The effect of atmospheric ammonia on ecosystems has been the subject of ongoing research. Its adverse effects as an air pollutant are well characterised, and may be even more widespread than previously thought (see Aber et al. 2003; Erisman et al. 2003; Krupa 2003; Purvis et al. 2003). The most important sources of NH_3 in Europe are agricultural activities, mainly crop fertilization and cattle management (Galloway et al. 2003; EPER 2004). Livestock housing facilities are recognised to be large point sources of NH_3 emissions. Close to such facilities, atmospheric NH_3 concentrations are very high, decreasing rapidly with distance over a few hundreds of meters to a few kilometres (Sutton et al. 1998). Measurement of atmospheric NH_3 in the vicinity of livestock housing include those by Pitcairn et al. (1998, 2003), with reported values in Scotland of 24–59 μg m^{-3} close to source, which declined to background values of 1.6–5 μg m^{-3} at 1 km. In order to assess the range of effects of NH_3 in natural ecosystems, that can be used for effective NH_3 mitigation policies (Dragosits et al. 2006), one can rely on two distinct approaches: (i) direct measurements of atmospheric NH_3 concentrations, which provide an estimate of dry NH_3-N deposition, but require intensive and costly operations; (ii) monitoring of effects on the biotic component. The latter approach should be carried out using groups of biota that are more sensitive to the pollutant of interest. Lichens have been reported as the most sensitive group to NH_3 emissions (e.g. Wolseley et al. 2006a; van Herk 1999). Lichens are symbiotic organisms widely used as biomonitors of environmental changes (e.g. Nimis et al. 1991; Vokou et al. 1999; Geebelen and Hoffmann 2001; Giordani et al. 2002; Pirintsos and Loppi 2003; Geiser and Neitlich 2007).

P. Pinho

Universidade de Lisboa, Faculdade de Ciências, Centro de Biologia Ambiental, Campo Grande, Edifício C2, Piso 5. 1749-016 Lisboa, Portugal

M. Sutton, S. Reis and S.M.H. Baker (eds), *Atmospheric Ammonia,*
© Springer Science+Business Media B.V. 2009

Monitoring atmospheric pollutants using lichens may be undertaken in three ways: (1) measuring variations in lichens diversity and/or abundance, (2) using variations in physiological parameters, and/or using lichens as accumulators of pollutants (Branquinho 2001), and (3) considering functional groups related to nutrients tolerance, such as the division between nitrophytic/oligotrophic (or nitrophytic/acidophitic) groups (see van Dobben and ter Braak 1999; Ruisi et al. 2005; Wolseley et al. 2006b).

For the Mediterranean region, a classification of lichen species based on their tolerance to environmental factors, namely eutrophication, is available (Nimis 2003), and was used for classifying lichen species in Portugal, under a similar climate. The mechanism by which NH_3 affects lichens is under debate. It has been suggested that changes in substrata pH play an important role (see van Herk 2001), a mechanism that may be confounded by the effects of dust, especially in Mediterranean regions (Loppi and Pirintsos 2000).

In this work our objective was to establish the critical levels of atmospheric NH_3 based on changes in epiphytic lichen communities in a Mediterranean climate.

10.2   Materials and Methods

Study area: The study area is a farm (called Malhada-de-Meias) with a cattle breeding site located in Portugal, 30 km East of Lisbon (Fig. 10.1c). Soil at the farm are sandy alluvium and eolic deposit, the average annual temperature is 17.5°C and the average annual precipitation is 600 mm (averages 1931–1960) (Instituto do Ambiente 2006). The farm is surrounded by intensive agriculture and livestock activities, and about 200 cows are permanently housed in a single barn measuring 800 m^2 (Fig. 10.1). To the south of this barn lies a cork-oak woodland (*Quercus suber L.*) grazed by cattle. Some areas of temporary pasture are also present in the woodland that is mainly for sheep. Ten kilometres west from the farm, a main motorway and urban areas are also located. The study area presents a Mediterranean climate (see Thompson 2005; Blondel and Aronson 2004), in the thermomediterranean climatic belt (Rivas-Martínez et al. 2004), characterized by dry hot summers and rainy mild winters (see climatogram in Fig. 10.1).

Ammonia sampling: Atmospheric NH_3 concentrations were measured over three continuous sampling periods in autumn 2006 (2-weekly measurements starting 3-Oct, 24-Oct and 6-Nov) using the high sensitivity ALPHA (_Adapted Low-cost Passive High Absorption_) passive diffusion samplers (Tang et al. 2001). Measurements were carried out at 22 sampling locations (2 × ALPHA samplers at each point) downwind of the cattle barn (Fig. 10.1a), with a larger proportion of sites within 200 m of the barn, since that was where the largest variation in concentrations were expected. Ten field blanks were randomly distributed amongst the sampling points. At four sampling locations, additional ALPHA samplers

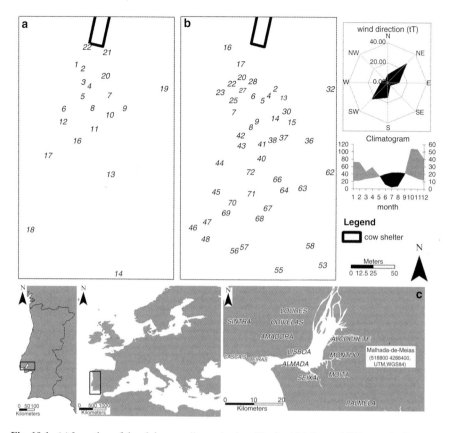

Fig. 10.1 (a) Location of the alpha sampling point (n = 22; site 15 is located 800 m to the SE, and is not shown), (b) location of the lichen diversity sampling plots (n = 74 trees). Wind direction represents the percentage of hours in which wind blows from the prevailing direction, during the study period. The climatogram refers to the average monthly precipitation (left) and average monthly temperature (right), from 1985 to 2006, the black area represents the xeric period according to Gausen, 1954 (in Blondel and Aronson 2004) (c) General location of the study area showing main cities and the location of the study site

provided by CEH were set up in parallel to those prepared by University of Lisbon for quality assurance purposes. Ammonium concentrations collected on the filter papers in the ALPHA samplers were determined by a modification of the Berthelot reaction (Cruz and Martins-Loução 2000). Results were compared with those obtained for the parallel CEH samples that were analyzed independently by CEH using the AMFIA (_AMmonium Flow Injection Analysis_) system (Sutton et al. 2001). Atmospheric ammonia concentrations determined with the ALPHA samplers were calibrated against measurements obtained with a CEH DELTA (_DEnuder for Long-Term Atmospheric sampling_) system. A good correlation was observed between the different methods used to determine atmospheric

NH$_3$ concentration in the air. The results presented are those obtained with the ALPHA samplers.

Lichen diversity: Lichen diversity on the main trunks of cork oak trees was measured using a sampling grid accordingly to the protocol described in Asta et al. (2002) and Scheidegger et al. (2002), but selecting trees differently. For each atmospheric NH$_3$ sampling point, the nearest three or four trees (in total, 74 trees) were selected for a survey (Fig. 10.1b). For each tree, a 50 × 10 cm grid, divided into five 10 × 10 cm subunits was placed on the tree trunk above the cork-harvest zone and used to determine lichen species frequencies, in each of the four main aspects. A lichen diversity value (LDV) was calculated for each atmospheric NH$_3$ sampling point according to Asta et al. (2002). LDV accounts both for species number and frequency. We calculated total LDV (using all species) and also the LDV of functional groups regarding nutrient preferences, according to Nimis (2003) and considered the highest classification for each lichen species. Groups were classified as: 4 and 5 as the nitrophytic group (LDV$_{nitro}$), 1 and 2 as the oligotrophic group (LDV$_{oligo}$), 5 as the strictly nitrophytic species group; and 3 as the indifferent group. Outliers were excluded from the analysis of critical levels.

10.3   Results and Discussion

10.3.1   Relating Lichen Diversity to Atmospheric Ammonia Concentration

The relationship between several lichen diversity indicators and atmospheric NH$_3$ concentration is depicted in Fig. 10.2. In general, at greater atmospheric NH$_3$ concentrations, lower total LDV values were found, showing that NH$_3$ has a significant impact in lichen communities. However, by looking at LDV divided into different nutrient functional groups, a much clear pattern is observed: for increasing NH$_3$ concentrations the oligotrophic lichens species decreased logarithmically from 18 to 2, whereas nitrophytic lichen species increased from 1 to a maximum of 20. Besides the two functional group indicators and because LDV$_{nitro}$ and LDV$_{oligo}$ have an opposing pattern with atmospheric NH$_3$ (Fig. 10.2), the ratio between LDV$_{oligo}$ and LDV$_{nitro}$ was also plotted. It was found that this indicator showed a significant correlation with NH$_3$, reflecting at the same time the pattern of the two different functional groups.

This relationship between the studied indicators and NH$_3$ was further interpreted by looking to the temporal changes of the atmospheric NH$_3$ represented by the standard deviations. Due to the fact that only three measurement periods were available, it was useful to evaluate this variability. Greater temporal changes are probably due to short term events, such as temporary pasture in a specific location, which temporally increase the NH$_3$ concentration but have a reduced effect in LDV. That

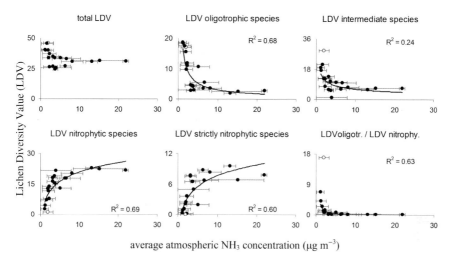

Fig. 10.2 Relationship between several lichen diversity indicators and atmospheric NH$_3$ concentration. Fitted line is a log function for downward curves and power function to upward curves. Site 15 is represented by an empty circle. Error bars represent the standard deviation of NH$_3$ concentration measured during the three sampling periods and were not considered in the adjustment of the fitted line

is probably the case of sampling location 15, which experienced periodic pasture by sheep and consequently higher NH$_3$ concentrations, but which is characterized by a high LDVoligo and a very low LDV nitro. This sample was excluded from the critical levels analysis. Another explanation for the greater standard deviations may be the existence of contrasting periods in the prevailing wind direction. For further analysis, the three indicators LDV$_{oligo}$, LDV$_{nitro}$ and LDV$_{oligo}$/LDV$_{nitro}$ ratio were chosen, since they showed the best correlation with NH$_3$. Given that most biodiversity changes occurred up to 4.5 µg m^{-3}, we focused further analysis at that level. Most authors have also reported that changes in lichens communities occurred at this level, such as Wolseley et al. (2006a) that detected changes in lichen diversity for concentrations above 3–4 µg m^{-3}.

10.3.2 Mapping of Lichen Diversity and Atmospheric Ammonia Concentration

The chosen indicators were interpolated using a simple interpolation function (inverse squared distance weighted) and mapped for the study area (Fig. 10.3). The observed pattern was a replacement of oligotrophic communities, more abundant at sites distant from the barn, by nitrophytic communities that become more abundant nearer the barn. This pattern is highlighted by the spatial distribution of NH$_3$ concentration that increases sharply near the barn, as predicted.

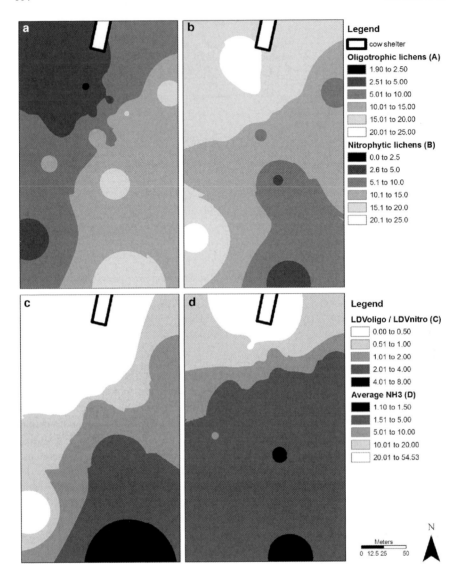

Fig. 10.3 (**a**) LDV of oligotrophic species (LDV$_{oligo}$); (**b**) LDV of nitrophytic species (LDV$_{nitro}$); (**c**) LDV$_{oligo}$/LDV$_{nitro}$ ratio; (**d**) average concentration of NH$_3$ (μg m^{-3}) for a 6 week period

10.3.3 LDV Levels in Undisturbed Sites

When defining critical levels for biodiversity, it must be assumed that any changes should be related to pristine and/or undisturbed sites. In this study, the NH$_3$ gradient was found always within the same farm, which has NH$_3$ sources other than the

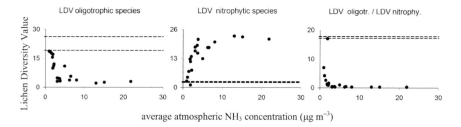

Fig. 10.4 Comparison between the distribution of lichen diversity indicators (LDV) of the farm site (represented by black dots) and the average of two control sites (represented by the two lines), located 5 km away and for which the same methodology was followed

cattle barn, such as: temporary pastures, soil mobilization, manure application, etc. To avoid the problem of not having a clear undisturbed site, the lichen diversity indicators at the farm were compared with indicators found at two distant control sites in the region: one without any livestock and with very low forestry activities and the other without livestock but with some forestry activities, such as selective tree cutting and cork harvesting. These two control sites were located 5 km north of the study site, and biodiversity data was collected using the same methodology. The distribution for the selected LDV indicators (Fig. 10.4) was compared, which shows that the biodiversity gradient at the farm can be considered to be disturbed, presenting on average lower LDV_{oligo} and higher LDV_{nitro}. Nevertheless, the maximum and minimum values of all three sites were within the same range, in particular the control site with some forest activities (lower line in LDV_{oligo} of Fig. 10.4) was similar to the more undisturbed sites in terms of NH_3 gradient.

10.3.4 Detecting Threshold Levels

In order to determine the critical levels in atmospheric NH_3 concentration for lichen diversity, the focus was on concentrations up to 4.5 μg m^{-3}. It is in this range of concentrations that most changes in lichen diversity occur and where the pattern between the two variables is linear (Fig. 10.5). In the case of LDV_{nitro}, the control values found in undisturbed sites were very similar to the lowest value found on the farm (Fig. 10.5). A critical level was calculated for NH_3 using an estimate with a probability level of 95% between NH_3 and LDV_{nitro} data fitted to a linear model (Fig. 10.5). Sampling points where the first visible changes in the LDV_{nitro} occurred were used. This was found to be 1.7 μg m^{-3}, for the LDV_{nitro} indicator. In the case of LDV_{oligo}, one of the controls calculated for undisturbed sites was much higher than the maximum level of this indicator, whereas the control with forest activities was within the expected range of the data. Using the methodology described above for LDV_{nitro}, the critical level was determined as 1.4 μg m^{-3} for LDV_{oligo} (Fig. 10.5). When applying the same methodology for the LDV_{oligo}/LDV_{nitro} ratio, a similar threshold was obtained (Fig. 10.5). Interestingly, these values show that oligotrophic lichens are slightly more

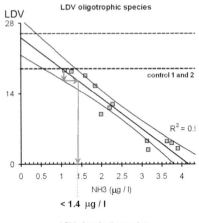

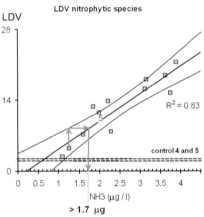

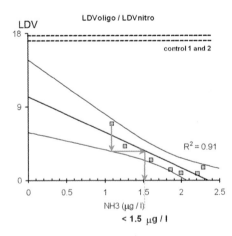

Fig. 10.5 Focus on the linear gradient range, showing the relationship between three lichen diversity indicators and concentrations of atmospheric NH_3. The highlighted NH_3 concentrations are those for which changes in lichen diversity have occurred

sensitive than nitrophytic ones which is in accordance to findings by Wolseley et al. (2006a). Surprisingly, the critical levels found here are very similar to values determined in other countries for the same organisms, namely in the UK. Wolseley et al. (2006a) have observed a loss of lichen diversity (measured in classes of abundance) at concentrations above $3-4 \mu g$ m^{-3}, but in Mediterranean climates that have higher temperatures during the summer, we would expect that lichen diversity critical levels would be different from those reported in colder climates.

The critical levels could be similar due to the fact that the same mechanism may be involved in establishing the changes in lichen communities, independent of climate. Atmospheric NH$_3$ is known to exhibit large seasonal variability in concentration (Sutton et al. 2001); the actual annual mean concentrations may therefore be either larger or smaller than the estimates from only three sets of measurements made over a 2 month period in autumn. An accurate assessment of the annual mean NH$_3$ concentrations would require continuous measurements for at least a year, which could then be used to provide a better estimate of critical levels for atmospheric NH$_3$.

10.3.5 Impact of Atmospheric NH$_3$ in Lichen Communities and Their Use as Biomonitors

It has been observed that with increasing air NH$_3$ concentrations, lichen communities changed from oligotrophic to nitrophytic species. The mechanism by which atmospheric NH$_3$ influences lichens is still under debate, but it has been suggested that both a direct impact and an indirect impact, by changing bark pH may occur (see van Herk 1999, 2001; Loppi and Pirintsos 2000; Frati et al. 2006; Wolseley et al. 2006a). Lichen diversity responds clearly to increases in NH$_3$ concentration. They are also long-living organisms able to integrate environmental changes over time. Therefore, lichen diversity is a suitable biomonitoring method, by providing an important measure of the direct cumulative dose-response to exposure to NH$_3$ emissions, which may vary rapidly over time (Jovan and McCune 2005).

By examining changes in lichen communities, specifically by using lichen indicators based on nitrogen-tolerance, an estimate of atmospheric NH$_3$ critical levels could be made for Mediterranean regions. This critical level, between 1 and 2 μg m^{-3} was found to be lower than the current limits (8 μg m^{-3}) but is in accordance with the concentrations found in other works using lichens in this volume. Because critical levels seem to be similar in regions with contrasting climate, the actual mechanisms involved in atmospheric NH$_3$ tolerance in lichens need further study, as well as an enlightenment of a possible role of climate in this sensitivity.

Acknowledgements This work was carried out with the support of the European Science Foundation NinE programme, COST Action 729, the EC NitroEurope Integrated Project as well as underpinning Portugese national funding. Pedro Pinho acknowledges Portugese Science and Technology Foundation (FCT-MCTES) for PhD grant SFRH/BD/17880/2004.

References

Aber J. D., Goodale C. L., Ollinger S. V., Smith M. L., Magill A. H., Martin M. E., Hallett R. A., Stoddard J. L. (2003) Is nitrogen deposition altering the nitrogen status of northeastern forests? Bioscience 53, 375–389.

Asta J., Erhardt W., Ferretti M., Fornasier F., Kirschbaum U., Nimis P. L., Purvis O. W., Pirintsos S., Scheidegger C., van Haluwyn C., Wirth V. (2002) Mapping lichen diversity as an indicator of environmental quality. In: Nimis, P. L., Scheidegger, C., Wolseley, P. A. (eds.), Monitoring with Lichens- Monitoring Lichens, Nato Science Program-IV, Kluwer, The Netherlands, volume IV, 273–279.

Blondel J., Aronson J. (2004) Biology and Wildlife of the Mediterranean Region. Oxford University Press, 2nd edition, New York, 328.

Branquinho C. (2001) Lichens. In: Prasad M. N. V. (eds.), Metals in the Environment: Analysis by Biodiversity, Marcel Dekker, New York, 117–158.

Cruz C., Martins-Loução M. A. (2000) Determination of ammonium concentrations in soils and plant extracts. In: Martins-Loução M. A., Lips S. H. Backhuys (eds.), Nitrogen in a Sustainable Ecosystem: From the Cell to the Plant.Leiden, The Netherlands, 291–297.

Dragosits U., Theobald M. R., Place C. J., ApSimon H. M., Sutton M. A. (2006) The potential for spatial planning at the landscape level to mitigate the effects of atmospheric ammonia deposition. Environmental Science & Policy 9, 626–638.

EPER (2004) European Pollutant Emission Register. Accessed on-line 15-February-2006 at http://www.eper.cec.eu.int/eper/

Erisman J. W., Grennfelt P., Sutton M. A. (2003) The European perspective on nitrogen emission and deposition. Environmental International 29, 331–325.

Frati L., Caprasecca E., Santoni S., Gaggi C., Guttova A., Gaudino S., Pati A., Rosamilia S., Pirintsos S. A., Loppi S. (2006) Effects of NO_2 and NH_3 from road traffic on epiphytic lichens. Environmental Pollution, 142, 58–64.

Galloway J. N., Aber J. D., Erisman J. W., Seitzinger S. P., Howarth R. W., Cowlingn E. B., Cosby B. J. (2003) The nitrogen cascade. Bioscience 53, 341–356.

Geebelen W., Hoffmann M. (2001) Evaluation of bio-indication methods using epiphytes by correlating with SO2-pollution parameters. Lichenologist 33, 249–260.

Geiser L. H., Neitlich P. N. (2007) Pollution and climate gradients in western Oregon and Washington indicated by epiphytic macrolichens. Environmental Pollution 145, 203–218.

Giordani P., Brunialti G., Alleteo D. (2002) Effects of atmospheric pollution on lichen biodiversity (LB) in a Mediterranean region (Liguria, northwest Italy). Environmental Pollution 118, 53–64.

Instituto do Ambiente (2006) Atlas do Ambiente Digital. Instituto do Ambiente, Amadora, Portugal. Accessed on line 15-February-2006 at http://www.iambiente.pt

Jovan S., McCune B. (2005) Air-quality bioindication in the greater central valley of California, with epiphytic macrolichen communities. Ecological Applications 15, 1712–1726.

Krupa S. V. (2003) Effects of atmospheric ammonia (NH_3) on terrestrial vegetation: a review. Environmental Pollution 124, 179–221.

Loppi S., Pirintsos S. A. (2000) Effect of dust on epiphytic lichen vegetation in the Mediterranean area (Italy and Greece). Israel Journal of Plant Sciences 48, 91–95.

Nimis P. L. (2003) TSB Lichen Herbarium 3.0, University of Trieste, Department of Biology, IH3.0 /02. Accessed on line 15-September-2005 at http://dbiodbs.univ.trieste.it/global/italic_tsb1

Nimis P. L., Lazzarin G., Gasparo D. (1991) Lichens as bioindicators of air pollution by SO2 in the Veneto region (NE Italy). Studia Geobotanica 11, 3–76.

Pirintsos S. A., Loppi S. (2003) Lichens as bioindicators of environmental quality in dry Mediterranean areas: a case study from northern Greece. Israel Journal of Plant Sciences 51, 143–151.

Pitcairn C. E. R., Leith I. D., Sheppard L. J., Sutton M. A., Fowler D., Munro R. C., Tang S., Wilson D. (1998) The relationship between nitrogen deposition, species composition and foliar nitrogen concentrations in woodland flora in the vicinity of livestock farms. Environmental Pollution 102, 41–48.

Pitcairn C. E. R., Fowler D., Leith I. D., Sheppard L. J., Sutton M. A., Kennedy V., Okello E. (2003) Bioindicators of enhanced nitrogen deposition. Environmental Pollution 126, 353–361.

Purvis O. W., Chimonides J., Din V., Erotokritou L., Jeffries T., Jones G. C., Louwhoff S., Read H., Spiro B. (2003) Which factors are responsbile for the changing lichen floras of London? Science of the Total Environment 310, 179–189.

Rivas-Martínez S., Penas A., Díaz T. E. (2004) Bioclimatic map of Europe- Thermoclimatic belts. University of León, Spain. Accessed on line 15-February-2006 at http://www.globalbioclimatics.org/form/tb_med.htm

Ruisi S., Zucconi L., Fornasier F., Paoli L., Frati L., Loppi S. (2005) Mapping environmental effects of agriculture with epiphytic lichens. Israel Journal of Plant Sciences 53, 115–124.

Scheidegger C., Groner U., Keller C., Stofer S. (2002) Biodiversity assessment tools-lichens. In: Nimis P. L., C. Sheidegger, P. A. Wolseley (eds.), Monitoring with Lichens, Monitoring Lichens. Kluwer, Dordrecht, The Netherlands, pp 359–365.

Sutton M. A., Milford C., Dragosits U., Place C. J., Singles R. J., Smith R. I., Pitcairn C. E. R., Fowler D., Hill J., ApSimon H. M., Ross C., Hill R., Jarvis S. C., Pain B. F. Phillips V. C., Harrison R., Moss D., Webb J., Espenhahn S. E., Lee D. S., Hornung M., Ullyett J., Bull K. R., Emmett B. A., Lowe J., Wyers G. P. (1998) Dispersion, deposition and impacts of atmospheric ammonia: quantifying local budgets and spatial variability. Environmental Pollution 102, 349–361.

Sutton M. A., Miners B., Tang Y. S., Milford C., Wyers G. P., Duyzer J. H., Fowler D. (2001) Comparison of low cost measurement techniques for long-term monitoring of atmospheric ammonia. Journal of environmental Monitoring 3, 446–453.

Tang Y. S., Cape J. N., Sutton M. A. (2001) Development and types of passive samplers for NH3 and NOx. The Scientific World 1, 513–529.

Thompson J. D. (2005) Plant Evolution in the Mediterranean. Oxford University Press, 1st edition, New York, 293.

van Dobben H. F., ter Braak C. J. F. (1999) Ranking of epiphytic lichen sensitivity to air pollution using survey data: a comparison of indicator scales. Lichenologist 31, 27–39.

van Herk C. M. (1999) Mapping of ammonia pollution with epiphytic lichens in the Netherlands. Lichenologist 31, 9–20.

van Herk C. M. (2001) Bark pH and susceptibility to toxic air pollutants as independent causes of changes in epiphytic lichen composition in space and time. Lichenologist 33, 419–441.

Vokou D., Pirintsos S. A., Loppi S. (1999) Lichens as bioindicators of temporal variations in air quality around Thessaloniki, northern Greece. Ecological Research 14, 89–96.

Wolseley P. A., James P. W., Theobald M. R., Sutton M. A. (2006a) Detecting changes in epiphytic lichen communities at sites affected by atmospheric ammonia from agricultural sources. Lichenologist 38, 161–176.

Wolseley P. A., Stofer S., Mitchell R., Truscott A. M., Vanbergen A., Chimonides J., Scheidegger C. (2006b) Variation of lichen communities with landuse in Aberdeenshire, UK. Lichenologist 38, 307–322.

Part II
Temporal Trends in Atmospheric Ammonia

Chapter 11
Linking Ammonia Emission Trends to Measured Concentrations and Deposition of Reduced Nitrogen at Different Scales

Albert Bleeker, Mark A. Sutton, Beat Acherman, Ana Alebic-Juretic, Viney P. Aneja, Thomas Ellermann, Jan Willem Erisman, David Fowler, Hilde Fagerli, Thomas Gauger, K.S. Harlen, Lars Robert Hole, Laszlo Horvath, Marta Mitosinkova, Ron I. Smith, Y. Sim Tang, and Addo van Pul

11.1 Summary

This document builds on the Bern Background Document (Sutton et al. 2003), which was used to facilitate the discussion about following emission trends by means of measurement data at the UN/ECE Ammonia Expert Group meeting in Bern (Switzerland) in 2000. It is now 6 years since the Bern Workshop and major new datasets on European NH_3 and NH_4^+ monitoring and their relationship to estimated NH_3 emissions have become available for the following countries: UK, Germany, Hungary, Switzerland, Denmark, the Netherlands, North Carolina, Slovak Republic, Norway and Croatia. Based on these datasets our current scientific understanding about the different issues is updated. In particular, input will be given to questions like: is there still an "Ammonia Gap" in the Netherlands, does such a gap exist in other countries, can we be confident of the effectiveness of ammonia mitigation policies and how can we best address the relationships between emission and deposition using atmospheric modelling and improved monitoring activities.

11.2 Introduction

At the end of the 1980s in different countries policies were developed to reduce NH_3 emissions. Targets were set and monitoring was initiated. In the late 1990s a discussion was started when it became clear that expected changes in NH_3 emissions where not matched by observed reductions of NH_3 concentrations in air and/or NH_4 concentrations in rain water. One example of this mismatch was a case study for the Netherlands, were extensive NH_3 emissions reduction policy was implemented and it was therefore surprising that by 1997, NH_3 concentrations were no smaller than in 1993, when the policy was initiated (Erisman et al. 1998a; Van Jaarsveld et al. 2000). The issue became

Albert Bleeker
Energy Research Centre of the Netherlands (ECN), Petten, The Netherlands

M. Sutton, S. Reis and S.M.H. Baker (eds), *Atmospheric Ammonia,*
© Springer Science+Business Media B.V. 2009

known as the "Ammonia Gap", raising questions regarding the cost effectiveness of the NH_3 abatement policy. Additionally, in Eastern Europe, following the crash in agricultural livestock populations and fertilizer usage after the political changes of 1989, it was curious that available monitoring in Hungary could also not detect the expected reductions in NH_3 emissions (Horvath and Sutton 1998). Since the emissions in east Europe must have decreased, due to reduced sector activity, this raised the question of whether there were non-linearities in the link between NH_3 emissions and atmospheric concentrations and deposition. These issues were reviewed at the Bern Workshop in 2000 (Sutton et al. 2003), which noted how interactions with changing SO_2 emissions, local spatial variability, short term meteorological variability and interactions with NH_3 compensation points were among the factors explaining the difficulty to make the links.

One of the key findings of the Bern workshop described in the Working Group Report from the Bern Workshop (Menzi and Acherman 2001) was the severe lack of NH_3 monitoring data across Europe. Recommendations were therefore made regarding the need to establish robust monitoring networks, especially with the ability to speciate between NH_3 gas and NH_4^+ aerosol, a finding which was re-enforced by the Oslo Workshop (2004) on monitoring strategies (Aas 2005).

It is now 6 years since the Bern Workshop and major new datasets on European NH_3 and NH_4^+ monitoring and their relationship to estimated NH_3 emissions have become available. Based on these datasets the findings of the previous workshop will be evaluated, hopefully updating our current scientific understanding about the different issues that were addressed there. In particular, input will be given to questions like: is there still an "Ammonia Gap" in the Netherlands, does such a gap exist in other countries, can we be confident of the effectiveness of ammonia mitigation policies and how can we best address the relationships between emission and deposition using atmospheric modelling and improved monitoring activities.

In this background document first the major findings of the Bern Workshop results are summarized, after which new and/or updated datasets for different countries are presented. Like in the previous document also information from the USA is included in the overview, complementing the overall picture by showing increasing trends of atmospheric NH_x. This in comparison with the assumed downward trends in atmospheric NH_x, which ought to be found in Europe according to the reported downward emission trends.

11.3   Important Findings from the Previous Review

The Bern background document (BBD, Sutton et al. 2003) addressed the link between NH_3 emission abatement and atmospheric measurements, while considering two clear challenges when doing so:

- To quantify the changes between NH_3 emission and monitored atmospheric NH_x in situations where emissions have definitely changed
- Bearing in mind the uncertainties in the previous challenge, to assess the effectiveness of NH_3 emission abatement policies

Case studies from Europe and America were presented to illustrate inherent uncertainties in linking the emissions to concentrations and/or depositions. Also examples from countries where NH$_3$ abatement policies have been implemented, using these examples to address the extent to which monitoring data can determine the effectiveness of abatement measures. The following sections summarise some of the case studies from the BBD, providing some information from that study in order to better understand the new information presented in the next chapter.

11.3.1 Case Studies: Linking Agricultural Sector Activity and Atmospheric NH$_x$

Since the agricultural sector activity is thought to have declined dramatically in Eastern European countries since the mid 1980s peak emissions (overall reduction of 46%), it should be possible to see a clear response of the monitoring data from these countries. A few examples were considered in the BBD, including Hungary, comparison of former East and West Germany, Slovakia and the former Soviet Union (FSU). Figure 11.1 shows the changes in estimated NH$_3$ emissions for the different European countries that were included in the BBD. This figure clearly shows that while the Western European countries only faced a minor reduction of the NH$_3$ emissions, the emissions declined dramatically in the Eastern European countries.

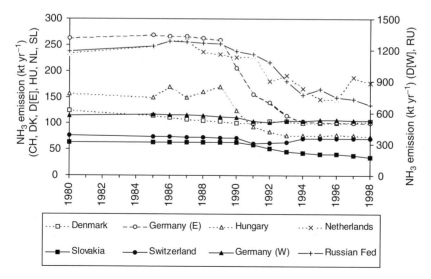

Fig. 11.1 Changes in estimated NH$_3$ emissions for European countries included in the BBD

For some of the Eastern European countries with the strong decline in emissions, examples are shown of the concentration and/or deposition trends, as presented in the BBD. The first example shows the trends of gaseous NH_3 and aerosol NH_4+ for the EMEP/GAW monitoring site K-puszta, located in the centre of Hungary (see Fig. 11.2). These data have been published by Horvath and Sutton (1998). While the emissions show a 53% reduction over the period 1980–1998, no trend was found in NH_3 and NH_4^+ over the same period. The Hungarian case did show a decrease in NH_4^+ in precipitation coupled to a decrease in SO_4^{2-} aerosol, which indicated an interaction with sulphur emissions and atmospheric chemistry. A possible explanation is that, with decreasing SO_2 emissions, a reduced rate of $(NH_4)_2SO_4$ aerosol formation is expected, resulting in less formation of NH_4^+ aerosol. This was also shown in other studies (see below).

Another example for a country with large reported reductions of NH_3 emissions was provided by Slovakia (EMEP station Chopok). The emission reduction was estimated to be 44% for the period 1990–1999, while the decrease in NH_4^+ concentrations is precipitation in that same period was around 20% (see Fig. 11.3).

The last Eastern European example is given here is for Russia. Figure 11.4 shows long-term data for four sites of the WMO *Global Atmospheric Watch* (Paromonov et al. 1999). These are sites within the European Territory of the Former Soviet

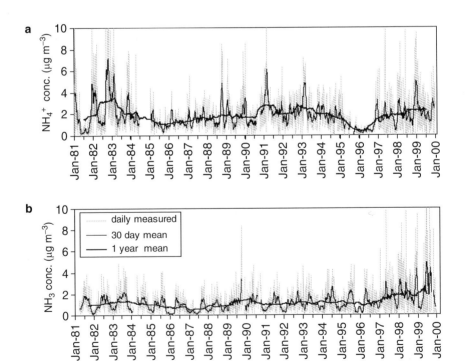

Fig. 11.2 Long-term record of gaseous NH_3 and aerosol NH_4^+ at K-puszta in Hungary (Sutton et al. 2003)

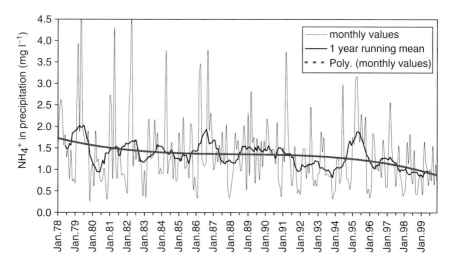

Fig. 11.3 NH_4^+ in precipitation at the EMEP/GAW station at Chopok, Slovakia (Sutton et al. 2003)

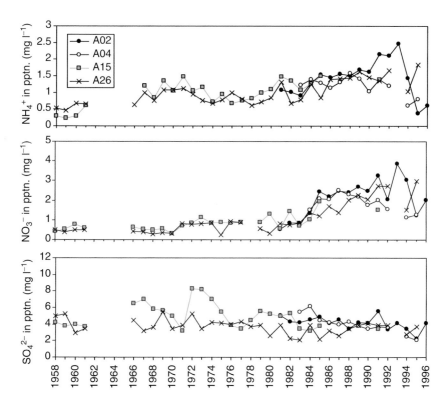

Fig. 11.4 Record of NH_4^+, NO_3^- and SO_4^{2-} precipitation concentrations for sites in the European Territory of the Former Soviet Union. A02: Berezina Biosphere Reserve (BR); A04: Oka-Terrase BR; A15: Central Forest BR; A26: Syktyvkar-1 (Sutton et al. 2003)

Union and are located within the region where NH_3 emissions are most likely to have occurred. Another point of attention is that also other countries in the region (see e.g. Hungary and Slovakia) show large reported reduction in NH_3 emissions, which suggests that a decrease in monitored NH_x data ought to be found.

When taking the average of three of the four sites, which show a decrease in NH_4^+ concentration in precipitation, the concentration decreased by approximately 40% in the period 1989–1995, being consistent with the national reduction in NH_3 emissions. However, like for Hungary, there is a simultaneous decrease of NO_3^- and SO_4^{2-} concentration in precipitation, possibly masking the effects of NH_3 emission trends. This implies an altered atmospheric transport distance of NH_x linked to changing NO_x and SO_2 emissions.

A contrasting case was found in the State of North Carolina (USA), where both NH_3 and NH_4^+ concentrations in air and precipitation respectively increased to the same degree as the NH_3 emissions (Fig. 11.5). Between 1985 and 1997 a major increase in NH_3 emission occurred of a factor 7, which reflects the rapid expanding pig sector in that region (Aneja et al. 2000; Walker et al. 2000). An advantage of this simultaneous change in emission and deposition is that the level of scatter in the measurement data can help indicate what would be the minimum detectable change or time-period to detect change. Walker et al. (2000) showed that a single site record of less than 5–7 year becomes increasingly less able to detect changes, even for such large changes shown in Fig. 11.5.

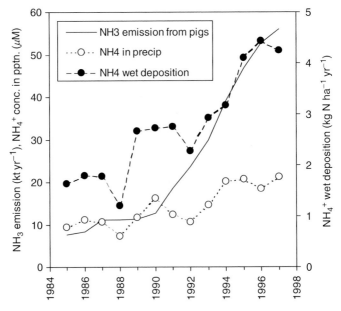

Fig. 11.5 Comparison of NH_3 emissions in coastal North Carolina, USA against NH_4^+ precipitation concentrations and wet deposition at the NADP monitoring site NC35 in Sampson County (from Aneja et al. 2000). Emissions are for the six counties of North Carolina surrounding NC35

11.3.2   Case Studies: Linking NH₃ Emission Abatement and Atmospheric NHₓ

When trying to link NH_3 emission abatement with measured concentrations and/or depositions of NH_x, the Netherlands forms a clear case of a country where a long history on this topic is available. In the BBD a description was given of the various studies that were (more or less) initiated because of the lack of measured trend following the implementation of different abatement measures. This lack of trend became known as the 'ammonia gap' and has been addressed by different authors (e.g. Erisman et al. 1998a, 2003; Boxman 1998; Van Jaarsveld et al. 2000).

The overall reported NH_3 emission reduction was 35%, being rather modest in comparison with e.g. Eastern European countries (see Fig. 11.1). A complicating factor is the lack of NH_3 concentration measurement data for a longer period at that moment (only 6 years, see Fig. 11.6). Information from the other case studies showed that it may turn out to be difficult to detect a 35% change with the available monitoring data. Especially for a small country like the Netherlands this is a real

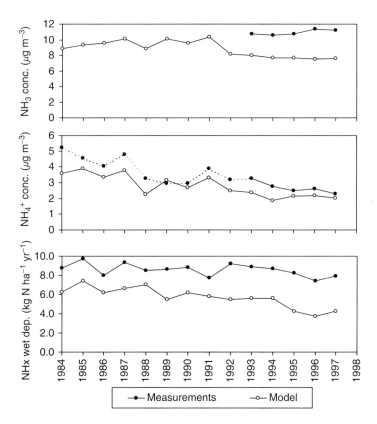

Fig. 11.6 Comparison of modelled and measured NH_x for the Netherlands (Sutton et al. 2003)

challenge, given the interactions with meteorology, other pollutant emissions and depositions and long-range transport, which puts high demands on the modelling and monitoring strategy designed for detecting these trends.

Besides the difference in trend for the different modelled and measured components, also the absolute difference was a topic for further investigations. Figure 11.6 shows the trends for both modelled and measured concentrations/deposition of NH_3 concentration, NH_4^+ aerosol concentration and NH_4^+ wet deposition. The NH_3 concentrations and NH_4^+ wet deposition show a systematic difference between modelled and measured values and the modelled NH_4^+ wet deposition decreased by 20% between 1993–1997, compared to a 10% reduction in the measurements. The opposite was true for the NH_4^+ aerosol concentrations, where the measurements showed a 29% decrease and the modelled values only a 14% decrease.

It was postulated that part of the observed differences were because of parallel changes in SO_2 and NO_x emissions over the same period (Erisman and Monteny 1998). This was assessed by Van Jaarsveld et al. (2000), by comparing modelled NH_x results with those estimated if SO_2 and NO_x emissions had remained at 1984 levels. The overall 'emission effect' is shown in Fig. 11.7, where less SO_2 and NO_x results in an increase in NH_3 concentrations and a simultaneous NH_4^+ wet deposition decrease. According to Van Jaarsveld et al. (2000) the higher NH_3 concentrations are a result of longer residence periods of gaseous NH_3 in the atmosphere due to a slower net rate of ammonium sulphate and nitrate aerosol formation. The lower wet deposition values are due to the decrease of NH_4^+ aerosol in the air. Since wet deposition of NH_4^+ is dominated by scavenging of NH_4^+ aerosol, the decreased levels of NH_4^+ aerosol in the air will thus result in decreased levels of wet deposition.

Based on the effect shown in Fig. 11.7 and less than average rain during the study period, Van Jaarveld et al. (2000) could explain part of the ammonia gap. Still there was not a complete explanation for the difference and it was concluded that only 45–70% of the foreseen reduction in emissions (due to implemented abatement measures) had been achieved (i.e. a national emission reduction of 16–25%).

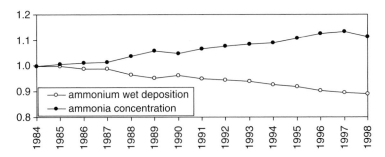

Fig. 11.7 The effect of changing SO_2 and NO_x emissions on NH_3 concentrations and the contribution of NH_4^+ scavenging to wet deposition, as assessed using the OPS model (van Jaarsveld et al. 2000)

11.3.3   Bern Background Document: Main Conclusions

From the different case studies presented in the BBD, the following main conclusions were drawn:

- It was clear that there are several difficulties and uncertainties in assessing the effectiveness of NH_3 abatement from monitoring networks.
- To do this requires sound monitoring methods implemented at sufficient sites and over a sufficiently long period.
- For NH_4^+ aerosol and NH_4^+ in rain, a modest number of sites can be used to indicate trends, whereas for NH_3 in source areas a high density of carefully selected sites is essential.
- In contrast to the need for many NH_3 sampling locations, is the requirement for high temporal resolution NH_3 concentration data at selected sites.
- Quantifying the interactions of NH_x, necessary to interpret long-term trends, also requires improved mechanistic understanding and modelling:
  - ○ Better generalization on the bi-directional controls on NH_3 exchange
  - ○ The chemical interactions that are recognized for atmospheric chemistry also need to be treated in relation to dry deposition
  - ○ Advancement of the regional-temporal modelling of NH_3 emissions in relation to environmental conditions
- It is important to retain caution in attributing changes in atmospheric NH_x to changes in NH_3 emission.
- There are clear difficulties trying to detect NH_3 emission changes even where these certainly occurred.
- In assessing the success of any abatement policy based on technical methods, a combination of appropriate modelling and sufficient measurements should be able to determine whether the abatement measures are broadly effective.
- However, where there is a gap between the monitoring response expected and that observed, this may be as much due to:
  - ○ Limitations in atmospheric process quantification and monitoring
  - ○ Ineffectiveness of the abatement techniques

11.4   Current Status of Studies for Verification

Since the BBD, 6 years have past and thus additional data have become available. Not only just by extending the measurement data with 6 additional years, but also because new studies investigating the relation between emission and concentration/deposition trends of reduced nitrogen. Some of these studies were initiated based the outcome and conclusions of the BBD.

This section summarizes the current status of these different verification studies. For this overview a distinction is made between country-specific case studies and a more general European overview on the measured trends. However, before discussing the concentration and/or deposition trends, updated information with respect to the European ammonia emissions is presented in the next section.

11.4.1 European Ammonia Emissions

For different European countries the changes in NH$_3$ emissions are shown in Fig. 11.8. The reported emissions to EMEP are shown here and the changes are presented for two periods:

- 1980–1998; corresponding to the presented trends in the BBD
- 1980–2003; the BBD information extended with emissions from an additional 5 years

In general the different countries show the same trends for both periods. However, some clear exceptions exist, like e.g. Spain, Cypress and Austria. Most countries have reduced their emissions since the BBD, but also there some exceptions exist (Czech Republic, Latvia, Cypress and Italy, see Fig. 11.9). The overall reductions of the ammonia emissions are still largest in the Eastern European countries, although the trend that was started during the 1980–1998 period clearly levelled off during the 1998–2003 period. These countries generally show a moderate emission reduction during this last period (Fig. 11.9). As in the BBD, these emission trends are

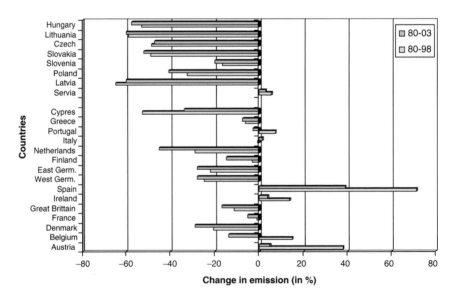

Fig. 11.8 EMEP ammonia emission changes for different European countries for two periods: 1980–1998 (according to BBD) and 1980–2003 (updated information)

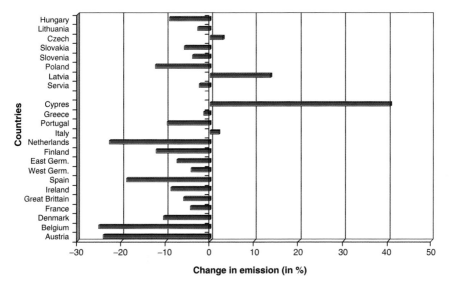

Fig. 11.9 EMEP ammonia emission changes for different European countries for the period 1998–2003

be evaluated by considering observed concentrations and/or deposition for different country-specific case studies.

11.4.2   Country Specific Case Studies

This section contains information from different countries, where activities have been going on in an attempt to better understand the relation between modelled and/or observed reduced nitrogen and the emission of NH_3. Other reasons for these studies are to verify reduction policies or to comply with e.g. EMEP activities. For some countries the presented information consists of an update of the BBD content (e.g. Hungary, Slovakia, North Carolina), while for other countries the results of some new studies are presented here.

11.4.2.1   The Netherlands

The discussion about the 'ammonia gap' has initiated different studies in the Netherlands, trying to close the gap between modelled emissions and those derived from atmospheric measurements. These studies focussed on different aspects, such as the emission calculations, the measurements and the concentration/deposition modelling. Some of the main results of these different studies are presented here.

The ammonia emission calculation procedures were evaluated, showing that some inconsistencies existed in the way these emissions were calculated. Especially issues like the transport of manure within the country, the use of fertilizer, the methods for applying manure to the field (and their ammonia reduction efficiencies) were studied in more detail and it turned out that improvements on some of these issues were needed. These improvements resulted in new emission estimates for the previous years, which are shown in Fig. 11.10. When comparing the emission trend with the trend in modelled and measured NH_3, it is clear that the different trends show the same pattern, except for the years 1997/1998. Until now one part of the gap was the deviation between the two trends (e.g. Erisman et al. 1998a), which seems to be solved now that the period is extended.

However, the rather large difference in absolute values of modelled and measured concentrations and/or depositions is still present (Erisman et al. 1998a; Van Jaarsveld et al. 2000; Van Pul et al. 2004). Figure 11.10 shows this more-or-less systematic difference of about 30% in the modelled and measured concentrations of NH_3. Since this underestimation could indicate an underestimation of the ammonia emissions, which might then cause problems in reaching the ammonia emission ceilings agreed in the UN-ECE and EU frameworks (Van Pul et al. 2004), special attention was paid to this aspect in different studies during the last few years.

One of the studies focussed on the representativity for trend detections of the eight stations of the Dutch Monitoring Network measuring ammonia concentrations at an hourly rate (Van Pul et al. 2004). The model that is normally used for performing the trends analyses (OPS, Van Jaarsveld 1995) is calibrated against the measured concentrations from the same eight measuring stations. The combination of these issues was a reason for launching a project measuring the ammonia concentrations in air in a large number of locations using passive samplers. The following goals of this project were:

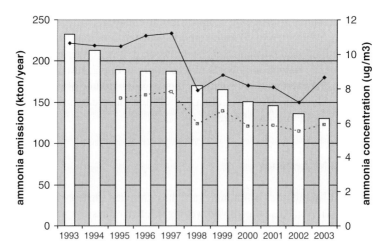

Fig. 11.10 Measured and calculated ammonia concentration at Dutch measurement sites (blue line = measured; red line = modelled) and ammonia emissions for the period 1993–2003 (bars) (From MNP/RIVM)

- To check the representativeness of the eight ammonia concentration measurement locations in the Dutch monitoring network for the Netherlands
- To discover the spatial pattern of the ammonia concentration in the Netherlands
- To determine the level of success in simulating the concentration pattern using the OPS model

Figure 11.11 shows the spatial distribution of the different passive sampler measuring locations (159 locations in total). During 1 year the ammonia concentrations in air were measured on a monthly basis (Duyzer and Weststrate 2002).

The country average concentrations resulting from the passive samplers were compared with those of the eight sites in the Dutch Monitoring Network, to investigate the representativeness of these eight sites. Figure 11.12 shows this comparison, using 155 of the passive sampler locations (four locations were in the direct neighbourhood of intensive animal farms and were therefore excluded for the analyses). Based on this comparison it was concluded that the measurements at the eight Dutch Monitoring Network locations represent the ammonia concentration level in the

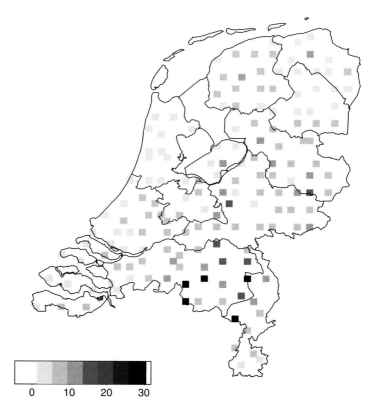

Fig. 11.11 Measuring locations for the passive samplers. In gray scale the annual average NH$_3$ concentration (in $\mu g/m^3$) is shown

NH$_3$ (µg/m^3)

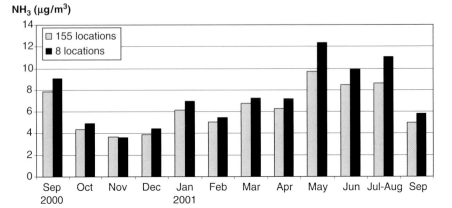

Fig. 11.12 Average ammonia concentrations in the Netherlands (in µg/m³). The average is calculated from 155 passive sampler sites (grey bar) and 8 passive sampler measurements at the locations of the Dutch Monitoring Network (black bar)

Netherlands reasonably well. The yearly average of these eight sites is 18% higher than the average based on the passive sampler sites. This confirms the careful design of the network (Erisman et al. 1998b; Mennen et al. 1996).

The third goal was to investigate the degree to which the OPS model is able to simulate the measured concentration pattern in the Netherlands. This was done by calculating the ammonia concentration at the 159 different measuring locations using emission data of different resolutions (5,000 × 5,000 m and 500 × 500 m). The modelled concentrations were compared with the measured ones and the result of this comparison is shown in Fig. 11.13. Using the 5,000 × 5,000 m emission data, 59% of the variance in the measurements could be explained, while for the 500 × 500 m resolution data this is 73%. Based on the 500 m resolution emissions, the underestimation of the measured concentrations amount to about 32%. Based on this information, it was concluded that the OPS model is able to describe the spatial pattern well. However, the measured concentrations are underestimated by about 30%, using the high resolution emission data. Van Pul et al. (2004) considered that the reason for this underestimation, although not clear yet, is most probably a combination of uncertainties in emission estimates and the parameterisation of the dry deposition in the OPS model.

Another study addressing the different uncertainties was the so-called 'Veld study' (Smits et al. 2005). In a 3 × 3 km area in the eastern part of the Netherlands, a detailed emission inventory was made. At the same time, NH$_3$ concentration measurements were performed using a combination of continuous measurements (every 5 min) at a central location in this area (focussing at the temporal variation) and 2-weekly passive sampler measurements at 50 locations in this area (focussing at the spatial distribution).

Using the area specific emission estimates, a 15% underestimation of the measured concentrations was found (see Fig. 11.14). Based on the available information it was concluded that there was no reason to assume a significant uncertainty in the local housing emissions, but that the main uncertainty was caused by an underestimation

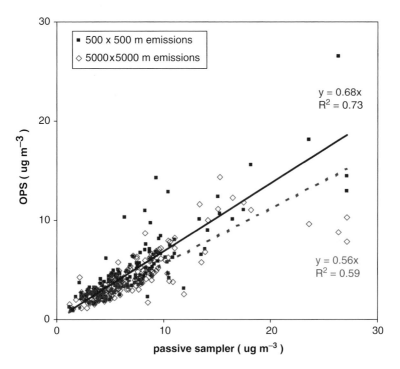

Fig. 11.13 Ammonia concentrations for 159 locations calculated with the OPS model, based on emissions on a 500 m resolution (black squares) and 5,000 m resolution (open diamonds) against ammonia concentrations measured with passive samplers

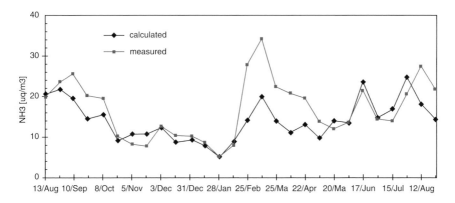

Fig. 11.14 Comparison of calculated and measured area averaged ammonia concentrations. The measurements are performed by means of passive samplers at 50 locations at 14 days time intervals

of the manure spreading emissions (especially during spring time). As a possible reason for this underestimation, the weather conditions during spreading were mentioned; under dry and sunny weather conditions more ammonia will volatilize than assumed in the emission calculations, using annual average emission factors

(Smits et al. 2005). When using emission estimates for the study area, calculated by means of the 'standard' national calculation procedure, an underestimation of the measurements of 30% was found.

One of the problems in the above mentioned 'Veld study' was that no actual emissions have been measured. Therefore, the conclusion that the underestimation of the concentrations was most likely due to underestimated spreading emissions was more or less indirect (although strong indications of the plausibility of this conclusion were available). To try to overcome this, a targeted study was started in 2005 to improve estimates of the spreading emissions based on measurements (using LIDAR and TDL measurements). At the same time, effort is put into the quality of the calculation of NH_3 concentrations by improving the dry deposition calculation procedures in the OPS model.

11.4.2.2 Switzerland

For Switzerland some new, but also updated, datasets are available. When looking at the reduced N compounds in Switzerland, monitoring data for NH_4+ in precipitation are available since 1985, for the sum of $NH_3 + NH_4^+$ (reduced N compounds in gas and aerosols) since 1993 and for gaseous NH_3 since 2000.

Figure 11.15 shows the monitored annual NH_4^+ wet deposition at several stations of the national monitoring network NABEL. The deposition values presented here expressed as precipitation-weighted loads of NH_4-N, with the analysis method being ion chromatography. In the BBD the Swiss data were not precipitation weighted and reflected only the measured concentrations in rain water. For a time-series analysis

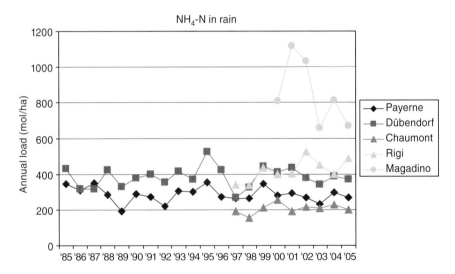

Fig. 11.15 NH_4^+ wet deposition measurements at sites of the NABEL monitoring network

with relevance concerning the impact on ecosystems and concerning the discussion of the relation between emission changes and air quality, the precipitation-weighted loads are thought to be more relevant. Figure 11.16 shows a map of the NABEL measuring locations, for easier reference.

Figure 11.17 shows the sum of $NH_3 + NH_4^+$ (reduced N compounds in gas and aerosol), monitored at the two NABEL stations Payerne and Rigi (see Fig. 11.16).

Fig. 11.16 Locations of the NABEL monitoring network sites

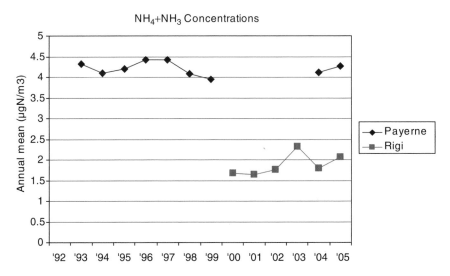

Fig. 11.17 Sum of $NH_4^+ + NH_3$ concentrations at two sites of the NABEL monitoring network (see Fig. 11.16 for locations)

Payerne was replaced by Rigi in 2000, but due to the high interest for these data, Payerne was introduced in 2004. The analysis is made with impregnated filters and ion chromatography. The Rigi station is ideally located to detect the influence of a major part of the Swiss plateau on aerosol formation. Only during wintertime it might occur that the station is sometimes above the inversion layer being formed over the Swiss plateau.

In Switzerland NH_3 concentrations are monitored at different stations mainly in the Swiss Plateau with the passive sampling technique and at some stations also with the denuder technique since the year 2000. Figure 11.18 shows a summary of the NH_3 concentration measurements at those sites which have been in operation since the year 2000 without interruption (16 sites in total). The annual data are shown in the form of percentiles covering the monitoring data of all the 16 sites together.

If the monitoring data presented above (precipitation, gaseous and aerosol chemistry) is considered, one can come to the conclusion that no significant trend can be seen towards lower levels or loads of reduced N compounds, since the beginning of the monitoring activities. Aerosol and gaseous concentrations even tend to show an increase since 2000. This is to a certain extent in contrast with NH_3 emission calculations obtained from agriculture, which show a decrease in NH_3 emissions of about 18% during the period 1995–2000 and with a further (slight) decrease of several percent from 2000 onwards. These emission calculations are currently under review. Serious doubts exist about these emissions since during the last 10 years more and more open housing systems and solid outdoor floors were constructed and the storage capacity for liquid manure increased by about 50–60% with mainly open silos. The review and assessment of these developments are ongoing and it is expected to have revised emission calculations available in 2007.

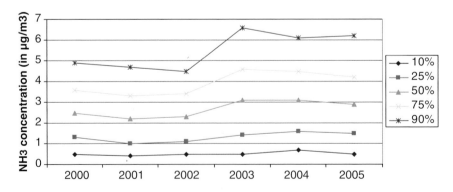

Fig. 11.18 NH_3 concentrations at 16 Swiss sites. Values are shown as percentiles for the 16 sites

11.4.2.3 United Kingdom

In the UK different studies have been undertaken in order to get more insight in the relation between emission and concentration trends of reduced nitrogen. This section gives an overview of some of the main activities. The first study presented here dealt with changing emissions after an outbreak of Foot and Mouth Disease and the way in which NH_3 concentrations responded to that. The second study focussed on long-term trends in concentrations, measured at the UK national ammonia monitoring network.

Foot and Mouth Disease study

A case study for investigating the link between changing NH_3 emissions and NH_3 concentrations in air was available after the outbreak of Foot and Mouth Disease (FMD) in the UK in 2001. The study was motivated by different previous studies showing the 'ammonia gap' between modelled and observed changes (e.g. Erisman et al. 1998a; Van Jaarsveld et al. 2000; Sutton et al. 2001 a,b, 2003). The outbreak of FMD provided the opportunity to assess whether future reductions in ammonia emissions would achieve the desired outcome of reduced air concentrations (Sutton et al. 2004, 2006). The FMD outbreak led to large regional reductions in animal numbers in some parts of the UK. Therefore, monitoring in these areas could be used to test whether NH_3 concentrations increase following subsequent animal restocking.

The basis for the measurement network was an initial modelling study, conducted to map the location of the FMD outbreak and its effect on emissions and atmospheric concentrations of e.g. NH_3 (Sutton et al. 2004). Figure 11.19 shows one of two regions studied, around Cumbria (Northern England). Analysis from Sutton et al. (2004) showed that the changes should be detectable for NH_3 and that these changes should also be larger than the inter-annual variability. Based on these first calculations a monitoring network was established for a study area centred on Cumbria and Devon (see Sutton et al. 2006). The monitoring network included about 15 sites in FMD-affected areas, with approximately five sites in surrounding areas little affected by FMD ("unaffected" sites), where measurements where done using triplicate passive "ALPHA" samplers (Tang et al. 2001) at a monthly interval. The spatial distribution of the different sites centred on Cumbria is shown in Fig. 11.19, overlain on the estimates of prior-modelled NH_3 concentration reductions.

The measured concentrations were compared with modelled values, based on calculations with the FRAME atmospheric dispersion model (Singles et al. 1998; Fournier et al. 2002), using monthly estimates of the NH_3 emissions (see Sutton et al. 2006 for more details).

Figure 11.20 shows the comparison between measured and modelled concentrations for the Cumbria area. The top graph shows the measured and modelled concentrations directly. From this graph it is not directly possible to detect a clear change between the period before and after restocking of the animals (October 2002). This is due to the large amount of within and between year variations. Only after plotting the values of the FMD-affected mean of normalized values as a percentage of the

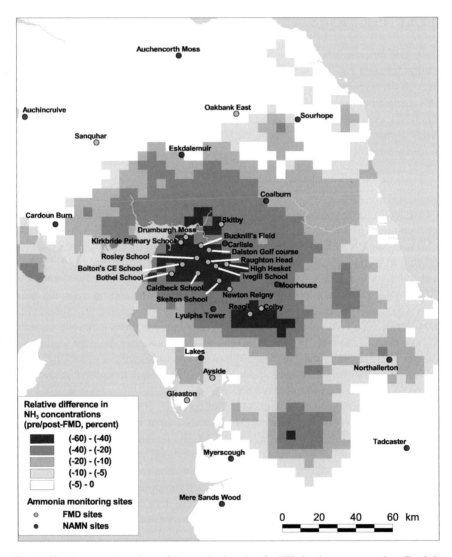

Fig. 11.19 Names and locations of the monitoring sites for NH$_3$ for the area centred on Cumbria. Sites shown in red are the existing nearby sites for NH$_3$ sampling under the National Ammonia Monitoring Network. The grid map shows the anticipated percentage of reduction in NH$_3$ concentrations after FMD (Sutton et al. 2006)

FMD-unaffected mean for both measured and modelled values did a clear change after restocking become visible (bottom graph of Fig. 11.20).

Sutton et al. (2006) considered that the FMD study has significant implications for rural air monitoring strategies, especially in the discussion about implementing:

- Low-frequency monthly measurements at very many sites
- Manual daily monitoring at key regional sites
- Advanced continuous multi-species monitoring at a few European 'super sites'

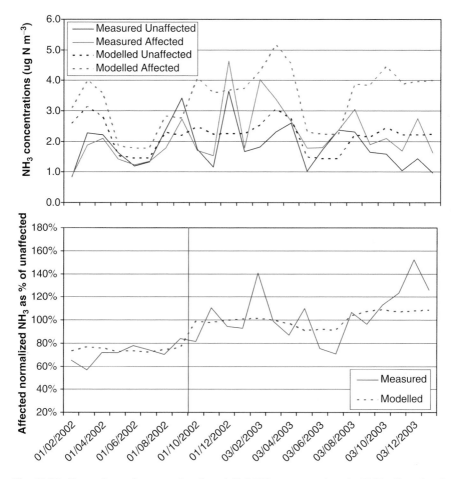

Fig. 11.20 Comparison of measured and modelled NH_3 concentrations in FMD-affected and unaffected areas of Cumbria: (top) mean concentration at sites, (bottom) mean of FMD-affected sites as a percentage of unaffected sites, from normalized data based on full recovery by October 2002 (Sutton et al. 2006)

Depending on the objectives with respect to rural air monitoring (e.g. detecting trends), one or more of these approaches are needed. The example of the FMD study showed that is was only possible to detect the trends in measured NH_3 in response to FMD by a paired comparison of FMD-affected and unaffected sites. In normal cases such a comparison is unlikely to be possible when assessing longer-term reductions in emissions, in the context of assessing the effectiveness of national abatement strategies. The report of Working Group 2 at Bern (Menzi and Achermann 2001) mentioned the need for longer data-series (e.g. 5–10 years) in order to detect significant trends. According to Sutton et al. (2006), monitoring efforts should ideally include parallel sites contrasting areas, in situations where regional or local abatement policies are implemented.

11.4.3 Long Term Trends

Data from the UK National Ammonia Monitoring Network (MAMN), already mentioned in the previous FMD section, can be used to analyse temporal trends in NH_3 and NH_4^+ concentrations (Tang et al. 2006). This also allows assessment of intra- and inter-annual trends between areas dominated by different ammonia emission source sectors (cattle, sheep, pigs etc.) to be made. As an example, the mean concentration from three background sites in Northern Scotland, which were measured using the same method over a period of 9 years are shown in Fig. 11.21. It appears that there is an upward trend in NH_3 concentrations, which may be due to the reduction in SO_2 emissions and aerosol NH_4^+ concentrations over the period. However, the trend is not statistically significant in the seasonally de-trended values, which is derived by dividing the concentrations by the national average seasonal cycle.

In contrast, NH_3 concentrations appear to have remained fairly constant at sites dominated by cattle emissions (Fig. 11.22). For the purpose of consistency in monitoring methods for analysing long-term trends, data in Fig. 11.22 are selected from 25 cattle sites that used the DELTA method over the entire period. In contrast, the simple seasonal detrending used for these analyses recognises the issue of seasonality confusing determination of longer term trends. However it is clear from the remote sites (Fig. 11.21) that a more variable seasonal pattern occurs due to for differences in weather patterns between years. At the cattle dominated sites, with almost no seasonal component the seasonal adjustment is minor.

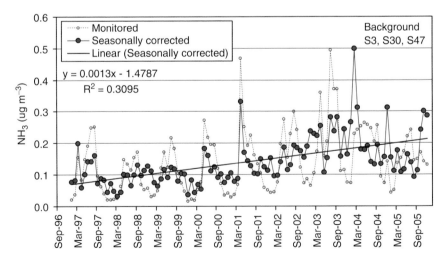

Fig. 11.21 Monitored and seasonally-detrended mean NH_3 concentration from three remote sites in the NAMN (S3 Inverpolly, S30 Strathvaich Dam and S47 Rum). All measurements are made using the DELTA system. The seasonal detrending was derived from the mean seasonal cycle for the whole period (Tang et al. 2009, see Section 3.3, this volume)

By contrast to the background and remote sites, overall there is some indication that NH$_3$ concentrations have decreased in pig and poultry areas (Fig. 11.23), although caution is needed since the passive sampling measurement methodology changes in 2000 from diffusion tubes to ALPHA samplers.

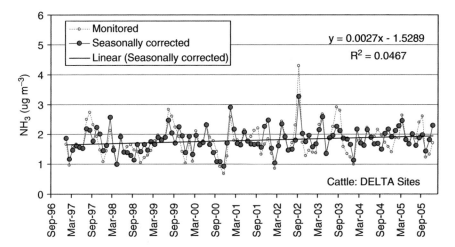

Fig. 11.22   Long term trend in mean monitored and seasonally-detrended NH$_3$ concentration from sites in grid squares dominated by emissions from cattle in the NAMN (Tang et al. 2009, see Section 3.3, this volume)

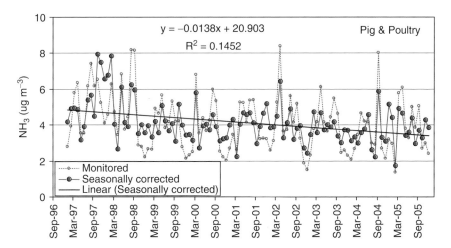

Fig. 11.23   Long term trend in mean monitored and seasonally-detrended NH$_3$ concentration from six sites in grid squares dominated by emissions from pigs and poultry in the NAMN (Tang et al. 2009, see Section 3.3, this volume)

Figure 11.24 shows the long term trends in aerosol ammonium according to the dominant source NH_3 sectors. Here the trends are plotted using seasonally corrected values, by comparison to the raw data. The difference between these shows clearly how NH_4^+ concentrations peak during spring. It is notable that while the overall dataset shows a decrease in NH_4^+ concentrations there are some differences between the different site groups. There is some indication that NH_4^+ aerosol concentrations may have increased in background, cattle and pig and poultry areas, although this is not significant.

The response of different regions to emission changes is also shown in the next few figures. Figure 11.25 gives the overall change in UK NH_3 emissions in the

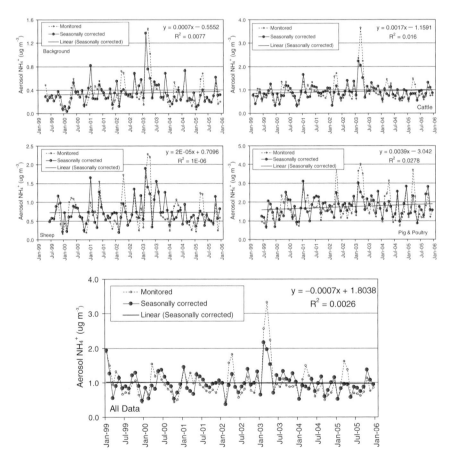

Fig. 11.24 Long term trend in mean monitored and seasonally-detrended NH_4^+ aerosol concentration from sites in grid squares classified as belonging to the following dominant emission categories, (a) background, mean data from eight sites, (b) cattle, mean data from 23 sites, (c) sheep, mean data from three sites, (d) pig and poultry, mean of data from two sites. The large graph shows data from all sites (Tang et al. 2009, see Section 3.3, this volume)

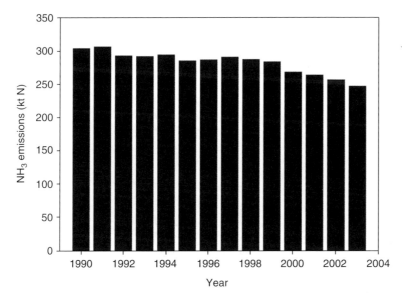

Fig. 11.25 NH$_3$ emission change (in kt N/year) in the UK in the period 1990–2003

period 1990–2003. The decrease of the national emission is about 15%. Figure 11.26 shows the division of wet deposition sites and 'acid water sites' in the UK in four different groups. These groups were identified through cluster analysis.

The response of the measurements to the changing emissions is shown in Fig. 11.27. Overall there is a clear (and significant at $\alpha = 0.05$ level) negative trend in measured values for Group 1, while there is a positive trend for Group 4. For Group 2 and 3 the changes are not very clear, although Group 2 seems to show a downward trend. Linking trends from Fig. 11.27 to the UK emission changes (Fig. 11.25) is difficult, since also long-range transport plays a role here, potentially influencing the observed trends for e.g. Group 1.

11.4.3.1 Germany

For Germany data are available from wet deposition monitoring networks of the individual German 'Bundeslander'. These data are compiled for Germany as a whole and used for different assessments. Figure 11.28 shows monitoring results for NH$_4^+$ wet deposition data, for the period 1987–2004 as averages for different regions in Germany and for the entire country. In general, the deposition is lowest in South West Germany, while being highest in the north-west. In the period 1987–1992 Eastern Germany shows a decline in wet deposition and then the trend is similar to the German average. For 2003 a clear depression in the depositions is visible for all regions, due to low precipitation amount and thus wet deposition.

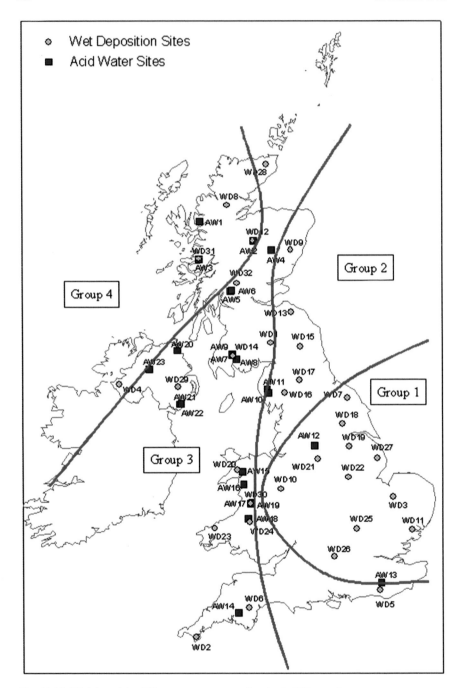

Fig. 11.26 Division of the different measurement sites in the UK over four groups

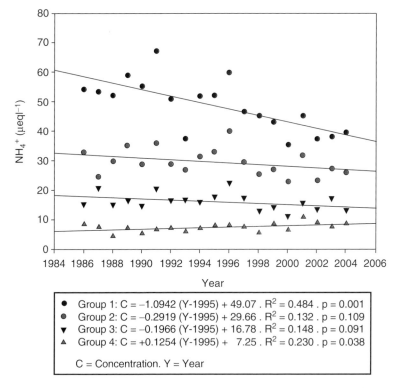

Group 1: C = −1.0942 (Y-1995) + 49.07 . R^2 = 0.484 . p = 0.001
Group 2: C = −0.2919 (Y-1995) + 29.66 . R^2 = 0.132 . p = 0.109
Group 3: C = −0.1966 (Y-1995) + 16.78 . R^2 = 0.148 . p = 0.091
Group 4: C = +0.1254 (Y-1995) + 7.25 . R^2 = 0.230 . p = 0.038

C = Concentration. Y = Year

Fig. 11.27 Response of measurements for the four different groups to national emission changes (see Fig. 11.25)

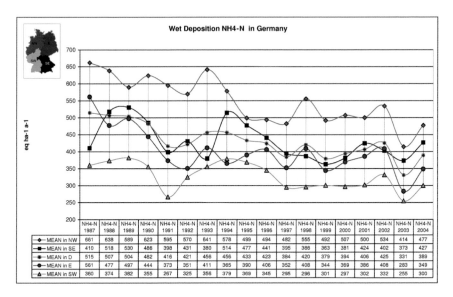

Fig. 11.28 Wet deposition of NH_4^+ in Germany in the period 1987–2004 for different regions in Germany and Germany as a whole

11.4.3.2 Slovakia

For Slovakia an update of an existing long term dataset is presented. Figure 11.3 already showed the in NH_4^+ concentration in precipitation for the EMEP site Chopok in Slovakia. Figure 11.29 now shows the extended dataset for this measuring station. In the period 1990–2004 a decrease in the concentrations can be observed of more than 50%, which seems to be corresponding with the decrease in emissions for the same period (see Fig. 11.30). However, the decrease in wet deposition (also shown in Fig. 11.29) for that period is less distinct and reaches about 30%.

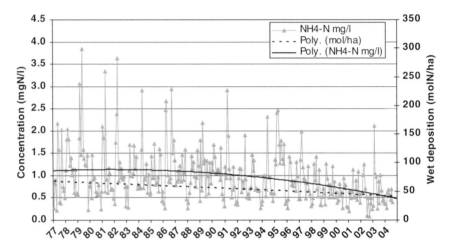

Fig. 11.29 NH_4^+ concentration in precipitation (in mg N/l) for the period 1977–2004 at the EMEP/GAW station at Chopok, Slovakia. Also shown are trend lines for both concentration and wet deposition. For clarity the data points for the latter are not shown here

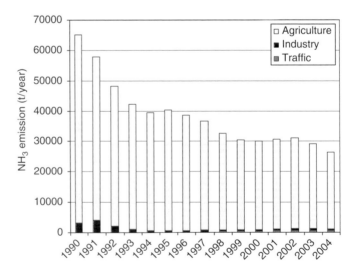

Fig. 11.30 NH_3 emission for Slovakia for the period 1980–2004 (in ton/year), with a distinction between agricultural, industrial and traffic emissions

11.4.3.3   Hungary

The data presented here originate from the EMEP/GAW station K-puszta in the centre of Hungary and extend the record presented in the BBD. Long-term daily records of gaseous NH_3, aerosol NH_4^+ and NH_4^+ wet deposition are available for this site.

Hungary showed a drastic decrease in NH_3 emission at the end of the 1980s. Since then, the emissions remained more-or-less the same (Fig. 11.31). Although estimated change in the NH_3 emissions is substantial, the trend is not followed in the measured NH_3 concentration in air (Fig. 11.32). On the contrary, the NH_3 air

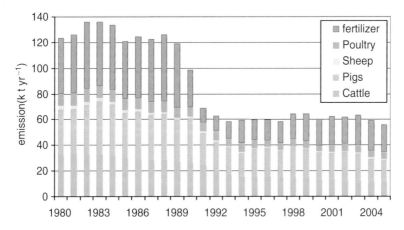

Fig. 11.31 NH_3 emission (in kt/yr) for Hungary, with a distinction between the different agricultural emission categories

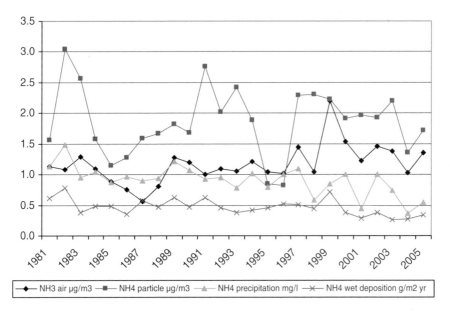

Fig. 11.32 Annual measured concentrations of NH_3 in air, NH_4^+ aerosols, NH_4^+ in precipitation and NH_4^+ wet deposition for the EMEP/GAW site K-puszta

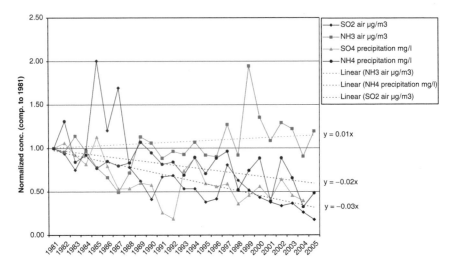

Fig. 11.33 Normalized concentrations/deposition values of NH$_3$, NH$_4^+$, SO$_2$ and SO$_4^{2-}$ (compared to 1981 levels)

concentration at the remote site shows a slight increase in the more recent years. For the NH$_4^+$ concentration in precipitation and the wet deposition of NH$_4^+$ there seems to be an indication of a small increasing trend during the last decade.

Figure 11.33 shows the normalized concentrations of NH$_3$ in air and NH$_4^+$ in precipitation (relative to 1981 values), together with these concentrations of SO$_2$/SO$_4^{2-}$. The SO$_2$ concentrations in air and SO$_4^{2-}$ concentration in precipitation both decreased in the period 1981–2004 at an annual rate of 3%. In the same period the NH$_3$ concentration in air increased with 1% annually, while the NH$_4^+$ concentration in precipitation reduced 2% annually. These observed trends in both SO$_x$ and NH$_x$, again indicate the relation between them, where reduces SO$_2$ emission hampers the formation of ammonium sulphate. Due to this reduced SO$_2$ in the air, the NH$_3$ concentration shows the increasing trend, masking a possible NH$_3$ emission trend at the same time, which seems to be only valid for background/remote areas. This interpretation is supported by the more detailed analysis of Horvath et al. (2009, see Section 3.2, this volume).

The message from Fig. 11.33 resembles the results from the modelling experiment from Van Jaarsveld et al. (2000) to a large degree (see Fig. 11.7). The increase of NH$_3$ concentration over time is almost equalled by a decrease of NH$_4^+$ concentration in precipitation, during decreasing SO$_2$ air concentrations.

11.4.3.4　Croatia

For Croatia measurement results are available for different sites in or around the City of Rijeka. The location of these sites is shown in Fig. 11.34. Listed below are the location names and distances from Rijeka:

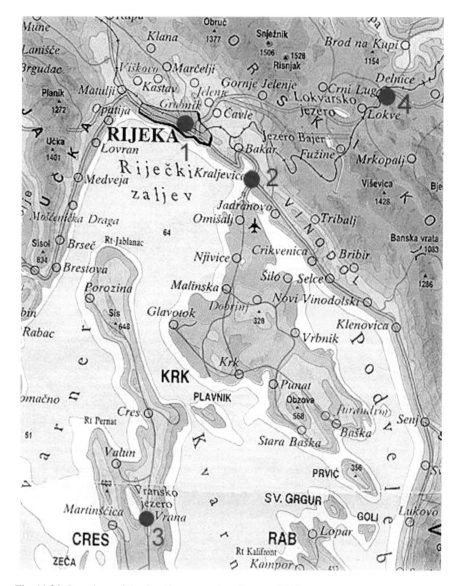

Fig. 11.34 Locations of the Croatian measuring sites near Rijeka

- Site 1: City centre of Rijeka
- Site 2: Kraljevica (20 km distance of Rijeka)
- Site 3: Island of Cres (80 km distance of Rijeka)
- Site 4: Delnice (50 km distance of Rijeka)

Figure 11.35 shows the wet deposition of SO_4^{2-}, NO_3^- and NH_4^+ for the period 1984–2005 (1996–2005 for NH_4^+). For SO_4^{2-} clear reduction were found for site 1

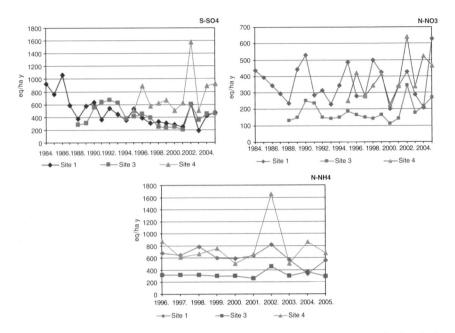

Fig. 11.35 Wet deposition of SO_4^{2-}, NO_3^- and NH_4^+ (in eq/ha/year) for three sites in Croatia for the period 1984–2005 (1996–2005 for NH_4^+)

and 3, while site 4 showed a contrasting trend. The very high values for 2002 (also found in NO_3^- and NH_4^+) are due to conditions of strong winds from the south, bringing Saharan dust into these regions.

For NO_3^- it is difficult to detect a trend, even when excluding the 2002 peak values. NH_4^+ shows a small decline for site 1 and 4 (after excluding the 2002 peak), while site 3 shows a small increase of the depositions. In general there are similar deposition patterns of NH_4^+ (and to a lesser extend NO_3^-) at sites 1 and 4. This is however for different reasons:

- Local washout of the atmosphere at site 1
- Higher precipitation depth at site 4

The decrease (although very small) in the ambient levels can also be found in the measured NH_3 concentrations. Figure 11.36 shows these concentrations for site 1 and 2 for the period 1980–2005, where the downward trend is to some extend visible at site 2.

11.4.3.5 Norway

For Norway data from seven wet deposition measurement sites are available. The location of these sites is shown in Fig. 11.37. Monthly measurement data

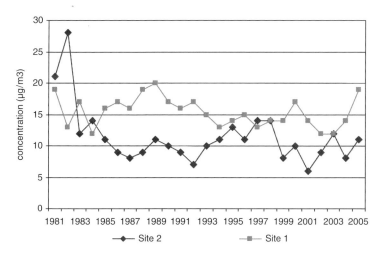

Fig. 11.36 NH₃ concentration (in μg/m³) at site 1 and 2 in Croatia (see Fig. 11.34) for the period 1980–2005

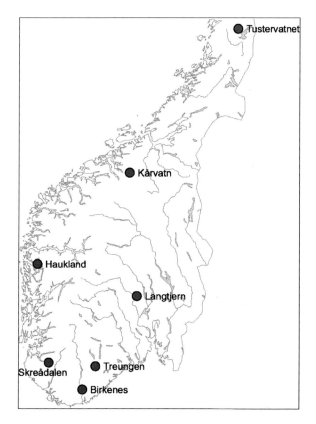

Fig. 11.37 Location of included NH₄⁺ wet deposition measuring sites in Norway

are shown in Fig. 11.38 for winter and summer months (Dec–Feb and June–Aug, respectively). When significant trends exist, they are included in the graphs, where dashed lines show 1980–2005 trends and solid lines show 1990–2005 trends. The measurements for the Northern Norway EMEP station Tustervatn is possibly influenced by local NH_3 farm emissions, therefore showing the somewhat contrasting trend compared to the other sites. However, there is no certainty of this, and it is curious that the increase in wet deposition of ammonium at this remote site matches the increase wet deposition of ammonium in the remote (Group 4) sites of the UK and of the gaseous NH_3 concentration at remote sites in North-West Scotland. Together, these give informative evidence of an increase in atmospheric transport distance of NH_x over recent years as SO_2 emissions have declined, and European NH_x chemistry becomes increasingly controlled by reversible reactions with nitrates.

For two of the stations (Treungen and Langtjern) a proper trend analyses could not be made because of too few measurement data. Two of the seven sites shown a significant decreasing trend for the period 1990–2005 (Birkenes and Haukeland), while for two other sites (Skreådalen and Kårvatn) no significant trend was found.

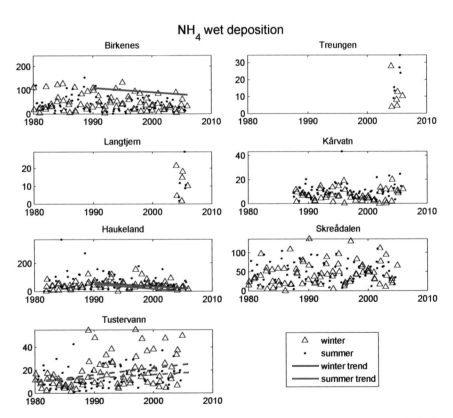

Fig. 11.38 NH_4^+ wet deposition (in mg/m²/month) for seven Norwegian sites for respectively the winter and summer months. For Birkenes, Haukeland and Tustervann also trend lines are shown

11.4.3.6   Denmark

Extensive monitoring of NH$_x$ is carried out in Denmark permitting assessment of the trends between 1989 and 2005 (Figs. 11.39, 11.40). The measurements of gaseous ammonia and particulate ammonium are made using the filter pack method.

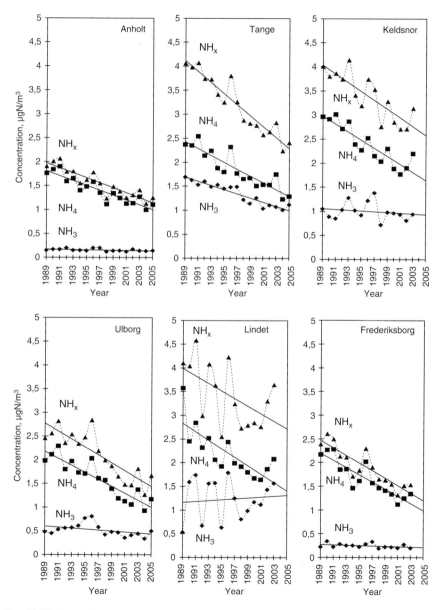

Fig. 11.39 Record of concentrations of total NH$_x$, gaseous NH$_3$ and particulate NH$_4^+$ at the Danish monitoring sites

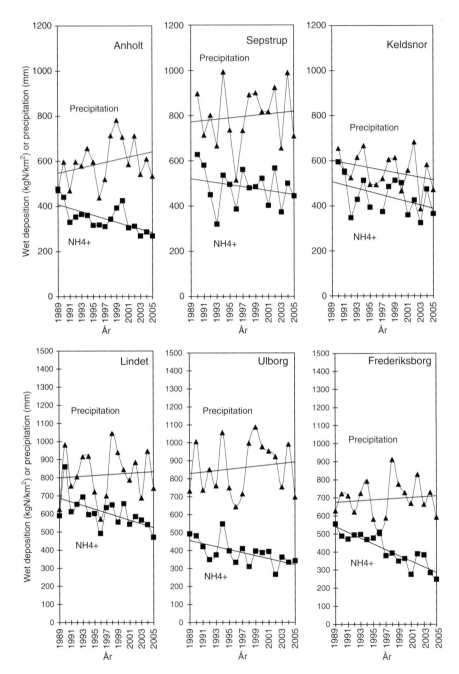

Fig. 11.40 Record of ammonium in wet deposition at selected monitoring sites in Denmark. The decreases are paralleled by reductions in SO_4^{2-} and NO_3^- depositions (data not shown)

This method may have interferences in the separation of the two phases. However, comparisons with measurements using denuders, which can separate the phases, have shown good agreement between the two methods (Andersen and Hovmand 1994; Ellermann et al. 2006). Under Danish conditions and with the Danish version of the filter pack method an acceptable separation of the two components is therefore achieved.

For particulate NH_4^+ a substantial and uniform decrease (45–55%) is seen on all the Danish stations for the period 1989–2005 (Table 11.1). For NH_3 there is a decrease (12–41%) on five of the stations. The decrease is less than for particulate NH_4^+ and the variations between the years and between the stations are considerable. The general picture with a large and uniform decrease of particulate NH_4^+ and a small and variable decrease of NH_3 in Denmark is therefore consistent with the previously reported observations (see last BBD). The different variability of ammonia and particulate NH_4^+ reflects that NH_3 is a primary and relatively short lived pollutant, while particulate NH_4^+ is a secondary and long lived pollutant. The main sources of NH_3 are therefore the local sources (agriculture) while the main part of particulate NH_4^+ is long range transport to Denmark. Model calculations using the Danish Eulerian Model (DEHM, Christensen 1997; Frohn et al. 2002, 2003) have shown that on average more than 95% of the ammonia in Denmark originates from national sources. Only 15% of the particulate NH_4^+ in Denmark originates from Danish sources.

The observed trends are in general in agreement with the trends for the emissions of ammonia. The Danish and European emissions of NH_3 have decreased with 27% and 17%, respectively, during the period 1990–2003 (NERI 2006; EMEP 2006). However, the decrease in atmospheric concentrations of the primary pollutant, NH_3, is on average less than the decrease in the emissions, while the decrease of the concentrations of the secondary pollutant, particulate NH_4^+, is higher. These findings support the suggestion that the trends in NH_3 and particulate NH_4^+ are impacted by

Table 11.1 Trends in concentrations of NH_x, NH_3 and particulate NH_4^+ and wet deposition of NH_4^+ in Denmark during the period 1989–2005. The trends are calculated on basis of linear regression lines and correspond to percentage change from 1989 to 2005

	NH_x	NH_3	Part. NH_4^+	Wet dep. NH_4^+
Anholt	−42	−19	−45	−31
Frederiksborg	−52	−22	−55	−47
Keldsnor	−36	−12	−45	−23
Lindet	−32	[a]	−50	−24
Tange/Sepstrup Sande[b]	−44	−41	−47	−13
Ulborg	−48	−27	−54	−30
Mean	−42	−24	−49	−28

[a]Not taken into account because it has moved location.
[b]NH_x, NH_3 and par. NH_4^+ in air is taken from Tange. Wet deposition of NH_4^+ is taken from Sepstrup Sande. The distance between these stations is about 30 km.
The wet deposition of NH_4^+ has also decreased (13–47%) on all of the stations. The variability between the years and between the stations can partly be explained by variations in the amount of precipitation at the stations. Model calculations using DEHM have shown that on average 33% of the wet deposition of NH_4^+ in Denmark originates from national sources.

changes in concentration levels of other pollutants like SO_2 and particulate SO_4^{2-}. The sulphur compounds have shown a significant decrease (more than 50%) in concentrations since 1989 (Ellermann et al. 2006).

Among the Danish stations, the reduction of the NH_3 concentration is lowest at Keldsnor, which is situated at a small island in Southern Denmark. This station is influenced by air coming from Germany and to some extend Poland. This low decrease may partly be explained by the larger decrease of emissions in Denmark compared to Germany (18% reduction, EMEP 2006) and EU25. The model calculations show that 100% of the NH_3 is the central part of Jutland originates from Danish sources, while the contribution from Danish sources are only about 71% in the southern and western part of Jutland.

It seems obvious that the efforts taken to reduce emissions of ammonia have had an impact on the long term trends of NH_x. Moreover, the reductions of the NH_3 emissions achieved by changes in the agricultural practise have shown an impact on the seasonal variation of the ammonia concentration in Denmark. Figure 11.41

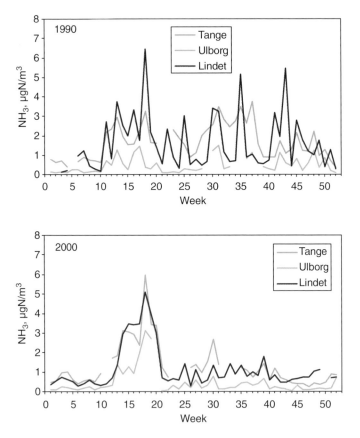

Fig. 11.41 Weekly average concentrations of NH_3 at the Danish sites of Tange, Ulborg and Lindet in 1990 and 2000

shows the seasonal pattern of NH_3 during the period 1990 and 2000. It is clearly seen that the seasonal pattern has changed from a more even distribution of NH_3 during spring-autumn to a pattern with much higher concentrations of NH_3 during spring. The Danish parameterisation of the seasonal variation of NH_3 emissions has shown that this change can be linked to the Danish regulations of the application of manure, which has caused a shift in emissions of NH_3 from autumn to spring and decreased the overall emissions from this source category (Ambelas et al. 2003).

11.4.3.7   USA

As already stated before, a contrasting situation with respect to the trends in reduced nitrogen can be found in the USA. Data were already shown in the BBD for the state of North Carolina, where a drastic increase both in NH_3 emissions and measurements was observed, due to increasing numbers of pigs. The following paragraphs show updated information for North Carolina, but also some overall information for the USA from the National Trends Network.

In the BBD this example from North Carolina was already presented, showing a dramatic simultaneous increase of both NH_3 emissions and atmospheric NH_4^+ levels. This was mainly caused by an increase of the number of pigs: factor of 7 increase between 1985 and 1997. Figure 11.42 shows an updated version of the previous North Carolina graph. It shows a clear change in the trends, both emission and concentration/deposition, caused by a moratorium on the number of pigs which started in 1996.

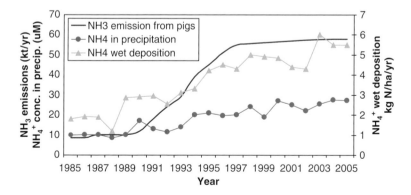

Fig. 11.42   Comparison of NH_3 emissions in coastal North Carolina (USA) against NH_4^+ precipitation concentrations and wet deposition at the NADP monitoring site NC35 in Sampson County (From Aneja et al. 2006)

The change in trend is again clearly shown in Fig. 11.43, where 4-weekly wet deposition concentrations are presented for Sampson County (or location NC35) for two distinct periods: before the moratorium on pigs and during the moratorium. The annual rate of increase in precipitation concentrations dropped from 9% to 4% per year, before and during the moratorium respectively.

Another dataset (containing the NC35 data for North Carolina presented above) is the one on wet deposition for the entire USA, from the National Atmospheric Deposition Program. For 258 sites nationwide, weekly measurements are available for some sites from 1978. These sites were selected to be regionally representative, which means that they avoid nearby pollution sources (like e.g. cities, power plants, major highways, cattle feedlots, etc.). Figure 11.44 shows the spatial distribution of the different sites over the USA.

At the different sites the following compounds are measured, SO_4^{2-}, NO_3^-, Cl^-, PO_4^{3-}, Ca_2^+, Mg_2^+, K^+, Na^+, pH and relative conductance.

Figure 11.45 shows the spatial distribution of the NH_4^+ wet deposition over the USA, after interpolation of the measured values. Maps are shown for 3 years, showing

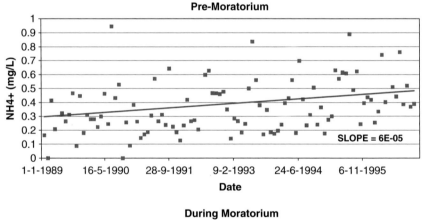

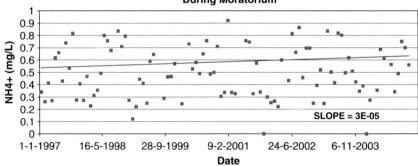

Fig. 11.43 Four-week Averaged NH_4^+ wet deposition concentrations at high emission site NC35 in the period 1989–1996 (above: pre-moratorium) and 1997–2004 (below: during moratorium)

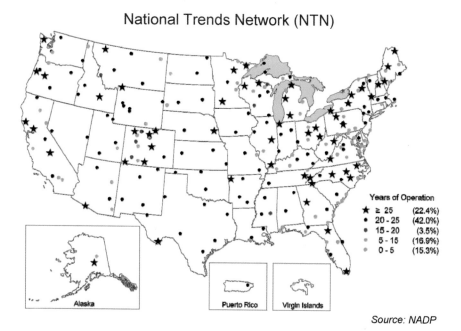

Fig. 11.44 Spatial distribution of the NTN monitoring sites, including information about the years of operation

the change of wet deposition in the period 1985–2003. Together with the maps for NH_4^+, maps for SO_4^{2-} and NO_3^- wet deposition are shown. Looking at the differences between the consecutive years for the three individual compounds, it seems that the increase in NH_4^+ coincides with decreasing loads for SO_4^{2-} and NO_3^-.

11.5 European Overview

Fagerli and Aas (2006) analysed trends of nitrogen compounds for the last decades at EMEP sites in Europe both from model calculations and observations. In general, the model was found to reproduce the trends in the measurements. They concluded that the emission estimates, their changes and the models response to the changes are reasonable.

For reduced nitrogen in precipitation, the largest decrease between the beginning of 1980 and 2003 was found in Eastern Europe (~40–60%). This is also the area were ammonia emissions have undergone the largest changes. In addition, sites in the Nordic countries show clear downward trends with magnitudes larger than the corresponding emission changes. These sites receive large portions of their pollution from Central and Central-East Europe, thus although ammonia emissions in this area have changed little, reduced nitrogen wet deposition in general declined.

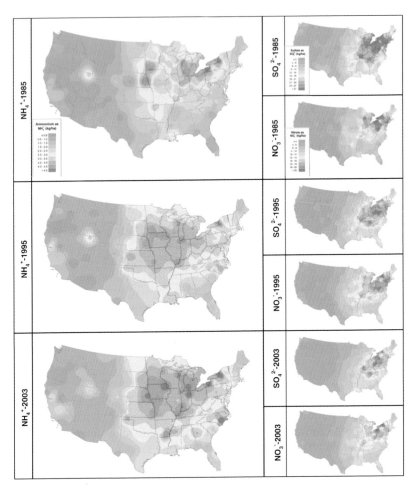

Fig. 11.45 Spatial distribution of NH$_4^+$, SO$_4^{2-}$ and NO$_3^-$ wet deposition (in kg/ha) over the USA for 1985, 1995 and 2003

A possible exception to this is ammonium in very remote areas, as it is shown here that concentrations and deposition increased in both North-West Scotland and North Norway, which may indicate an increasing atmospheric transport distance for NH$_3$ emissions.

On a European scale, the overall changes in ammonia and NO$_x$ emissions have been similar (around 25% from 1980–2004). However, the reductions in NO$_x$ emissions have been more uniform over Europe (Fig. 11.46). As a result, oxidized nitrogen in precipitation has decreased by 20–35% in most European countries with statistically significant declines at most sites. In Ireland and some South European countries (e.g. Spain, Portugal and Greece) the level of NO$_x$ emissions are at the same level or slightly

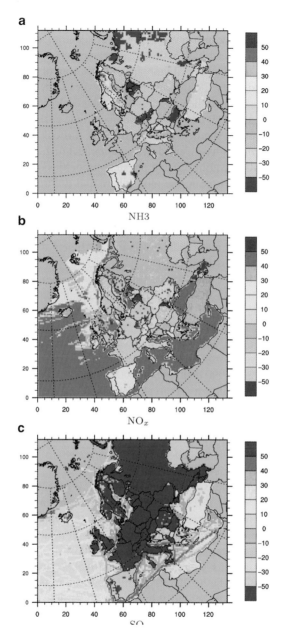

Fig. 11.46 Changes (%) in emissions from 1990 to 2004 (relative to 1990) (From EMEP)

higher at present compared to 1980, and in these countries no statistically significant trends were found at the EMEP sites for wet deposition of oxidized nitrogen.

On a European scale, the trend in the observations (and the EMEP model results) for wet deposited nitrogen correlate with the trend in the emissions.

For air concentrations, the picture is more complex. Unfortunately, much less information is available for measurements of air concentrations; most of the EMEP sites did not start to measure TIA (Total inorganic NH_x, sum of ammonia and ammonium aerosol) and TIN (Total inorganic xNO_3^-, sum of nitric acid and nitrate aerosol) until the end of the 1980s and only a few sites (~20) have reported measurements continuously since then. Because the meteorological variability is large (~20%, van Loon et al. 2005) and of the same magnitude as the change of the emissions in the same period, the detection of trends is difficult. Moreover, since the gas and particulate phases have very different chemical (e.g. their role in the NH_4^+ -NH_3-HNO_3-NO_3^--SO_4^{2-} equilibria) and physical properties (e.g. the aerosols have a much longer residence time in the atmosphere and are transported over longer distances) the trend in the gas and particulate phase may be different. Fagerli and Aas (2006) concluded that in general the trend in TIA in air followed the trend in ammonia emission. However, both model calculations and measurements indicated that in some areas the decrease in TIA was more efficient than the corresponding decrease in ammonia emissions.

For TIN in air, few of the ~20 EMEP sites with continuous measurements from around 1990 to 2003 show statistically significant declines (in observations or model results), despite that NO_x emissions have been reduced by as much as 30–50% in some areas during this period.

A problem for Fagerli and Aas (2006) was that model calculations were not available for all the years of interest, e.g. in the analysis of air concentrations trends from 1990 to 2003 only 1990 and 1995 to 2003 were available. In order to average out meteorological variability, a set of calculations have been made with the EMEP Unified model using the same meteorological year (2004) and three different sets of emissions; (1) 1990, (2) 2004 and (3) 2004, but with SO_x emissions as in 1990. The model version and its setup for these calculations are the same as in Fagerli and Aas. In Figs. 11.47 and 11.48 the modelled changes from 1990 to 2004 are presented for the different nitrogen species.

Whilst the concentrations of reduced nitrogen in precipitation are predicted to have changed by about the same amount as ammonia emissions, and with a similar spatial pattern (except for the reductions in the Nordic countries), NH_x in air is reduced somewhat less in some areas and somewhat more in other areas. For instance, in Czech Republic NH_3 emissions have declined by around 55% whilst NH_x concentrations decrease only by 40–50%. In Germany, however, NH_x concentrations decline by 20–30% whilst emission reductions are reported to be 10–20%. The reason behind this pattern appears to be a combination of a less efficient formation of ammonium aerosol (due to decreasing SO_x emissions) and less efficient dry deposition of NH_3 due to less acidic surfaces. Both these effects lead to a shift towards gaseous ammonia relative to particulate ammonium. For instance, model calculations predict that, despite the reduction in ammonia emissions over much of Europe, ammonia concentrations in background air are estimated to increase in many areas (Fig. 11.47). By contrast, simulated ammonium aerosol concentrations decrease everywhere, also in areas

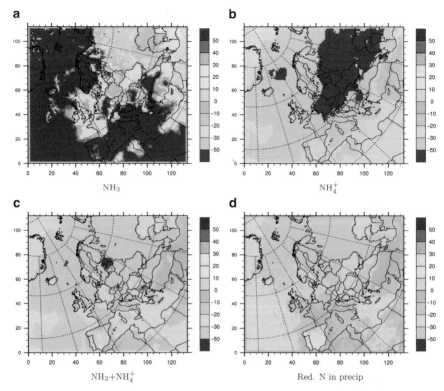

Fig. 11.47 Modelled changes (%) in reduced nitrogen from 1990 to 2004 (relative to 1990)

where ammonia emissions have increased (e.g. in Spain). If SO_x emissions had not been reduced (Fig. 11.49), model calculations predict that the reductions in the reduced nitrogen species would have followed the changes in ammonia emissions more closely.

EMEP model calculations predict that the largest decline in oxidized nitrogen should be seen for nitric acid (Fig. 11.48), with reductions similar or larger to the reductions in NO_x emissions. Nitrate aerosol concentration is suggested to decrease in most of Europe, but not to the same extent as NO_x emissions. Small declines, or even increases are seen in the areas where SO_x emission reductions have been largest. The sum of the two, which is the most commonly measured, change by 0–30% in most of Europe, with the largest changes in the areas where both NO_x and NH_3 emissions have been reduced. With a dataset containing only few years, such a small change may be masked by the meteorological variability. In contrast, model calculations predict that if SO_x emissions had remained at the 1990 level (Fig. 11.50), TIN would have decreased by 30–50% over most of Europe, which should have been easily detectable in measurements.

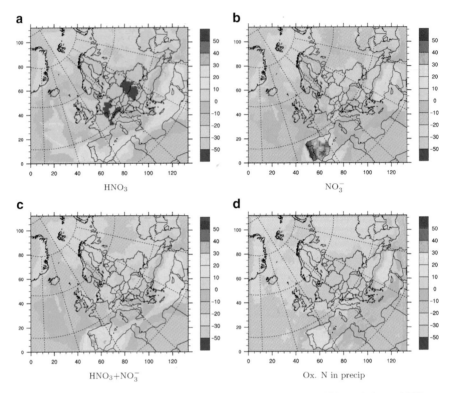

Fig. 11.48 Modelled changes (%) in oxidized nitrogen from 1990 to 2004 (relative to 1990)

In a more generalized way the EMEP emission and concentration data can also be investigated (Fowler et al. 2006). Here the EMEP emission changes were compared with the measured NH_4^+ concentrations in precipitation for five different regions. Figure 11.51 shows an overview of these regions, while Table 11.2 gives some information about the import/export status of these regions for the year 2000 for respectively NH_x, SO_x and NO_y.

Overall there is a reduction of the NH_3 emissions in Europe of 23% in the period 1980–2000. However, as was already showed in Fig. 11.8, there are large differences over Europe with respect to these changes. Trying to detect these emission changes by measurements on a European scale is only possible using measurements of NH_4^+ in precipitation, since a European wide measurement network for NH_3 is not available at the moment. Countries like UK, the Netherlands and Denmark have an operational NH_3 monitoring network. However, differences exist with respect to measurement techniques, etc., hampering a proper comparison of the measurement results. In particular, it is evident that measurements of TIA (which do not separate gaseous NH_3 and aerosol NH_4^+) are inadequate to analyze the changes.

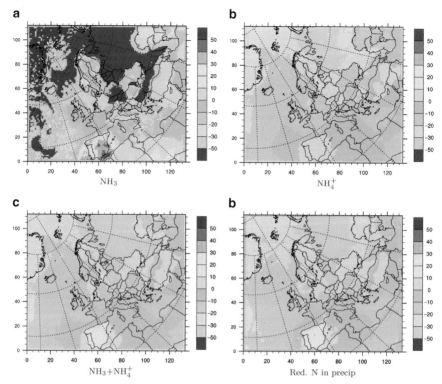

Fig. 11.49 Modelled changes (%) in reduced nitrogen from 1990 to 2004 (relative to 1990) if SO_x emissions had remained at the 1990 level

When looking at the changes of NH_4^+ concentration in precipitation, clear differences between the different regions can be seen (Table 11.3):

- Monitoring NH_4^+ in aerosol and rain provides effective integration at the regional scale and reveals the trends in emissions.
- Concentrations in most of Europe are declining, while in remote regions concentrations and deposition are increasing.
- There has been a change in the chemistry of ammonia as a consequence of sulphur emission reductions.

11.6   Discussion and Conclusions

The discussion about the different issues presented in the previous sections focus here on two main items: (1) what conclusions can we draw from the new information and do they differ from those of the first assessment (BBD)? and (2) are

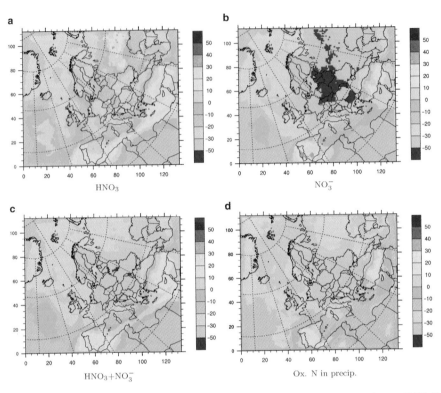

Fig. 11.50 Modelled changes (%) in oxidized nitrogen from 1990 to 2004 (relative to 1990) if SOx emissions had remained at the 1990 level

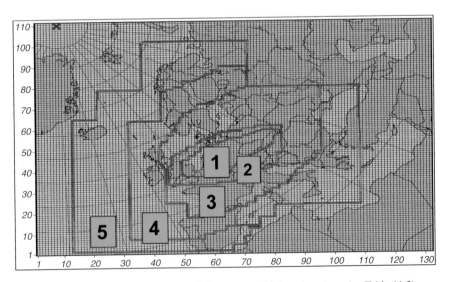

Fig. 11.51 Division of Europe over five different source/sink regions (see also Table 11.2)

Table 11.2 Overview of the five regions and the amount of NH_x, NO_y and SO_x being exported or imported (in %)

	Type	NH_x (%)	NO_y (%)	SO_x (%)
Region 1	Source – export	20	50	20
Region 2	Source – export	10	60	30
Region 3	Sink – import	10	10	20
Region 4	Sink – import	60	10	20
Region 5	Sink – import	70	50	80

Table 11.3 Changes (in %) of NH_4^+ in precipitation and NH_3 emission in the period 1980–2000 for the five regions (Data taken from EMEP)

	ΔNH_4^+ (%)	ΔEmission (%)
Region 1	−28	−29
Region 2	−41	−22
Region 3	−26	−28
Region 4	+7	−6
Region 5	+43	+10

the present measurements adequate for assessing reductions of reduced nitrogen emissions? Related to the second question, it can be asked: Is it possible to improve the overall European measurement strategy?

11.6.1 Have the Conclusions Changed Between the First Assessment (BBD) and the Results Presented Here?

Here the different conclusions from the BBD (Sutton et al. 2003) are listed again (in italics), adding information based on the new datasets.

It was clear that there are several difficulties and uncertainties in assessing the effectiveness of NH_3 abatement from monitoring networks.

Assessing the effectiveness of NH_3 abatement is still a difficult issue, although substantial progress was made in the last few years in understanding the problems related to this. A good example is the FMD study from the UK, presented in Section 2.2.3. The major conclusions from that study can be taken forward in the discussion about the implementation of different monitoring strategies (i.e. low-frequency at many sites, daily at key regional sites, advanced at a few European 'super sites'). The FMD study showed that, in order to detect the response of NH_3 concentrations to abatement measures, monitoring efforts should include parallel sites in both affected and unaffected areas. Although this should be taken into account in an ideal situation, such a comparison is unlikely to be possible in normal situations, when assessing longer-term reductions in emissions in the context of looking at the effectiveness of national abatement strategies. There is no clear and quantitative monitoring strategy aiming at an assessment of the effects of policies/measures on different scales.

Assessing this effectiveness requires sound monitoring methods implemented at sufficient sites and over a sufficiently long period.

The definition of 'sufficient' will largely depend on the actual situation. Again, the UK FMD study provided valuable input to this discussion. Multiple low-cost sites in two contrasting regions where needed to detect the changes in a period of half a year. However, wet deposition results from one site during 10–15 years were needed to show the effect of the moratorium on animal numbers in North Carolina.

Conditions like e.g. other relevant NH_3 sources will determine the final monitoring strategy. However, this requires a good insight in the situation with respect to the expected effect of the measures, which can be evaluated by means of modelling exercises. This was also shown for the FMD study: monitoring activities where pointed towards those regions where the largest effects were foreseen. However, this assumes the availability of adequate modelling tools, capable of describing the present situation and the expected changes due to the different measures (see also below).

For NH_4^+ aerosol and NH_4^+ in rain, a modest number of sites can be used to indicate trends, whereas for NH_3 in source areas a high density of sites is essential.

When only looking at these measurements in terms of detecting trends, it might be discussed whether indeed the assumed difference in measurement strategy between NH_4^+ (both aerosol and rain) and NH_3 is valid. The Dutch study presented in before showed that only eight intensive monitoring sites distributed over the Netherlands are able to detect the overall trend. Comparing the average concentrations based on these eight sites with those from 155 sites, showed that the eight sites provided a good estimate of the average concentrations in the Netherlands. It should be taken into account, however, that the locations of these eight monitoring sites are well chosen and based on thorough research about the representativeness of these sites with respect to covering the different concentration situations in the Netherlands. A high density of NH_3 measurement sites is essential for e.g. calibrating transport and deposition models. The 155 measuring sites in the Netherlands were also used for this purpose and it showed that the OPS model was capable of modelling the spatial distribution of the NH_3 concentrations rather well. However, this was only possible in the presence of good quality emission data.

From some of the studies presented before, it became clear that the measurements for NH_4^+ aerosol and wet deposition show the required integration over space and time for evaluating the trends in NH_3 emissions on a European scale. For more regional/local trends studies the focus should be more on NH_3 concentrations, since these reflect the local NH_3 situation better.

In contrast to the need for many NH_3 sampling locations, is the requirement for high temporal resolution NH_3 concentration data at selected sites.

The availability of concentration data on a high temporal resolution is important for different reasons. An obvious reason is to fill the gap in time when e.g. only low-cost samplers are available, like passive sampling tubes, which are normally used for time periods of 2–4 weeks. By using high temporal resolution data, a better understanding of the linkage between different sources and the resulting concentrations can be acquired. This is done by combining the measured air concentrations with meteorological data (wind direction and speed being most important). By doing so sources of NH_3 can be evaluated by means of their contribution to the measured concentration at the high temporal resolution sites. Mosquera et al. (2000) showed an example of such an evaluation, where emission factor for cattle were evaluated using this kind of measurement information.

Quantifying the interactions of NH_x, necessary to interpret long-term trends, also requires improved mechanistic understanding and modelling

- better generalization on the bi-directional controls on NH_3 exchange
- the chemical interactions that are recognized for atmospheric chemistry also need to be treated in relation to dry deposition
- advancement of the regional-temporal modelling of NH_3 emissions in relation to environmental conditions

The need for improvements on these topics was again shown in some case studies, like the one for the Netherlands. Uncertainties with respect to e.g. the bi-directional NH_3 exchange in the modelling are thought to be one of the reasons why the 'ammonia gap' is still existing. Further research is therefore ongoing to reduce these uncertainties. Also the availability of good quality emission data is important for a better understanding of the observed concentrations. But this is also important for developing an adequate monitoring strategy, based on good modelling tools in combination with emission data at relevant temporal and spatial resolutions.

It is important to retain caution in attributing changes in atmospheric NH_x to changes in NH_3 emissions.

Again it was shown for different studies that the simultaneous changes in SO_2, NO_x and NH_3 emissions cause problems when investigating the changes in measured NH_x. It was already presented in the BBD that the change in SO_2 concentration causes an increase of NH_3 air concentrations and a decrease of NH_4^+ wet deposition. The European overview also showed these parallel changes in air concentrations and wet deposition, due to changing SO_2 emissions over the years. This was based on modelling studies, but also measurements from Hungary showed this simultaneous increase of air concentrations and decrease of wet deposition corresponding with lower SO_2 emissions over the period 1981–2004.

The degree to which SO_2 and NO_x emission changes result in increasing or decreasing concentration/deposition levels should ideally be taken into account when evaluating the overall effectiveness of abatement measures with respect to NH_3 emissions trends. Long term measurements like those in Hungary might provide the necessary input to this kind of investigations. It is however, not clear to what extend that can be done for Europe.

There are clear difficulties trying to detect NH_3 emission changes even where these certainly occurred.

Some clear advances in this field were made by e.g. the UK FMD study and by the longer term datasets now available, particularly from the Netherlands and Denmark. From the FMD study it became clear that there is a need for contrasting areas when trying to investigate the 'certain' NH_3 emission changes. The FMD situation provided the possibility to study these contrasting areas and indeed 'discover' the changing concentration levels after populating the cattle farms again. Attention should be paid to finding these contrasting areas when starting these type of studies, even though it might prove to find them within a proper distance of the study area. Where such comparison is not possible, the only alternative is for long data series (ideally longer than 10 years) that speciate the separate NH_x components.

In assessing the success of any abatement policy based on technical measures, a combination of appropriate modelling and sufficient measurements should be able to determine whether the measures are broadly effective.

Again, the UK FMD provided a good example of this. Modelling showed the areas where changes were most likely to occur, while the measurements proved it. In this case the availability of low cost measurement techniques (passive sampler tubes) at a monthly time resolution was enough to give a proper representation of the changes in NH_3 emissions within the study area. Similarly, the Dutch and Danish examples reflect situations where estimated NH_3 emission reductions are related to abatement policies. In both these cases, the atmospheric data suggest broadly that the expected emission reductions were achieved, although questions remain regarding the absolute magnitude of the NH_3 emissions.

However, where there is a gap between the monitoring response expected and that observed, this may be as much due to

- limitations in atmospheric process quantification and monitoring
- ineffectiveness of the abatement techniques

Given the different considerations on the previous BBD conclusions, this last conclusion is still valid. However, there has been some clear advancement in closing this gap in the last few years, where we do get to a better understanding of the reasons behind it. Further work on the different monitoring methods has to be continued to close this gap. However, joint efforts are needed to facilitate this, focussing both on the description of the atmospheric processes by means of models and the measurement techniques used to evaluate them.

11.6.1.1 Are the Measurements Adequate to Assess the Emission Reductions of Reduced Nitrogen?

The deposition of reduced nitrogen is one of the important drivers in international policies to decrease nitrogen emissions. It is clear that for individual ecosystems concentration, exposure and deposition in general, and dry deposition in particular, cannot be quantified with sufficient accuracy using deposition models. The various methods have different advantages and drawbacks and the choice of a certain method for estimation of the flux of a specific pollutant to a specific ecosystem may in many cases depend on the purpose of the study and on requirements on accuracy and costs. For the time being, it is impossible to obtain an accurate annual average concentration and deposition map of Europe based on current deposition measurements and therewith is difficult to target emission reductions based on reduced critical level and deposition/critical load exceedances. Dry, cloud and fog deposition show very strong horizontal gradients in ambient concentrations due to variations in land use, in surface conditions and meteorology. Deposition maps are generated based on a combination of models and measurements (e.g. Van Pul et al. 1995; Erisman et al. 2001). Regarding spatial and temporal scales, measurements are supplementary to models in such a method. Furthermore, measurements are used for developing process descriptions and for evaluation of model results. Finally, measurements can act as an independent tool for assessing policy targets through trend detection. These issues, outlined below, require a combination of different measuring, monitoring and modelling approaches.

11.6.2 Process-Oriented Studies

Process-oriented studies are primarily used to provide insight into emission and deposition processes, and to obtain process descriptions and parameters to be used in models. Micrometeorological methods provide the best methods for these purposes. Three super sites equipped within the LIFE project fulfilled the role for this purpose (Erisman et al. 2001). In cases where micrometeorological methods cannot be used, such as complex terrain and within forest stands, the throughfall method is the only one currently available, even though the uncertainty in NH_x fluxes is large due to canopy exchange processes. Process-oriented studies can be used to test or verify simple/low cost measuring methods, which might be used for other purposes such as monitoring.

11.6.3 Evaluation of Models

For evaluation or validation of model results, preferably simple and low cost monitoring methods are desired. In general, monthly to annual average concentrations and fluxes are used for validation. The uncertainty in results obtained by these monitoring methods should be within acceptable limits, and the community needs to agree quantitative values of these limits. Furthermore, results should be representative for areas used as receptor areas in the model. Validation of long-range transport model results can be done using area representative measurements of wet deposition and of ambient concentrations. Low-cost micrometeorological measurements suitable for monitoring or super site data might be used for evaluating model concentrations and dry deposition fluxes. Throughfall measurements might be used as a validation method for spatial variability in dry (and total) deposition, provided that several criteria on the method and site are met and the measurements are corrected for canopy exchange (e.g. Draaijers et al. 1996). In Europe about 400 throughfall sites have been operational since 1995 (De Vries et al. 2001). It is advisable to equip several monitoring locations in Europe with dry deposition monitoring systems, wet-only sensors, and cloud and fog deposition measuring methods, which act as reference stations for testing of low-cost equipment and which can serve to derive surface exchange parameterisations used in deposition models. The locations with so-called 'intensive monitoring methods' should be selected on the basis of pollution climates and type of vegetation. Furthermore, the surroundings should be homogeneous and no significant sources should be near the site.

11.6.4   Detection of Trends

If the purpose of measurements is trend detection, annual averages must be measured as accurately as the magnitude of the trends. Ambient concentration and wet deposition measurements such as those of the EMEP monitoring network can be used for trend detection. The trend in precipitation concentrations is representative for the dry deposition trend, which cannot be measured accurately enough at present. The disadvantage of using only concentration measurements is that a change in dry deposition due to ecosystem response (as a result of reduced loads or climatic change) or due to changes in surface conditions (interaction with other gases, etc.) cannot be detected. Extensive deposition and concentration monitoring (see previous section) might be useful for trend detection, especially where larger emission reductions have occurred, otherwise intensive methods should be applied.

11.6.5   Modelling

An essential component is the use of process-based models to complete emission, transport and dry deposition inputs from existing air concentration monitoring networks. These models will be applied to quantify ecosystem specific inputs of these components. The core dry deposition monitoring stations and the low cost deposition monitoring network will provide the validation data to test and refine models for pollution climates and land uses in Europe.

11.6.6   Monitoring and Modelling Strategy

The EDACS model (or its successor IDEM) is an example of a method to estimate small scale deposition fluxes and critical load exceedances in Europe (e.g. Erisman and Draaijers 1995). The resolution, determined by the land use maps, is good enough to estimate ecosystem specific inputs. Surface resistance parameterisations should be more detailed to describe the complex surface exchange of gases. The monitoring stations used in e.g. the LIFE project provide and now established as Level 3 "super sites" under NitroEurope, provide detailed data to evaluate models and to improve parameterisations under a range of climates and conditions. The data from these stations are, however, not representative for the total range of ecosystems, climates and conditions in Europe. It is therefore necessary to extend the sites, in order to cover more of the ranges. This for example matches to the objective under NitroEurope to establish a series of Level 2 "regional sites" for nitrogen flux measurements.

In general it can be stated that ecosystem type, site management, roughness characteristics and the surface conditions (wetness, snow cover, etc.) are all important in controlling the deposition rates of reduced nitrogen., This shows that both the major ecosystem/management types need to be considered as well as major differences in environmental conditions (dry weather, cold, wet, etc.), to verify whether the model assumptions are correct.

11.6.7   *European Monitoring Strategy*

A European monitoring strategy for transboundary air pollution has been extensively discussed in the last 5 years for the period 2004–2009, under the frame of the UNECE Task Force on Measurement and Modelling (e.g. Aas 2005). It is therefore relevant to review the progress in implementing this strategy in relation to reduced nitrogen, and address the most critical limitations, in the light of the new datasets presented and discussed in this document.

Acknowledgements This review was conducted as a contribution to the Verification component of the NitroEurope IP, and with travel financing from COST 729. We are grateful to the European Commission and the large number of national funding bodies for underpinning support of the measurements and modelling.

References

Aas W. (2005) Workshop on the implementation of the EMEP monitoring strategy (Ed.) (Oslo 22–24 November 2004) EMEP/CCC Report 2-2005. NILU, Kjeller, Norway. www.nilu.no/projects/ccc/reports/cccr2-2005.pdf

Ambelas S.C., Hertel O., Gyldenkærne S., Ellermann T. (2003) Implementing a dynamilcal ammonia emission parameterisation in the large-scale air pollution model ACDEP. Journal of Geophysical Research – Atmospheres 109, 1–13.

Andersen H.V., Hovmand M.F. (1994) Measurements of ammonia and ammonium by denuder and filter pack. Atmospheric Environment 28, 3495–3512.

Aneja V.P., Chauhan J.P., Walker J.T. (2000) Characterization of atmospheric ammonia emissions from swine waste storage and treatment lagoons. Journal of Geophysical Research 105 (D9), 11535–11545.

Boxman D. (1998) Effects of changing nitrogen deposition on coniferous forests; comparison of European NITREX locations. KUN report, Nijmegen, The Netherlands (in Dutch).

Christensen J. (1997) The Danish Eulerian hemispheric model - a three dimensional air pollution model used for the Arctic. Atmospheric Environment 31, 4169–4191.

De Vries W., Reinds G.J., Van der Salm C., Draaijers G.P.J., Bleeker A., Erisman J.W., Auee J., Gundersen P., Kristensen H.L., Van Dobben H., De Zwart D., Derome J., Voogd J.C.H., Vel E.M. (2001) Intensive monitoring of forest ecosystems in Europe. Technical Report 2001. UN/ECE, EC, Forest Intensive Monitoring Coordinating Institute, Geneva and Brussels, 177.

Draaijers G.P.J., Erisman J.W., Spranger T., Wyers G.P. (1996) The application of throughfall measurements for atmospheric deposition monitoring. Atmospheric Environment 30, 3349–3361.

Duyzer J.H., Weststrate H. (2002) Mapping the spatial distribution of ammonia over the Netherlands. TNO Report no 2002/074, TNO, Apeldoorn, The Netherlands (in Dutch).

Ellermann T., Andersen H.V., Bossi R., Brandt J., Christensen J., Frohn L.M., Geels C., Kemp K., Løfstrøm P., Mogensen B.B., Monies C. (2006) Atmosfærisk deposition 2005. NOVANA (in English: Atmospheric deposition 2005. NOVAN). NERI Technical report No. 595: 66. National Environmental Research Institute, Roskilde Denmark.

EMEP (2006) http://webdab.emep.int/

Erisman J.W., Draaijers G.P.J. (1995) Atmospheric deposition in relation to acidification and eutrophication. Studies in Environmental Research, vol. 63. Elsevier, The Netherlands.

Erisman J.W., Bleeker A., Van Jaarsveld J.A. (1998a) Evaluation of the effectiveness of the ammonia policy using measurements and model results. Environmental Pollution 102, 269–274.

Erisman J.W., Bleeker A., Van Jaarsveld, J.A. (1998b) Atmospheric deposition of ammonia to semi-natural vegetation in the Netherlands: methods for mapping and evaluation. Atmospheric Environment 32, 481–489.

Erisman J.W., Hensen A., Fowler D., Flechard C., Grüner A., Spindler G., Duyzer J., Weststrate H., Römer F., Vonk A.W., Van Jaarsveld H. (2001) Dry deposition monitoring in Europe. Water Air and Soil Pollution 1 (5/6), 17–27.

Erisman J.W., Grennfelt P., Sutton M.A. (2003) The European perspective on nitrogen emission and deposition. Environment International 29, 311–325.

Erisman J.W., Monteny G.J. (1998) Consequences of new scientific findings for future abatement of ammonia emissions. Environmental Pollution 102, 275–282.

Fagerli H., Aas W. (2006) Trends of nitrogen in air and precipitation: model results and observations at EMEP sites in Europe, 1980–2003 (submitted for publication).

Fournier N., Pais V.A., Sutton M.A., Weston K.J., Dragosits U., Tang Y.S., Aherne J. (2002) Parallelisation and application of a multi-layer atmospheric transport model to quantify dispersion and deposition of ammonia over the Britisch Isles. Environmental Pollution 116 (1), 95–107.

Fowler D., Muller J., Tang Y.S., Dore T., Vieno M., Smith R.I., Nemitz E., Sutton M., Erisman J.W. (2006) Measuring and modeling gaseous NH3 and aerosol NH4 at the regional scale: how does ambient concentration respond to emission controls? Presentation at 'Workshop on Agricultural Air Quality: State of the Science'. Potomac, Maryland, June 5–8, 2006.

Frohn L.M., Christensen J.H., Brandt J. (2002) Development of a high resolution nested air pollution model – the numerical approach. Journal of Computational Physics 179, 68–94.

Frohn L.M., Christensen J.H., Brandt J., Geels C., Hansen K.M. (2003) Validation of a 3-D hemispheric nested air pollution model. Atmospheric Chemistry and Physics Discussions 3, 3543–3588.

Horvath L., Sutton M.A. (1998) Long term record of ammonia and ammonium concentrations at K-puszta Hungary. Atmospheric Environment 32, 339–344.

Mennen M.G., Van Elzakker B.G., Van Putten E.M., Uiterwijk J.W., Regts T.A., Van Hellemond J., Wyers G.P., Otjes R.P., Verhage A.J.L., Wouters L.W., Heffels C.J.G., Romer F.G., Van den Beld L., Tetteroo, J.E.H. (1996) Evaluation of automatic ammonia monitors for application in an air quality monitoring network. Atmospheric Environment 30, 3239–3256.

Menzi H., Achermann B. (eds.) (2001) Proceedings of the UNECE Ammonia Expert Group. Swiss Agency for Environment, Forest and Landscape (SAEFL), Bern, 18–20 September 2000.

Mosquera J., Hensen A., Van den Bulk W.C.M., Vermeulen A.T., Erisman J.W., Möls J.J. (2000) NH_3 flux measurements at Schagerbrug and Oostvaardersplassen, the Dutch contribution to the GRAMINAE experiment. ECN Report no. ECN-C–00–079, ECN, Petten, The Netherlands.

NERI (2006) http://www2.dmu.dk/1_Viden/2_miljoe-tilstand/3_luft/4_adaei/default_en.asp

Paromonov S., Ryaboshapko A., Gromov S., Granat L., Rodhe H. (1999) Sulphur and nitrogen compounds in air and precipitation over the former Soviet Union in 1980–95. Report GM-95. Department of Meteorology, Stockholm University.

Singles R.J., Sutton M.A., Weston K.J. (1998) A multi-layer model to describe the atmospheric transport and deposition of ammonia in Great Britain. Atmospheric Environment 32, 393–399.

Smits M.C.J., Van Jaarsveld J.A., Mokveld L.J., Vellinga O., Stolk A., Van der Hoek K.W., Van Pul W.A.J. (2005) The 'Veld' project: a detailed inventory of ammonia emissions and concentrations in an agricultural area. A&F Report no. 429. Agrotechnology & Food Innovations, Wageningen, The Netherlands (in Dutch).

Sutton M.A., Tang Y.S., Dragosits U., Fournier N., Dore T., Smith R.I., Weston K.J., Fowler D. (2001a) A spatial analysis of atmospheric ammonia and ammonium in the UK. The Scientific World 1 (S2), 275–286.

Sutton M.A., Tang Y.S., Miners B., Fowler D. (2001b) A new diffusion denuder system for long-term, regional monitoring of atmospheric ammonia and ammonium. Water Air and Soil Pollution: Focus 1, 145–156.

Sutton M.A., Asman W.A.H., Ellerman T., Van Jaarsveld J.A., Acker K., Aneja V., Duyzer J.H., Horvath L., Paramonov S., Mitosinkova M., Tang Y.S., Achermann B., Gauger T., Bartnicki J., Neftel A., Erisman J.W. (2003) Establishing the link between ammonia emission control and measurements of reduced nitrogen concentrations and deposition. Environmental Monitoring and Assessment 82 (2), 149–185 [Revised version earlier published in: UNECE Ammonia Expert Group (Berne 18–20 Sept 2000) Proceedings (eds: Menzi H. and Achermann B.) pp 57–84. Swiss Agency for Environment, Forest and Landscape (SAEFL), Bern, 2001]

Sutton M.A., Dragosits U., Dore A.J., McDonald A.G., Tang Y.S., Van Dijk N., Bantock T. Hargreaves K.J., Skiba U., Simmons I., Fowler D., Williams J., Brown L., Hobbs P., Misselbrook T. (2004) The potential of NH_3, N_2O and CH_4 measurements following the 2001 outbreak of Foot and Mouth Disease in Great Britain to reduce the uncertainties in agricultural emissions abatement. Journal of Environmental Science and Policy 7, 177–194.

Sutton M.A., Dragosits U., Simmons I., Tang Y.S., Hellsten S., Love L., Vieno M., Skiba U., Di Marco C., Storeton-West R.L., Fowler D., Williams J., North P., Hobbs P., Misselbrook T. (2006) Monitoring and modelling trace-gas changes following the 2001 outbreak of Foot and Mouth Disease to reduce the uncertainties in agricultural emissions abatement. Environmental Science & Policy 9, 407–422.

Tang Y.S., Cape J.N., Sutton M.A. (2001) Development and types of passive samples for NH_3 and NO_x. The Scientific World 1, 513–529.

Tang Y.S., Love L., Van Dijk N., Simmons I., Storeton-West R., Smith R.I., Dore A.J., Vieno M., Dragosits U., Theobald M.R., Fowler D., Sutton, M.A. (2006) Ammonia in the United Kingdom: spatial patterns and temporal trends 1996–2005. Report of the National Ammonia Monitoring Network. Centre for Ecology and Hydrology (Edinburgh Research Station), Penicuik, UK.

Van Jaarsveld J.A., Bleeker A., Hoogervorst N.J.P. (2000) Evaluatie ammoniak emissieredukties met behulp van metingen en modelberekeningen. RIVM Report 722108025. RIVM, Bilthoven, The Netherlands (in Dutch).

Van Jaarsveld J.A. (1995) Modelling the long-term atmospheric behaviour of pollutants on various spatial scales, PhD Thesis, University of Utrecht, The Netherlands.

Van Loon M., Wind P., Tarrason L. (2005) Meteorological variability in source allocation: transboundary contributions across Europe. In: Transboundary Acidification, Eutrophication and Ground Level Ozone in Europe. EMEP Status Report 1/2005, The Norwegian Meteorological Institute, Oslo, Norway, 89–107.

Van Pul A., Van Jaarsveld H., Van der Meulen T., Velders G. (2004) Ammonia concentrations in the Netherlands: spatially detailed measurements and model calculations. Atmospheric Environment 38, 4045–4055.

Van Pul W.A.J., Potma C.J.M., Van Leeuwen E.P., Draaijers G.P.J., Erisman J.W. (1995) EDACS: European deposition maps of acidifying components on a small scale. Model description and Preliminary Results. RIVM Report no. 722401005, RIVM, Bilthoven, The Netherlands.

Walker J.T., Aneja V.P., Dickey D.A. (2000) Atmospheric transport and wet deposition of ammonium in North Carolina. Atmospheric Environment 34, 3407–3418.

Chapter 12
Long-Term Record (1981–2005) of Ammonia and Ammonium Concentrations at K-Puszta Hungary and the Effect of Sulphur Dioxide Emission Change on Measured and Modelled Concentrations

Laszlo Horvath, Hilde Fagerli, and Mark A. Sutton

12.1 Introduction

In the 3 years of political and economical changes in Hungary (1989–1991) the number of animals and the use of fertilizers highly decreased with parallel decrease of calculated ammonia emission (Fig. 12.1). In an earlier study (Horvath and Sutton 1998) long-term trend of background ammonia/ammonium concentrations were analysed between 1981 and 1995 at Hungarian K-puszta station (longitude 19′33″, latitude 46′58″, h=125 m a.s.l.). Despite the reduction in NH_3 emission, the concentration of ammonia remained at the same level. Only small reductions were observed both for particle ammonium concentration and wet ammonium deposition. It was suggested that the reasons for the discrepancy included changing SO_2 emissions as well as buffering of the atmospheric NH_3 concentration by vegetation compensation points.

12.2 Recalculation of Ammonia/Ammonium Trend for Longer Period

Trends in concentrations were recalculated using additional data for 1996–2005 (Fig. 12.2). The 25 year long record of measured ammonia concentration did not show any decrease even in the period of large ammonia emission reduction (from 1989), moreover, a small increase can be observed. Concentration of particle ammonium and the wet deposition decreased by 10–20 and 40–50%, respectively.

Laszlo Horvath
Hungarian Meteorological Service, Gilice 39, 1118 Budapest, Hungary

Hilde Fagerli
Norwegian Meteorological Institute, Niels Henrik Abels vei 40, 0313 Oslo, Norway

M. Sutton, S. Reis and S.M.H. Baker (eds), *Atmospheric Ammonia,*
© Springer Science + Business Media B.V. 2009

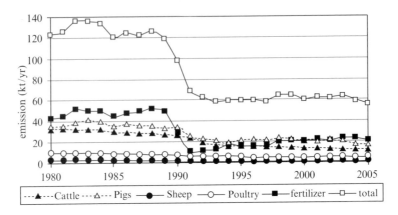

Fig. 12.1 Trend of ammonia emission in Hungary

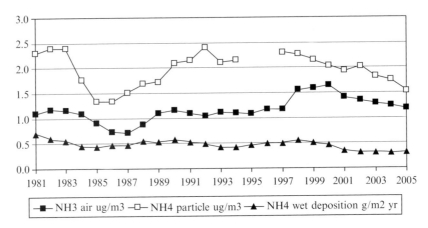

Fig. 12.2 Trend of concentrations of reduced nitrogen compounds in air and precipitation (3 year moving average)

12.3 Possible Reason for the Lack of Reduction in Ammonia Concentration

One possible explanation for the lack of reduction in ammonia concentration is the decrease of SO_2 emission and concentration by 60–70% since 1989, as it can be seen in Fig. 12.3. The co-deposition of NH_3 and SO_2 and the effect of chemistry on the concentrations of both gases have earlier been demonstrated (Flechard et al. 1999; Fowler et al. 2001; Erisman et al. 2001). Reduced SO_2 concentration decreases the rate of conversion of gaseous NH_3 to ammonium (hydrogen) sulphate particles.

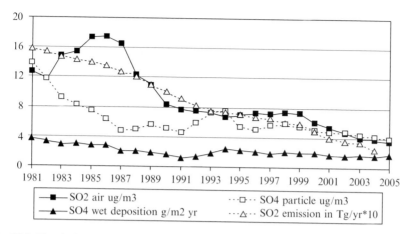

Fig. 12.3 Trend of sulphur emission in Hungary and concentrations of sulphur compounds in air and precipitation (3 year moving average)

12.4   Modelling of the Effect of Sulphur Dioxide Level on the Predicted Ammonia Concentrations

To investigate the role of sulphur dioxide in the budget of ammonia, the EMEP model was applied (Simpson et al. 2003; Fagerli et al. 2004). Two different sets of model calculations for the period 1990, 1995–2004 were set up; set (1) used the best available emissions for each year and set (2) used the same emissions as set (1) except that SO_2 emissions were kept at the 1990 level for all the years. In Figs. 12.4 and 12.5 the results from the two sets of model simulations are compared for ammonia. Note that the modelled ammonia concentrations are scaled to the 1990 measured ammonia concentrations. Using the set (1) calculations, the measured and modelled trend of ammonia concentrations agree well. In accordance with the measurements, the model simulations predict ammonia concentrations to increase. In contrast, model simulations from set (2) show decreasing ammonia concentrations in the period 1990–2004.

The model simulations predict lower ammonia levels in the period between 1995–2004 if the sulphur dioxide emission remained the 1990 level than in the calculations with the most realistic SO_2 emissions.

The agreement between measured and modelled particle NH_4^+ concentration is good if the model runs with the real (set 1) sulphur emission (Fig. 12.6). In contrast, the EMEP model predicts higher particle ammonium level if sulphur dioxide emission remained the same as for 1990 in the period between 1995–2004. These results demonstrate the important role of sulphur dioxide in the ammonia/ammonium gas to particle conversion.

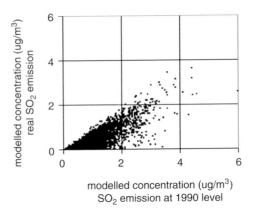

Fig. 12.4 Modelled daily ammonia concentrations calculated with real and high sulphur emission

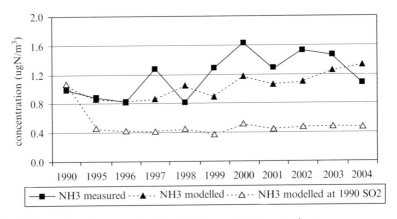

Fig. 12.5 Comparison of measured and modelled ammonia concentrations

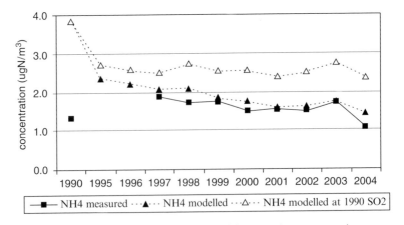

Fig. 12.6 Comparison of measured and modelled particle ammonium concentrations

12.5   Conclusions

The ammonia emission has been considerably reduced in Hungary since the 1989–1991 while the gaseous ammonia level did not follow the emission pattern. We have demonstrated that this discrepancy at least partly may be attributed to reductions in sulphur dioxide emissions. One possible explanation is that the sulphur dioxide to sulphate (sulphuric acid) conversion promotes the ammonia to ammonium (hydrogen) sulphate conversion (neutralizing in particle phase). Moreover, the decrease of SO_2 emissions leads to less acidic surfaces and thereby a less efficient dry deposition of ammonia. Both these processes lead to higher ambient ammonia concentrations with decreasing SO_2 concentrations.

When model simulations are carried out with a constant (1990) sulphur dioxide level for 1995–2004, the ambient level of ammonia follows the change of ammonia emission pattern, whilst an increase in ammonia levels are found if model simulations with more realistic emissions are performed. Thus, the EMEP model simulations support the hypothesis that the lack of reductions in NH_3 concentrations following 1989–1990 largely can be explained by the decrease in SO_2 concentrations throughout the 1990s.

A lack of higher particulate NH_4^+ concentrations prior to 1989, when SO_2 concentrations were even higher, may be due to not all of the sulphate being neutralized by ammonia, i.e. indicating the presence H_2SO_4 and NH_4HSO_4 containing aerosol during the 1980s. Further analysis is required to investigate this problem.

Acknowledgements   We are grateful for the financial support from COST Action 729 and the NitroEurope Integrated Project.

References

Erisman J.W., Hensen A., Fowler D., Flechard C., Grüner A., Spindler G., Duyzer J., Westsrate H., Römer F., Vonk A., Jaarsveld H. (2001) Dry deposition monitoring in Europe. Water Air and Soil Pollution: Focus, 1, 17–27.

Fagerli H., Simpson D., Tsyro S. (2004) Unified EMEP Model: Updates EMEP Report 1/2004, Transboundary Acidification, Eutrophication and Ground Level Ozone in Europe. Status Report 1/2004. www.emep.int

Flechard C.R., Fowler D., Sutton M.A., Cape J.N. (1999) A dynamic chemical model of bi-directional ammonia exchange between semi-natural vegetation and the atmosphere. Quarterly Journal of the Royal Meteorological Society, 125, 2611–2641.

Fowler D., Sutton M.A., Flechard C.R., Cape J.N., Storeton-West R., Coyle M., Smith R.I. (2001) The control of SO_2 dry deposition on to natural surfaces and its effects on regional deposition. Water Air and Soil Pollution: Focus, 1, 39–48.

Horvath L., Sutton M.A. (1998) Long term record of ammonia and ammonium concentrations at K-puszta, Hungary. Atmospheric Environment, 32, 339–344.

Simpson D., Fagerli H., Jonson J.E., Tsyro S., Wind P., Tuovinen J.-P. (2003) The EMEP Unified Eulerian Model. Model Description. EMEP MSC-W Report 1/2003. www.emep.int

Chapter 13
Assessment of Ammonia and Ammonium Trends and Relationship to Critical Levels in the UK National Ammonia Monitoring Network (NAMN)

Y. Sim Tang, Ulrike Dragosits, Netty van Dijk, Linda Love, Ivan Simmons, and Mark A. Sutton

13.1 Introduction

The UK National Ammonia Monitoring Network (NAMN, Sutton et al. 2001a) was established in 1996 to quantify the spatial distribution and long-term trends in concentrations of atmospheric ammonia (NH_3) and also aerosol ammonium (NH_4^+) (since 1999). There are currently 94 sites. At 59 of these sites, the CEH DELTA methodology (Sutton et al. 2001b) is used to provide the spatial and temporal patterns of NH_3 and aerosol NH_4^+ across the UK, while passive diffusion samplers (Tang et al. 2001) are used to assess regional and local scale variability in air NH_3 concentrations in source regions. Monitoring is on a monthly timescale, which is optimal to provide information on seasonality and for estimating annual mean in air concentrations.

Emissions of NH_3 in the UK have fallen by 12% between 1990 and 2004 (http://www.naei.org.uk). The long-term dataset from the UK NAMN, which comprises 9 years of gaseous NH_3 data (since September 1996) and 7 years of aerosol NH_4^+ data (since 1999) may therefore be analysed to assess trends in air concentrations. The data can also be used to compare with critical levels of NH_3 concentrations set in the UK to protect vegetation and ecosystem.

13.2 Trends in Ammonia Concentrations

Trends in atmospheric NH_3 are not easily detectable at individual sites due to significant inter-annual variability that are linked to meteorological differences and periodic local influences. Therefore, a trend assessment was based on the results from multiple sites. In Figs. 13.1–13.4, the long-term trend in mean monitored and seasonally-detrended

Y. Sim Tang
Centre for Ecology & Hydrology, Bush Estate, Penicuik, Midlothian, EH26 0QB, United Kingdom

M. Sutton, S. Reis and S.M.H. Baker (eds), *Atmospheric Ammonia,*
© Springer Science + Business Media B.V. 2009

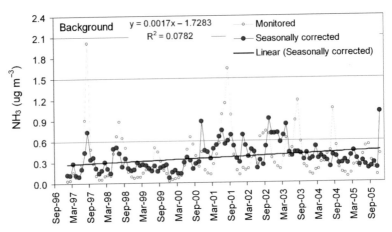

Fig. 13.1 Long-term trend in mean monitored (–o–) and seasonally-detrended mean NH$_3$ concentration (–•–) from six background sites (defined by 5 km grid average emissions < 1 kg N ha^{-1} year^{-1}) in the NAMN. All measurements are made using the DELTA system throughout

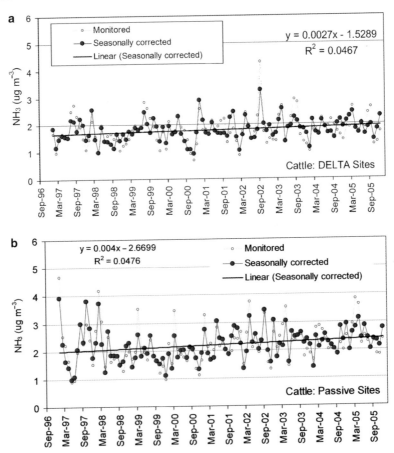

Fig. 13.2 Long-term trend in mean monitored (–o–) and seasonally-detrended mean NH$_3$ concentration (–•–) from sites in grid squares dominated by emissions from cattle in the NAMN. (**a**) data from 20 sites where measurements are made using the DELTA system throughout, and (**b**) data from 17 sites where all measurements are made using the Diffusion Tube (Sep96–Apr00) and subsequently the ALPHA sampler (from May00)

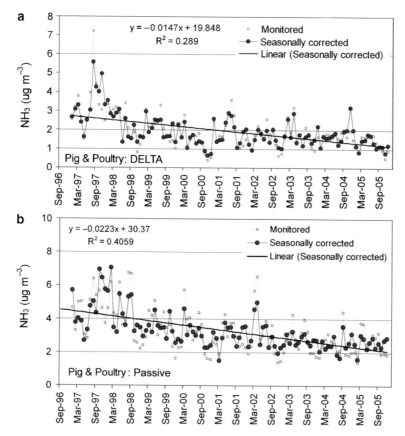

Fig. 13.3 Long-term trend in mean monitored (–o–) and seasonally-detrended mean NH₃ concentration (–•–) from sites in grid squares dominated by emissions from pig and poultry in the NAMN. (**a**) data from four sites where measurements are made using the DELTA system throughout, and (**b**) data from 12 sites where all measurements are made using the Diffusion Tube (Sep96–Apr00) and subsequently the ALPHA sampler (from May00). Since the passive methods are subject to a correction derived from their regression against the DELTA system, caution needs to be exercised in interpreting long-term trends where the sampling methodologies are changed

mean NH₃ concentrations of sites grouped into four different emission source sectors are compared (sites are classified according to 2003 NH₃ emission sector data/map by Dragosits and Sutton 2005). The seasonal detrending was derived from the mean seasonal cycle for the whole period normalized to 1, and then multiplying each monthly value by the appropriate value.

13.2.1   Emission Sector: Background

The mean concentration from six background sites (S3 Inverpolly, S7B Gulabin Lodge, S12 Halladale, S22 Moorhouse, S30 Strathvaich Dam, S41 Lagganlia), which were measured using the same method (DELTA) throughout the measurement period in the UK NAMN are presented in Fig. 13.1.

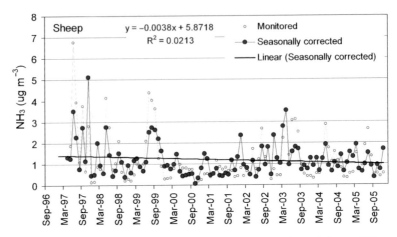

Fig. 13.4 Long term trend in mean monitored (–o–) and seasonally-detrended mean NH₃ concentration (–•–) from two sites in grid squares dominated by emissions from sheep in the NAMN (S70 Cwmystwyth, S93 Llynn Brianne). Measurements at Cwmystwyth were made using the DELTA system throughout, whilst ALPHA samplers were used at Llynn Brianne

The temporal patterns are clearly reproducible between years, with monitored mean monthly concentrations in the range of $0.1–1\,\mu g\,NH_3\,m^{-3}$. There appears to be an upward trend in NH_3 concentrations. However, the trend is not statistically significant in the seasonally de-trended values, which is derived by dividing the concentrations by the national average seasonal cycle. The apparent increase may be a feature of the reduction in SO_2 emissions over the same period, leading to a longer atmospheric lifetime of NH_3, thereby increasing NH_3 concentrations in remote areas.

13.2.2 Emission Sector: Cattle

In cattle dominated areas, there is a slight (non-significant) increase in NH_3 concentrations (Fig. 13.2), consistent (but not as large as) with the estimated increase in emissions from this sector since 2002. For the purpose of consistency in monitoring methods for analysing long-term trends, data in Fig. 13.2a are selected from 20 cattle sites that used the DELTA method over the entire period, whereas the mean data of 17 passive sampling sites were used in Fig. 13.2b. In both graphs, the temporal patterns are clearly reproducible between years, with monitored mean monthly concentrations in the range of $0.9–4.5\,\mu g\,NH_3\,m^{-3}$.

13.2.3 Emission Sector: Pig and Poultry

There is some indication that NH_3 concentrations have decreased in pig/poultry areas (Fig. 13.3), which is consistent (but not as large as) with the estimated reduction in NH_3 emissions from these sectors. Again, for the purpose of consistency in monitoring methods for analysing long-term trends, data in Fig. 13.3a are selected from four pig and poultry sites that used the DELTA method over the entire period, whereas the mean data of 12 passive sampling sites were used in Fig. 13.3b. In both graphs, the temporal patterns are clearly reproducible between years, with monitored mean monthly concentrations in the range of 0.4–7 μg NH_3 m^{-3}. The larger NH_3 concentrations at passive sites are due to the deployment of passive samplers in source regions to explore regional and local scale variability.

13.2.4   Emission Sector: Sheep

There are only two sites in the UK NAMN that are classed as being in sheep emission grid squares (according to 2003 emission data). Overall there is a slight downward trend in NH_3 concentrations (non-significant) at these two sites (Fig. 13.4). There is however a degree of uncertainty in the disaggregation of emission data for sheep, and subsequent classification of 5 km grid squares in the AENEID emission model, particularly in Wales. This is currently under further investigation.

13.3   Trends in Aerosol Ammonium Concentrations

For NH_4^+ aerosol, it has already been shown that this is a regional product, and there is therefore less relationship to the dominant ammonia source sectors in the data. This is confirmed by very similar long-term trends in mean monitored and seasonally-detrended NH_4^+ concentrations derived for sites grouped into different emission source sectors (data not shown). Analysis was therefore made using all measurement site data (Fig. 13.5). The overall dataset shows no detectable trend in NH_4^+ concentrations, indicating that a longer measurement period is needed before trends can be detected.

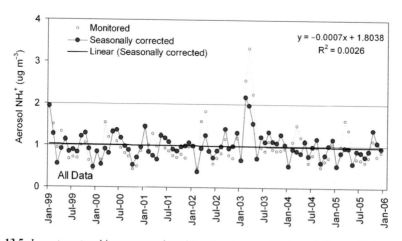

Fig. 13.5 Long-term trend in mean monitored (–o–) and seasonally-detrended mean NH_4^+ aerosol concentration (–•–) from all sites in the UK NAMN. The seasonal detrending was derived from the mean seasonal cycle for the whole period normalized to 1, and then multiplying each monthly value by the appropriate value

13.4   Critical Levels Assessment

Policy assessment of N deposition impacts at a UK scale uses the "critical loads" and "levels" approach (NEGTAP 2001). The latter is used to assess the direct effects of ammonia using "critical levels" of concentrations, which is complementary to the critical loads approach.

For Ammonia, critical levels (CLE) of concentration for the protection of vegetation and ecosystem have been set for the following reference periods:

- Twenty-three μg NH$_3$ m^{-3} for a 1-month average period
- Eight μg NH$_3$ m^{-3} for a 1-year average period

National interpolated maps of NH$_3$ concentrations derived from the UK NAMN may be used to give an initial estimate of critical levels exceedance over the country, e.g. at specific designated nature conservation sites (e.g. SSSIs). From these maps, exceedance of the 1-year average CLE of 8 μg NH$_3$ m^{-3} occurred at two sites in East Anglia, and at one site in Oxfordshire (Fig. 13.6a). Exceedance of the 1-month average CLE of 23 μg NH$_3$ m^{-3} occurred in very small areas of the UK (Fig. 13.6b). By contrast, a critical loads exceedance map based on the monitoring data identifies large areas of the UK at risk (data not shown). Therefore the critical levels are perhaps set too high to protect sensitive receptors.

For each site in the network, the ratio of the highest monthly NH$_3$ concentration to the highest annual NH$_3$ concentrations recorded for that site was calculated and then mapped (Fig. 13.6c). The ratio of the monthly CLE (23 μg NH$_3$ m^{-3}) to annual CLE (8 μg NH$_3$ m^{-3}) is 2.9. Sites (5 km grid squares) that have ratios higher than 2.9

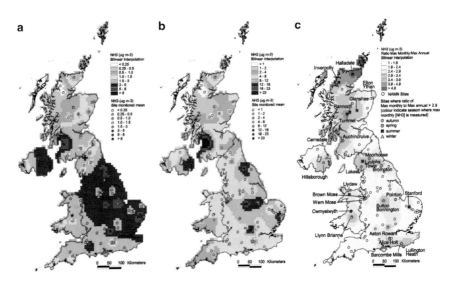

Fig. 13.6 Interpolated map showing (**a**) highest annual mean for each site over the period 1997–2006, (**b**) highest monthly value for each site over the period 1996–2006, and (**c**) ratio of 'max monthly [NH$_3$]: max annual [NH$_3$]' at each site

are then either (a) background sites with strong seasonality (i.e. summer maxima, with winter minima in NH$_3$ concentrations, e.g. Halladale and Inverpolly), or (b) agricultural sites with occasional spike in NH$_3$ concentrations due to for example, manure spreading activities in spring (e.g. S76 Pointon) or autumn (e.g. S40 Sutton Bonnington). By plotting the frequency distribution, 25% of sites have ratios that are greater than 2.9 (Fig. 13.7).

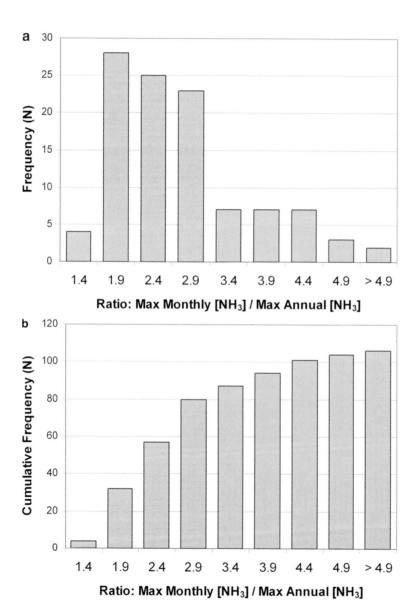

Fig. 13.7 (**a**) Frequency and (**b**) cumulative frequency distribution of site (max monthly [NH$_3$]: max annual [NH$_3$]) ratios

Acknowledgement This work was carried out with funding from the Department for Environment, Food and Rural Affairs (DEFRA), and from supporting CEH programmes. The assistance of the network of site operators is also gratefully acknowledged.

References

Dragosits U., Sutton M.A. (2005) Maps of Ammonia emissions from agriculture, waste, nature and other miscellaneous sources for the NAEI 2003. Report to AEAT/DEFRA.

NEGTAP (2001) Transboundary air pollution: acidification, eutrophication and ground level ozone in the UK. Report of the National Expert Group on Transboundary Air Pollution, DEFRA, London.

Sutton M.A., Tang Y.S., Dragosits U., Fournier N., Dore T., Smith R.I., Weston K.J., Fowler D. (2001a) A spatial analysis of atmospheric ammonia and ammonium in the UK. The Scientific World 1(S2), 275–286.

Sutton M.A., Tang Y.S., Miners B., Fowler D. (2001b) A new diffusion denuder system for long-term, regional monitoring of atmospheric ammonia and ammonium. WASP: Focus 1, 145–156.

Tang Y.S., Cape J.N., Sutton M.A. (2001) Development and types of passive samplers for NH_3 and NO_x. The Scientific World 1, 513–529.

Chapter 14
Review of Published Studies Estimating the Abatement Efficacy of Reduced-Emission Slurry Spreading Techniques

J. Webb, Brigitte Eurich-Menden, Ulrich Dämmgen, and Francesco Agostini

14.1 Introduction

The Gothenburg Protocol to reduce atmospheric pollution drawn up by the United Nations Economic Commission for Europe (UNECE) Convention on Long-Range Transboundary Air Pollution (LRTAP) limits national emissions of ammonia (NH_3). To help meet this agreement, the EU has agreed a National Emissions Ceilings Directive (NECD), under which the UK target for NH_3 emission is a maximum of 297 $\times 10^3$ t NH_3. By 2007, large pig and poultry units in the EU will be required to reduce NH_3 emission in consequence of the EU Directive on Integrated Pollution Prevention and Control (EEC 1996). Webb and Misselbrook (2004) estimated that c. 27×10^3 t NH_3-N, c. 12% of UK NH_3-N emissions from livestock production, arose following the application of slurries to land. Much effort has been directed toward reducing NH_3 emission following slurry application, however, few studies of reduced-emission slurry spreading techniques have been carried out using field-scale equipment and few balanced comparisons have been made of more than one abatement technique. In consequence it was considered useful to review published data.

14.2 Methods

The work was carried out as part of a UK project (Defra AM0123, A Collation of Ammonia Research) hence an initial review was carried out of studies undertaken in the UK. Results were analysed separately for studies carried out at plot or field scale and also for those carried out on arable or grassland. Results were also sorted according to the type of slurry used and the time of year the measurements took place. A separate examination was then carried out of studies reported from elsewhere in Europe. The reduced emission techniques considered were as

J. Webb

AEA, Gemini Building, Harwell Business Centre, Didcot, Oxfordshire OX11 0QR, United Kingdom

M. Sutton, S. Reis and S.M.H. Baker (eds), *Atmospheric Ammonia,*
© Springer Science + Business Media B.V. 2009

follows. Application of slurry by: trailing hose (TH); trailing shoe (TS); open-slot injection (OSI); closed slot injection (CSI). Rapid incorporation of slurry by: plough (P), disc (D) or tined implement (T). Due to the limited number of studies published, the differences in experiment design, equipment used and times of slurry application, the data were not suitable for robust statistical analysis. Results quoted are means, weighted according to the number of individual experiments carried out in each study.

14.3 Results and Discussion

14.3.1 Reduced Emission Slurry Application Techniques, All Studies

14.3.1.1 Trailing Hose

Ten datasets were available from UK studies, six on grassland, four on arable. The results (Table 14.1) suggested abatement may be a little greater on grassland (c. 35%) than arable land (c. 30%). The greater abatement reported by Smith and Misselbrook (2000) may have been due to the slurry being applied to a growing

Table 14.1 Abatement efficiency of reduced emission spreading techniques, percentage reduction from surface applications

	Trailing			Injection		Plough	Disc	Tines
	Hose	**Shoe**	**Slot**	**Shallow**	**Deep**			
Cattle slurry, grass								
UK data	38 (5)	60 (6)	64 (2)	63 (4)				
Other data	33 (13)	60 (23)	82 (10)	91 (7)				
Mean	*37 (18)*	*59 (29)*	*77 (12)*	*81 (11)*				
Weighted mean	*34*	*60*	*79*	*81*				
Cattle slurry, arable								
UK data	29 (4)	46 (3)	−5 (2)	81 (4)	86 (2)			
Other data	18 (7)	41 (1)	47 (2)	84 (3)				
Mean	*22 (11)*	*44 (4)*	*21 (4)*	*83 (7)*	*86 (2)*			
Weighted Mean	*22*	*45*	*21*	*82*	*86*			
Pig slurry, grass								
UK data	24 (1)							
Other data	56 (3)	66 (5)		89 (1)		89 (1)		25 (1)
Mean	*40 (4)*	*66 (5)*		*89 (1)*				
Weighted Mean	*48*	*66*		*89*				
Pig slurry, arable								
Other data	29 (2)	78 (3)	97 (1)			98 (1)	74 (8)	64 (9)
Mean	*29 (2)*	*78 (3)*	*97 (1)*					
Weighted mean	*29*	*78*	*97*					
Overall weighted mean	***32 (35)***	***60 (41)***	***67 (17)***	***82 (19)***	***86 (2)***	***94 (2)***	***74 (8)***	***60 (10)***

crop, whereas the results reported by Misselbrook et al. (2002) were following application of slurry to cereal stubbles. Data from other European countries appear to confirm a difference with average abatement of c. 39% on grassland, but only c. 21% on arable land.

There was considerable variation in the results of individual studies, ranging from 11–82% on grass and 4–50% on arable. The overall weighted abatement means were c. 37% and 23% for grassland and arable respectively. The very small abatement of 11% reported by Mannheim et al. (1995) was attributed to the slurry being applied to a 'short' crop. Malgeryd (1998) followed application of slurry by TH with immediate irrigation, and hence this result should be discounted when assessing the effectiveness of the TH machine. UK results covered all the year, except for January. Data from other European countries is almost all from February to August.

The current estimate of 30% abatement used in the UNECE Guidance Document (Anon 2006) seems reasonable, although this should perhaps be increased to 35% for grassland.

14.3.1.2 Trailing Shoe

Nine datasets were available from the UK, six on grassland, three on arable. Average abatement measured on grassland was 60% but only 46% on arable. The lesser effectiveness on arable land is not surprising as the technique was developed for use on grassland. Results from other European countries were similar for cattle slurry with average abatement of 60% and 45% for grassland and arable respectively. Data from the Netherlands were only available for pig slurry applications. Based on these data, the current estimate of 60% abatement used in the UNECE Guidance document appears well-founded. UK data cover every month except December and January. Results from outside the UK were confined to the period March to August.

14.3.1.3 Open Slot Injection

There were only four datasets available from the UK, two on grassland two on arable. Mean abatement of 60% obtained on grassland (64%) was less than obtained elsewhere in Europe (73–84%). The overall mean of 79% was greater than that currently used in the Guidance document (70%). However, the two UK studies carried on an arable land did not appear to achieve any reductions in emission. This appears due to very heavy rainfall after application maintaining the moisture content of the slurry surface exposed to the air and hence facilitating transport of NH_4^+ ions to the emitting surface. The UK data for OSI on grassland covers the months February to September. It would be useful to have measurements from October to December. Results from other European countries were confined to the period April to June.

14.3.1.4 Shallow, Closed Slot Injection

This approach has been superseded, for grassland at least, by OSI. The latter does much less damage to grass swards and is reported to be nearly as effective as CSI. Results of the two UK studies produced on average abatement efficiency of c. 63%, no more than obtained using OSI. Two German studies gave an average abatement efficiency of 87% while Dutch studies gave a mean abatement of 92%. In contrast to the results obtained using OSI, four UK studies of CSI on arable land produced mean abatement of 81%, similar to results reported from Germany (84%). Seasonal coverage for the UK was good, with data obtained in every month except January. Data reported from elsewhere in Europe covered the periods March to September.

14.3.1.5 Rapid Incorporation of Slurry into Arable Land by Tillage

There were very few UK data for this. In three experiments in Defra Project NT2001, NH_3 emissions following incorporation of pig slurry by disc were reduced by c. 25%, while incorporation of cattle slurry (two experiments) within 1 h by plough reduced emissions by c. 90%. Four experiments in which pig slurry was incorporated by rotavator (Defra WA0632) reduced emissions by c. 70%. One publication in English reported abatement efficiencies of 89% and 25% for incorporation by plough and by harrow respectively, similar to the estimates used in NARSES (Webb and Misselbrook 2004) of 90% and 30%. More data were available from the Netherlands. One study of incorporation by ploughing gave an abatement of 98%. Incorporation by discs and by tines gave abatement efficiencies of c. 75% and 65% respectively. This work was carried out either in March or April or September.

14.3.2 Slurry, Field-Scale Studies Only

Given the likely differences in efficiency between abatement carried out on the plot scale with purpose-built machinery, we thought it would be worthwhile examing the results of work carried out at the field scale. Studies referred to as field scale are those where commercially-available machines were used to apply slurry to large plots; at least $100 m^2$ and up to $8,000 m^2$. Measurements of NH_3 emissions were made using micrometeorological techniques.

Some projects made direct comparisons of methods within the same experiment, and the results of these are reported first. Such comparisons may be objected to as a single time of application will not be optimal for all techniques. However, such comparisons reduce the amount of confounding, and so it is better to begin by examining the results of orthogonal comparisons before looking at averages produced under a range of different experimental conditions. Moreover, in practice farmers and even contractors are unlikely to be in a position to have a choice of machines to suit the conditions.

Malgeryd (1998) reported a comparison of NH_3 emissions following application of pig slurry to a barley crop by means of TH, TS, and OSI with emissions following surface broadcasting. However, the TS applicator is described as 'trenching with sliding foot' and hence suggests that the TS placed the slurry under the soil. If this is so it may explain why it produced abatement similar to that of the OSI (90%). The TH reduced NH_3 emissions by 40%.

Mattila (1998) made two paired comparisons of cattle slurry to grass by TH and injection to 8 cm. Subsequent results (personal communication) indicate that TH reduced emissions by an average of 23%, shallow injection by 99%. No explanation was given for the relatively poor performance of the TH, except that emissions, compared with surface application, were only reduced over the first day after application. The large application rates (35 and 42 m^3 ha^{-1}) may also have been a factor.

Mulder and Huijsmans (1994) reported a large set of data which include four paired experiments. Averages for the small number of paired results are in reasonable agreement with the overall average abatement efficiencies (grassland TH, 41%; TS, 61%; SLI, 89%: arable BS, 36%; TS, 73%; I, 91%).

Misselbrook et al. (2002) also reported a large number of mainly pair-wise comparisons between surface broadcast and either TH, TS or OSI. Overall average reductions (with the number of experiments in brackets) are given in Table 14.2. Overall the abatement efficiencies on grassland were similar to those reported by Mulder and Huijsmans (1994). On arable land efficiencies for both TS and OSI were substantially less. The results of five 3-way comparisons are also reported in Table 14.2.

Table 14.2 Results of studies carried out at the field scale

Authors	Grass			Arable			
	Trailing hose	Trailing shoe	Slot injector	Trailing hose	Trailing shoe	Slot injector	Other injector
Smith and Misselbrook (2000)							
Overall	43 (10)	48 (10)	55 (10)	32 (5)	36 (4)	78 (5)	
Dry soil	37	44	46				
Moist soil	48	51	64				
Lorenz and Steffens (1997)	30 (24)	70 (24)	89 (24)				
Misselbrook et al. (1996)			60(3)				
Rubæk et al. (1996)	*0 (2)		* 60 (2)				
Dosch and Gutser 1996)				48 (1)			90 (1)
Döhler (1990)				30 (1)			
Simple mean	37	59	72	40	36	78	90
Weighted mean	34 (3)	64 (2)	79 (4)	35 (3)	36 (1)	78 (1)	90

* relative to TH application and not included in means.

These comparisons may be distorted because:

1. The comparison of TH and OSI for grassland included the two (of seven) experiments that gave the smaller reduction for injection (although TH results were similar to the average for the whole dataset).
2. The comparison of TH and OSI also included an abatement efficiency from OSI was less than the average for the whole dataset, while those for Band-spreading were more representative.

These inconsistencies with the overall dataset perhaps will illustrate the difficulties of basing conclusions on three- or four-way comparisons based on only one experiment.

14.3.2.1   Factors Affecting Abatement Efficiencies

Trailing Hose

Sommer et al. (1997) concluded efficiency of TH to cereal crops increased with increasing crop height and density. No reduction in NH_3 emission was measured when slurry was applied to a wheat crop 10 cm tall but with a leaf area index of only 0.3. Bless et al. (1991) found TH reduced NH_3 emissions by only 15% from slurry applied to oilseed rape in October, although emissions from the broadcast treatments were less than expected, possibly as a result of rain washing slurry off the rape leaves. Malgeryd (1998) found that TH reduced NH_3 losses when applied to a growing crop but not when applied to a bare soil, although Misselbrook et al. (2002) did measure reductions from TH-spread slurry applied to cereal stubbles. Sommer et al. (1997) also found TH to be less efficient on a wet soil when infiltration was reduced (but see below). Mulder and Huijsmans (1994) showed abatement efficiency to increase with increasing height of the grass sward, but to decrease with increasing application rate.

Results were examined to see if any trends in abatement efficiency could be seen in respect of time of year of application. No consistent differences were seen for grassland. Other factors such as grass height, soil moisture status and application rate, none of which would be consistently related to the time of year, are likely to be the cause of variation in results. While within series of experiments the effectiveness of Band-spreading could be quite well related to crop height, average efficiencies from application to cereal stubbles by Mulder and Huijsmans (1994) and Misselbrook et al. (2002) at 32% were not much less than the average efficiency of applications to a growing crop (39%).

Trailing Shoe

Misselbrook et al. (2002) found regression on grass height explained 47% of the variance in NH_3 abatement. The technique was ineffective (0% reduction) when slurry was applied at $40 m^3$ ha^{-1} to a recently grazed sward. Mulder and Huijsmans (1994) found the TS to be less effective when slurry was applied at $16\,m^3\,ha^{-1}$ than when it was applied at $8\,m^3\,ha^{-1}$. This was attributed to more soiling of grass at the greater application rate. They also considered long grass to increase the effectiveness of the technique. There is perhaps a potential conflict here since, in the UK at least, farmers like to apply slurry to fairly short aftermaths to reduce the risk of silage taint

The TS appeared to work less well in February and March than between May and December in the study of Misselbrook et al. (2002), but Mulder and Huijsmans' (1994) results over the same period show no differences in efficiency.

Shallow Open-Slot Injection

On average OSI reduced NH_3 losses by 70% or more, but the UK studies produced smaller average efficiencies. Pain and Misselbrook (1997) found rainfall after application reduced the effectiveness of shallow injection. However, treatments were applied at similar times of the year in most studies and there is no obvious explanation for the lesser efficiency measured in the UK. The results of Mulder and Huijsmans (1994) and Misselbrook et al. (2002) showed no difference according to season. The smaller average efficiencies measured by Pain and Misselbrook (1997) were not due to applications being made at different times of year to the other studies.

14.4 Conclusions

Overall average results for Trailing Hose at c. 30% for grassland and for arable are in reasonable agreement with overall mean of all studies presented in Table 14.2. Slot injection for grassland and arable are slightly greater at 70% (compared with 67% in Table 14.1). The efficiencies for Trailing Shoe on grassland at 60% were the same as the overall average in Table 14.1, while the efficiency of 70% on arable was somewhat greater. Averages are also shown weighted by the number of experiments in each study. These are similar to the unweighted averages and do not alter the conclusions above.

References

Anon (2006) Control Techniques for Preventing and Abating Emissions of Ammonia (EB.AIR/ WG.5). UN Economic Commission for Europe Executive Body for the Convention on Long-Range Transboundary Air Pollution.

Bless H.-G., Beinhaur R., Sattelmacher B. (1991) Ammonia emission from slurry applied to wheat stubble and rape in north Germany. Journal of Agricultural Science, Cambridge 117, 225–231.

Döhler H. (1990) Laboratory and field experiments for estimating ammonia losses from pig and cattle slurry following application. In: Nielsen V.C., Voorburg J.H., L'Hermite, P. (eds.) Odour and Ammonia Emissons from Livestock Farming. Proceedings of a seminar in Silsoe, 26–28 March, UK, 132–140.

Dosch P., Gutser R. (1996) Reducing N losses (NH_3, N_2O, N_2) and immobilization from slurry through optimized application techniques. Fertilizer Research 43, 165–171.

EEC (1996) Concerning integrated pollution prevention and control. EEC/96/61. Official Journal No. L257 (10 Oct. 1996). Brussels, Belgium: EEC.

Lorenz F., Steffens G. (1997) Effect of application techniques on ammonia losses and herbage yield following slurry application to grassland. 287–292 In: Jarvis J.C., Pain B.F. (eds.) Gaseous Nitrogen Emissions from Grasslands. CAB International, Wallingford, UK.

Malgeryd J. (1998) Technical measures to reduce ammonia losses after spreading of animal manure. Nutrient Cycling in Agroecosystems 51, 51–57.

Mannheim T., Braschkat J., Marschner H. (1995) Reduktion von Ammoniakemission nach Ausbringung von Rinderflüssigmist auf Acker- und Grünlandstandorten: Vergleichende Untersuchungen mit Prallteller, Schleppschlauch und Injektion. Zeitschrift für Pflanzenernähr. Bodenkunde 158, 535–542.

Mattila P.K. (1998) Ammonia volatilization from cattle slurry applied to grassland as affected by slurry treatment and application techniques – first year results. Nutrient Cycling in Agroecosystems 51, 47–50.

Misselbrook T.H., Shepherd M.A., Pain B.F. (1996) Sewage-sludge applications to grassland - influence of sludge type, time and method of application on nitrate leaching and herbage yield. Journal of Agricultural Science, 126, 343–352.

Misselbrook T.H., Smith K.A., Johnson R.A., Pain B.F. (2002) Slurry application techniques to reduce ammonia emissions: results of some UK field-scale experiments. Biosystems Engineering 81, 313–321.

Mulder E.M., Huijsmans J.F.M. (1994) Restricting ammonia emissions in the application of animal wastes. Overview of measurements by DLO field measurement team 1990–1993. IMAG-DLO – Wageningen ISSN 0926–7085.

Pain B.F., Misselbrook T.H. (1997) Sources of variation in ammonia emission factorsfor manure applications to grasland. In: Jarvis S.C., Pain B.F. (eds.) Gaseous Nitrogen Emissions from Grasslands, (293–301. CAB International, Wallingford, UK.

Rubæk G.H., Henriksen K., Petersen J., Rasmussen B., Sommer S.G. (1996) Effects of application technique and anaerobic digestion on gaseous nitrogen loss from animal slurry applied to ryegrass (Lolium perenne). Journal of Agricultural Science, Cambridge 126, 481–492.

Smith K.A., Misselbrook T.H. (2000) Optimising the performance of low-trajectory slurry application techniques on grassland and arable land. MAFF Project report WAO 650.

Sommer S.G., Friis E., Bach A., Schørring J.K. (1997) Ammonia volatilization from pig slurry applied with trail hoses or broadspread to winter wheat: effects of crop developmental stage, microclimate, and leaf ammonia absorption. Journal of Environmental Quality 26, 1153–1160.

Webb J., Misselbrook T.H. (2004) A mass-flow model of ammonia emissions from U.K. livestock production. Atmospheric Environment 38(14), 2163–2176.

Part III
Analysis of Ammonia Hotspots

Chapter 15
Ammonia Deposition Near Hot Spots: Processes, Models and Monitoring Methods

Benjamin Loubet[*], Willem A.H. Asman, Mark R. Theobald, Ole Hertel, Y. Sim Tang, Paul Robin, Mélynda Hassouna, Ulrich Dämmgen, Sophie Genermont, Pierre Cellier, and Mark A. Sutton

15.1 Introduction

Atmospheric reduced nitrogen (NH_x) mainly originates from hot spots, which can be considered as intensive area or point sources. A large fraction of the emitted NH_x may be recaptured by the surrounding vegetation, hence reducing the contribution of these hot spots to long-range transport of NH_x. This paper reviews the processes leading to local recapture of NH_x near hot spots, as well as existing models and monitoring methods. The existing models range from research models to more operational models that can be coupled with long-range transport model provided the necessary information on emissions is available. Local recapture of NH_3 ranges from 2% to 60% within 2 km of a hot-spot and it is sensitive to source height, atmospheric stability, wind speed, structure of the surrounding canopies, as well as stomatal absorption, which mainly depends on green leaf area index and stomatal NH_3 compensation point of vegetation, and finally, cuticular deposition, which depends primarily on vegetation wetness. The main uncertainties and limitations on NH_x recapture models and monitoring techniques are discussed.

Due to the decrease of sulphur and nitrogen oxides emissions under a series of UNECE protocols, reduced nitrogen (NH_x), has become the dominant pollutant in Western Europe contributing to acidification of ecosystems (e.g. Vestreng and Støren 2000). At the global scale NH_x and NO_x emissions are comparable, although large uncertainties exist on NH_x emissions (Dentener and Crutzen 1994; Bouwman et al. 1997). Moreover, NH_x deposition, with other nitrogen (N) deposition, leads to eutrophication and changes in the biodiversity of semi-natural ecosystems (Van Breemen and van Dijk 1988; Roelofs et al. 1985; Fangmeier et al. 1994; Krupa 2003; EEA 2003). Although atmospheric ammonia (NH_3) is not a greenhouse gas (GHG), deposition of NH_x may lead to increased GHG emissions (N_2O) (Melillo et al. 1989) or reduced consumption of CH_4. Additionally, ammonium sulphate aerosols ($(NH_4)_2SO_4$, contribute to half of the negative radiative forcing of the

B. Loubet,
Institut National de la Recherche Agronomique (INRA), Unité Environnement et Grandes Cultures, 78850 Thiverval-Grignon, France

atmosphere due to aerosols (Houghton et al. 2001; Adams et al. 2001), as well as contributing to impacts of secondary aerosol on human health.

As it has been known since the end of the 19th century (Eriksson 1952), atmospheric ammonia originates particularly from livestock (Bouwman et al. 1997; Oudendag and Luesink 1998; Misselbrook et al. 2000; Dämmgen and Erisman 2005). The main NH_x sources are housing and waste storage (Jarvis and Pain 1990; Bussink and Oenema 1998; Pain et al. 1998; Döhler et al. 2002), and land spread manure (Génermont and Cellier 1997; Asman et al. 2004). Hence the main NH_x emissions are "hot spots" sources in the sense that they are intense and either spatially small (point sources, such as animal houses and manure storage) or temporally short (application of manure). The emitted NH_x is either (i) dry-deposited as gaseous NH_3 by stomatal absorption and non-stomatal adsorption to canopy surfaces (e.g. Sutton et al. 1993a, b, 1995a), (ii) dry-deposited as particulate ammonium (NH_4^+) essentially by Brownian diffusion (particle size < 1 µm), (iii) wet-deposited as ion NH_4^+, or (iv) transformed by chemical reactions with other gases or aerosols (Dlugi et al. 1997; Nemitz et al. 2002, 2004a, b; Nemitz and Sutton 2004). But NH_3 can also be emitted by the plants themselves, which can either act as sinks or sources of NH_3 depending on their nitrogen (N) nutrition status and the atmospheric NH_3 concentration (Farquhar et al. 1980; Sutton et al. 1995a, b, 2001a; Schjoerring et al. 1998; Andersen et al. 1999; Milford et al. 2001a; Hill et al. 2001). Moreover, non-stomatal adsorption of NH_3 is influenced by the load of acidic pollutants to the surface (Erisman and Wyers 1993; Sutton et al. 1993c; Fléchard et al. 1999).

The combination of hot spots sources and effective deposition processes leads to sources and sinks of NH_x being spatially heterogeneous at a scale of a square kilometre (Sutton et al. 1998a; Dragosits et al. 1998, 2002; Hutchings et al. 2001a, b). Direct measurement of NH_x deposition near hot spots is challenging due to local advection (Loubet et al. 2001, 2003; Hensen et al. 2008a). Indirect estimates using mass balance, ^{15}N labelling, SF_6 to NH_3 ratio methods, as well as modelling studies, have estimated that the fraction recaptured within 2 km downwind from the source of NH_3 emitted ranges between 2% and 60% (Asman 1998; Loubet and Cellier 2001; Sommer and Jensen 1991; Theobald et al. 2001; Loubet et al. 2006). The large variability of NH_x deposition near sources is known to depend critically on the canopy structure surrounding the source (roughness, side fluxes) (Klaassen 1991; Draaijers et al. 1994; De Jong and Klaassen 1997; Theobald et al. 2001; Loubet et al. 2006), the NH_3 emissions from the canopy (Schjoerring et al. 1998; Riedo et al. 2002), the litter (Nemitz et al. 2000a, b), or from the soil (Génermont and Cellier 1997), as well as the non-stomatal NH_3 fluxes (van Hove et al. 1989; Erisman and Wyers 1993; Sutton et al. 1995a; Fléchard et al. 1999; Loubet and Cellier 2001). Less known, are wet deposition fluxes and chemical transformations of NH_x near intensive sources, as well as direct emissions of particulate NH_4^+ (McCulloch et al. 1998). Despite this knowledge on local deposition, its quantitative assessment within regional atmospheric models is still a challenge at European scale, and probably in all regions having large livestock populations. As a result, options related to spatial interactions in hot spots have until now been little considered within mitigation strategies to reduce trans-boundary NH_x pollution.

The complexity of the processes involved and the variability of the deposition fluxes near hot spots have led to the use of models and monitoring techniques to evaluate the fraction of NH_x re-deposited locally.

15.1.1 Models

There are few models for NH_3 deposition near hot spots that deal with the complexity of the processes noted above. Some existing models treat within-canopy vertical transfer and leaf-scale exchange (Baldocchi 1988, 1992; Harper et al. 2000), other models that treat dispersion and deposition of NH_3 above the canopy, DEPO1 (Asman 1998), LADD (Hill 1998; Fowler et al. 1998), and FIDES (Loubet et al. 2001), or at larger scales (Asman and Janssen 1987; Singles et al. 1998; Fournier et al. 2002; van Pul et al. 2009, this volume). There are also models for the dispersion of tracers within the canopy, although without exchange processes (e.g. Wilson and Sawford 1996). However, there are few models addressing within- and above-canopy dispersion together with ammonia exchange within the canopy and at the ground, as done by MODDAS-2D (Loubet et al. 2006). Existing overviews of modelling local deposition of NH_x have been provided by Asman (2002) and Hertel et al. (2006). The main limitations of current NH_x short-range deposition models are:

1. A good knowledge of the emission from the hot spot, in both time and space, at both the daily and the yearly time scale for a diversity of animal species, building types, and effluent management practices
2. The parameterisation/modelling of the NH_3 emissions from the plants, which is known to depend on the plant, the N and water supply (and therefore on N deposition if it is intensive), the variations in global radiation, and to physiological changes throughout the season (Schjoerring 1997; Milford et al.2001a; Loubet et al. 2002)
3. The non-stomatal NH_3 deposition, which depends on leaf surface wetness but also on the ionic composition of these water films, and hence on the NH_x load, but also on the interaction with other chemical compounds (Fléchard et al. 1999)
4. The use of a detailed turbulence model to precisely evaluate the sensitivity of modelled NH_3 deposition on turbulence in complex situation, such as downwind of buildings and within tree-belts
5. The wet deposition and throughfall component near hot spots, which amount should be evaluated for a range of climatic conditions with contrasted precipitation frequency and temperatures
6. The amount of other nitrogen compounds emissions from hot spots (fine aerosols, dusts, amines) and their subsequent nearby deposition
7. The parameterisation of chemical reactions between NH_3 and acidic compounds, especially in "rural-urban" zones

15.1.2 Monitoring and Effect Assessment

As measuring deposition of NH_3 with strong local advection is still challenging (Hensen et al. 2008a), alternative methods have been developed consisting in monitoring NH_3 concentration and nitrogen (N) enrichment in plants around hot spots (Sommer and Jensen 1991; Pitcairn et al. 1998, 2002; Sutton et al. 2001b; Tang et al. 2001; Dämmgen et al. 2005). The monitoring methods are based on low cost techniques (diffusion denuders, filter packs, passive samplers, batch samples). Specific experiments have also been conducted to assess the deposition to tree belts (Theobald et al. 2001). The main limitations of the NH_3 monitoring and effect assessment methods are:

- To accurately measure NH_3 concentrations over long periods under fluctuating conditions, in such a way that results are representative of larger areas
- To relate atmospheric NH_3 concentration, NH_x concentration in plants and NH_x deposition, taking account of saturation effects at very high deposition rates
- To be able to estimate local NH_3 deposition rates to sensitive receptors (e.g. nature areas) in the vicinity of hot spots with sufficient accuracy for regulatory screening and detailed assessment,while accounting for corrections due to local advection effects, when necessary (e.g. Loubet et al. 2001; Milford et al. 2001b)
- The challenge to provide basic level screening assessments of the effects of NH_3 hot spots on adjacent nature areas on a routine basis at modest costs

15.2 Ammonia Emissions from Hot Spots

Ammonia is mainly emitted from animal housings, manure storage, and land-spread manure, and to a smaller extent from mineral fertiliser application and grazing (Bouwman et al. 1997; Misselbrook et al. 2000). Hence, in intensive agricultural areas, most NH_3 sources are concentrated in small areas surrounding the farms, which may be defined as a hot spot. It is well known that NH_3 concentration above a solution containing NH_4^+ increases exponentially with temperature and pH of the solution (e.g. Génermont and Cellier 1997). This feature, as well as turbulent exchange, is among most important processes involved in emissions of NH_3.

Emissions of NH_3 from housings depend mainly on the number of animals, and the feeding quantity, the construction of the floors, but also on the type of management (Sommer et al. 2006), and the pH of the litter. The ventilation rate (Seedorf et al. 1998a) and the temperature inside the stables (Seedorf et al. 1998b; Wathes et al. 1998) are also essential factors regulating the emissions. In naturally ventilated stables, the rate of ventilation results from a combination of free and forced convection and hence depends on wind and outdoor and indoor temperatures (Hensen et al. 2008b). However, the emissions from animal housings do not only depend on the number of animals, but also partly on the manure handling system (liquid manure/solid manure,

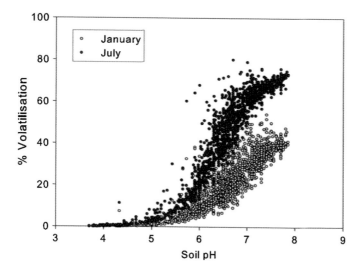

Fig. 15.1 Percentage NH$_3$ volatilised from land-spread manure as a function of soil pH, as output from the VOLT'AIR model, for January and July 2000, over England and Wales (Theobald et al. 2005)

slatted floor, partially slatted floor, deep litter etc.) and storage system (open tanks, tanks with a crust, tanks with a cover) (Hutchings et al. 2001a).

Emission of NH$_3$ after fertilisation depends on the nature of the fertiliser (liquid or solid manure, mineral fertilizer, etc.), the application method and farming practices (Hutchings et al. 2001b), the location and time where the application takes place, soil properties (essentially pH and buffer capacity), as well as the meteorological conditions (temperature and wind speed). The time-span between application and incorporation, and the precipitation also plays a role (Génermont and Cellier 1997; Huijsmans et al. 2003; Rosnoblet et al. 2007). Emissions of NH$_3$ from storage facilities mainly depend on pH, temperature, wind speed and the presence of crust (see e.g. Olesen and Sommer 1993). Figure 15.1 illustrates the effect of soil pH on NH$_3$ volatilisation from land-spread manure.

15.2.1 *Temporal Variation of NH$_3$ Emissions*

Seasonal variability of NH$_3$ emissions is due to a combination of both management practices and meteorological conditions. The application of manure and mineral fertilizer occurs predominantly during spring, but to a lesser extent in autumn. Similarly animals are grazing outdoor only part of the year. The temperature and wind speed dependence of NH$_3$ emissions also explains seasonal variations, but also induce diurnal variations (Asman 1992; Battye et al. 2003; Gilliland et al. 2003; Anderson et al. 2003; Aneja et al. 2003). The diurnal variation of NH$_3$ emissions from a naturally ventilated housing is illustrated in Fig. 15.2, while Fig. 15.3 illustrates modelled seasonal variability of NH$_3$ at a national scale in Denmark.

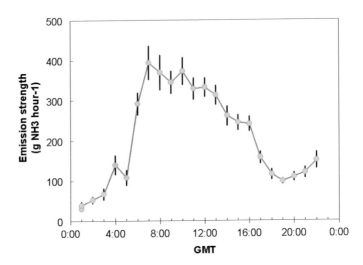

Fig. 15.2 Daily variability of NH₃ emissions from naturally ventilated farm buildings containing 550 animals (2/3 cattle and 1/3 pigs) in Braunschweig, Germany. The estimate was obtained by an inversion method, using measured NH₃ concentration at 230 m and a local dispersion model (FIDES-2D, Loubet et al. 2001; Hensen et al. 2008b). The error-bars are ± standard errors

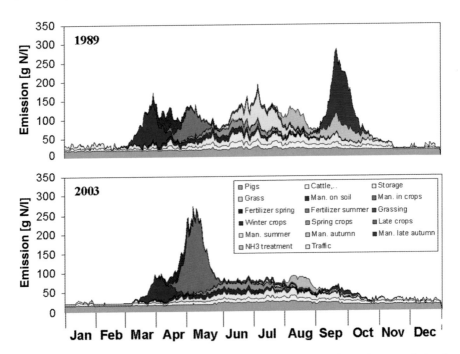

Fig. 15.3 Seasonal variations in modelled Danish ammonia emissions in 1989 (upper plot) and 2003 (lower plot) simulated with a crop growth model combined with local meteorological data. Differences between the 2 years are due to changes in local legislation regarding application of manure on the fields in Denmark that have taken place over the period (Hertel et al. 2006)

15.2.2   Modelling NH₃ Emissions from Hot Spots

Many transport-chemistry models in the past have not taken the seasonal or diurnal variations in NH₃ emissions into account. One reason was the difficulty to get detailed information to model those variations (Hutchings et al. 2001b). However, improvement has been made, from constant emission models (Singles et al. 1998), and sine functions; which was first derived for TREND (Asman 1992), and have also been applied in various of the early versions of the EMEP model (Hov et al. 1994; Olendrzynski et al. 2000), and the ACDEP model (Hertel 1995; Skjøth et al. 2002). Recently, with information on agricultural practices being more accessible, NH₃ emissions are often modelled in a more dynamic way (Génermont and Cellier 1997; Søgaard et al. 2002; Van Jaarsveld 2004; Pinder et al. 2004, 2006; Sommer et al. 2006). The current versions of TREND and OPS use detailed parameterizations of the diurnal and seasonal variations in the NH₃ emission rate that are based on dynamic models (Van Jaarsveld 2004). Models that are applied to smaller areas, where more detailed information has been collected, can take additional factors into account (Smits et al. 2005).

Emission inventories with high spatial and temporal resolution have been implemented in a couple of studies. One example is the American study by Pinder et al. (2006). They combined a model for housing activity (Pinder et al. 2004) with a redistribution method (Gilliland et al. 2003) to obtain a high-resolution inventory for application in their model system. Another example is the Danish process based emission inventory (Skjøth et al. 2004; Gyldenkærne et al. 2005a), which accounts for climatic conditions, agricultural practice and meteorology. The emission inventories obtained for the site of one of the Danish monitoring stations Tange are shown in Fig. 15.3 for the years 1989 and 2003. The differences between these 2 years are due to changes in Danish legislation that have been applied over this period: Danish farmers are now only allowed to apply manure and mineral fertilizer during the growth season. Model calculations based on these inventories are performed on routine basis under the Danish Background Air Quality Monitoring Programme (Ellermann et al. 2006). The model applied in this programme has been the Lagrangian long-range transport model ACDEP (Hertel 1995), which was also used for the first testing of the inventory. Recently ACDEP has been substituted with the Eulerian model DEHM-REGINA, which has been shown to make a better description of the transport processes.

15.2.3   Uncertainties in NH₃ Emissions Estimates

When modelling the deposition around a single farm, there is usually rather detailed information available about the housing and manure handling system. This is usually not the case when the deposition is modelled for a larger area. The spatially most detailed information that is publicly available in very few countries is the number of animals present belonging to each farm at a given date (e.g. in a husbandry register). Estimates then have to be made about the housing, storage and applicatoin practices. The emission belonging to one animal category can e.g. vary by a factor of 2, depending on the housing system (Hutchings et al. 2001a, b). This means that emissions in transport models are rather sensitive to assumptions regarding the housing and manure handling system. Housing and manure handling practices may be variable over countries or within a country. There is also some uncertainty in the location of the emissions, since in local databases the farm location is usually that of the farmer's house, instead of the animal housings and storage facilities.

15.2.4   Emission of Other Nitrogen Compounds

Although nitrogen emissions from farms are mainly in the form of gaseous NH_3, other forms may contribute as well: N_2O, NO and N_2 are emitted, in particular from solid systems (Dämmgen and Hutchings 2008, in press). Primary aerosols (skin and feather particles) also contain nitrogen, and some nitrogen containing volatile organic compounds may also be emitted. Schade and Crutzen (1995) found considerable emissions of methylamine, mainly in the form of trimethylamine-N, which possibly can react to N_2O and HCN. The deposition velocities of these volatile organic species have however hardly been studied. Nitrous oxide (N_2O) and N_2 are also common in composting elements (Fukumoto et al. 2003).

15.2.5   Emission of Other Reactive Species

Other reactive species may interact with NH_x in the atmosphere surrounding the farm as well as modifying deposition rate of NH_3 near the farm (Erisman and Wyers 1993).

A first estimate of the emissions of volatile organic compounds from animal husbandry reveals a considerable emission of sulfur species, mainly as dimethyl sulfide (DMS) from mammals (two thirds of the total) and dimethyl disulfide (DMDS) from poultry (one third of the total) (Smith D. et al. 2000; Chavez et al. 2004), as well as hydrogen sulfide (H_2S), from pig units (Lim et al. 2003). The overall emissions of these species are so large that their chemical fate needs to be considered. In general, all S species apart from sulphur dioxide (SO_2) and sulphuric acid or sulfates (H_2SO_4, SO_4^{2-}) are unstable under atmospheric conditions. However, SO_2 concentrations in ambient air indicate that animal husbandry is not a large indirect source of SO_2.

Once released from the animal house, DMS and DMDS are likely to be oxidized (Möller 2003; Sørensen et al. 1996; Saltelli and Hjorth 1995). However, their reaction pathways and products differ:

- Approximately two thirds of the DMS released react with OH radicals forming methane sulfonic acid (MSA) CH_3-$S(O_2)$-OH and dimethyl sulfone ($DMSO_2$) CH_3-$S(O_2)$-CH_3. These reactions do not result in the formation of SO_2. Nevertheless, the reaction products, MSA in particular, play a role in the formation of condensation nuclei and should be removed from the atmosphere primarily by wet deposition.
- One third of the DMS reacts with OH radicals to form sulfur dioxide (SO_2) and sulfuric acid (H_2SO_4). Both species are deposited dry and wet. A considerable increase of SO_2 concentrations is not likely to occur (Shon et al. 2005). However, both species will contribute to acidification. The mean atmospheric lifetime of DMS is in the order of magnitude of a day.
- Within the atmosphere, DMDS reacts much faster than DMS. Its mean lifetime is minutes. The reactions with OH radicals result in the formation of methane sulfenic acid (CH_3-S-OH) and methyl sulfide radicals (CH_3-S). Methane sulfenic acid should react to form MSA (Finlayson-Pitts and Pitts 1986); the reaction products of the radical should be SO_2 and H_2SO_4.

Sulfur compounds emitted from hot spots may no interact with NH_3 in the gas phase, but they may increase the potential for NH_3 recapture downwind from hot spots (Durenkamp and De Kok 2002) by modifying the pH of the receptors (vegetation surfaces, ground). There is however little information to get a clear view of potential interactions between NH_3 and sulfuric compounds deposition.

15.3 Modelling NH_x Local Deposition

15.3.1 Atmospheric Diffusion

The speed at which substances released into the atmosphere are dispersed depends on the wind speed, the turbulence and atmospheric stability (e.g. Seinfeld and Pandis 1998). Moreover, vertical mixing in the atmosphere can be limited above a certain height, called the mixing height. This mixing height shows diurnal and seasonal variations and depends on turbulence and the atmospheric stability. The existence of a mixing height is especially important for dispersion and transport at regional scales, but during nighttime the mixing height may be so low that it also influences local dispersion. The turbulence and the wind field do not only depend on meteorological conditions, but are influenced by the presence of buildings and by the surface roughness. The wind speed increases with height and this will have an influence on the dispersion away from sources.

Deflection and disturbance of the wind field by a structure (housing, storage tank etc.) has an influence on the dispersion of the pollutant released from this structure (see e.g. Bjerg et al. 2004). In the upwind displacement zone the approaching airflow is deflected around the structure. Immediately leeward of the structure there is a zone that is relatively isolated from the main flow, and further downstream there is a highly disturbed wake. If the pollutant emitted very close to the top of the structure and the exit velocity of any stack small, then the pollutant will be transported downward by "downwash" on the leeward side of the building. This will often be the case for animal houses with outlets on the roof. If an exit stack is relatively high, the pollutant will not be transported downward. The dimensions of the structure (height, width, orientation to the wind, inclination of the roof etc.) all have an influence on the airflow.

The influence of building on the wind field (and hence deposition) can be neglected at distances 5–10 times the emission height and at least as large downwind, as the buildings reach upwind (Irvine et al. 1997; Flesch et al. 2005). This means in practice often of the order of 50–150 m downwind.

A description of the local scale transport with the option of simulating the complex flow around buildings may be provided with application of Computational Fluid Dynamics (CFD) models. These models provide a very high spatial resolution, but are in general very demanding with respect to computer resources. An inter-comparison of European CFD models has shown that they provide similar flows even for the more complex building configurations (Ketzel et al. 2002). However, there may be discrepancies between the CFD models concerning where the peak-values appear. The description of the flow very close to buildings may have considerable impact on the calculated concentrations and thereby also on the obtained deposition values further away from the source. CFD models are usually only able to generate wind/turbulence fields and concentration fields, but not calculate deposition. Some atmospheric transport and deposition models, such as the Lagrangian stochastic dispersion models may use the output of the wind/turbulence fields of the CFD-models as an input (e.g. Bouvet et al. 2006).

15.4 Chemical Reactions

15.4.1 Photochemical Reactions

NH_3 reacts with OH radicals ($NH_3 + OH \rightarrow NH_2 + H_2O$). The rate of this reaction has been estimated at $3.3 \ (\pm 1) \times 10^{-12} \exp(-933 \ (\pm 100)/T \ (K) \ cm^3 \ mol^{-1} \ s^{-1}$ (Diau et al. 1990). Assuming $[OH] = 5 \times 10^5 \ mol \ cm^{-3}$ the lifetime of NH_3 is about 110 days, which is much longer than the actual lifetime of NH_3. This means that this reaction is not an important sink for NH_3 in the atmosphere.

15.4.2 Reactions with Acids

Ammonia reacts with acids: H_2SO_4-containing aerosol (almost all H_2SO_4 is in the particulate form because it has a very low vapour pressure) and gaseous HNO_3 and

HCl. $(NH_4)_2SO_4$ as a solid or in aqueous solution is the preferred form of SO_4^{2-}. Once formed it does not evaporate again. Two regimes can now be distinguished (Nenes et al. 1998; Seinfeld and Pandis 1998):

- **An NH_3-poor atmosphere**. In this case there not enough NH_3 to neutralize all H_2SO_4 and the aerosol will for that reason be acidic and the vapour pressure of NH_3 remain low.
- **An NH_3-rich atmosphere**. In this case all H_2SO_4 reacts with NH_3 and the remaining NH_3 can then react with HNO_3 or HCl to form particulate NH_4NO_3 or NH_4Cl. NH_4NO_3 and NH_4Cl may exist as a solid or as an aqueous solution of NH_4^+, NO_3^- and Cl^-, depending on the relative humidity, temperature as well as the presence of other inorganic salts (Stelson and Seinfeld 1982). In an ammonia-rich atmosphere NH_4NO_3 or NH_4Cl in solid or dissolved form is generated if the concentration products in the gas phase, $[NH_3][HNO_3]$ or $[NH_3][HCl]$, exceed threshold values. These threshold values decrease with temperature and decrease with increasing humidity, i.e. at lower temperatures and higher humidities NH_4NO_3 formation is enhanced.

The time scales to achieve gas-aerosol equilibrium vary from 1 h to several hours for fine particles (0.1 μm < diameter < 1 μm). For coarse particles (diameter > 1 μm) the time scale is so long that NH_3 generally is not in equilibrium with these particles.

The reaction rate of NH_3 with acids depends on the concentrations of the acid species present in the atmosphere and shows temporal and spatial variations and is also dependent on the height above ground level. In most regions of Europe, more acid precursors in the form of SO_2 and NO_x are present than the emitted NH_3 can neutralise. Since NH_3 emissions occur at or near ground level, the fraction of acid neutralised by NH_3 decreases with height (Erisman et al. 1988). It is likely that acids are often fully neutralised at ground level in large parts of Europe, though not during the whole year. However, in some places with low NH_3 emissions, as in northern Scandinavia neutralisation of acids might not be reached even at ground-level (H. Fagerli 2006, EMEP, Norway, personal communication).

In practice the gas-to-particle conversion factor of NH_3 can be determined from field experiments, based on many assumptions (Erisman et al. 1988; Harrison and Kitto 1992), or can be inferred from fitting transport models with measured concentrations of gases and aerosols. Using this technique Asman and Janssen (1987) found a pseudo-first order reaction rate of 8×10^5 s^{-1} (about 30% h^{-1}), a rate that might have been reduced since 1987 due to reduced emissions of SO_2 in Europe (van Jaarsveld et al. 2000). The reaction rate is so low that it can be assumed that all released NH_3 has not reacted within a few kilometres from sources, but at larger distances the reaction should be taken into account. This can either be done by having a separate chemistry model to parameterise the reaction rate as a function of the NH_3, SO_2 and NO_2 concentrations and then to apply this relation to find the reaction rate of NH_3 to particulate NH_4^+ in an atmospheric transport and deposition model (van Jaarsveld et al. 2000) or by modelling the reaction rate within the atmospheric transport model. For the last option, emissions of involved compounds (NH_3, SO_2, NO_x) are needed, preferably on the same scale. Moreover, information is also needed on many reaction rates, which will slow down the calculations considerably.

The transformation of NH_3 to particulate NH_4^+ produces sub-micron aerosols, which have relatively small deposition velocities (they are too small to settle by gravity and too large to diffuse by Brownian motion), hence favouring long-range transport of NH_x. For that reason, gas-to-particle conversion of NH_3 is of great importance for trans-boundary transport of NH_x, although remains an issue of secondary importance in detailed analysis of NH_3 hot spots.

15.5 Dry Deposition

15.5.1 General

Although this section focuses on ammonia, sometimes some information is given on particles or other gaseous compounds, so that the reader gets an impression of the difference in properties. Some general information can be found in Sutton et al. (1995a, 1998b, 2000).

Transport to and from the surface occurs by turbulent and molecular diffusion. In the atmosphere turbulent diffusion (transport by eddies of different sizes) is responsible for the transport, and can be considered the same for gases and particles smaller than ~10 μm in diameter (Wilson 2000). For larger particles gravitational settling and cross-trajectory effects (due to gravitational forces and inertia, heavy particles do not follow the fluid trajectories exactly) cannot be neglected (Sawford and Guest 1991). However, ammonium-containing particles are not that large, except maybe for dusts ejected from housings, which may contain nitrogen.

Close to the surface, in the so called "laminar boundary layer", transport takes place by molecular diffusion (gases) or Brownian motion (particles). For particles with a diameter > 0.4 μm interception plays also a role close to the surface especially if the surface has sharp edges or is covered with hair-like objects. For particles with a diameter > 1 μm, impaction increases deposition due to inertia (see e.g. Seinfeld and Pandis 1998).

Gases are absorbed by diffusion through stomata, but can also be deposited onto the leaves/needles especially if the gases are soluble in water and the leaves are covered by a water-layer. Stomata are almost closed when it is dark (no photosynthesis occurs then), when the leaves are CO_2 saturated, and during water stress. Particles diffuse too slowly and can therefore not be transported through the stomata. For that reason deposition of particles occurs mainly onto leaves/needles cuticules. Once deposited to vegetation, particles are not easily released back to the atmosphere, apart from re-suspension for large particles.

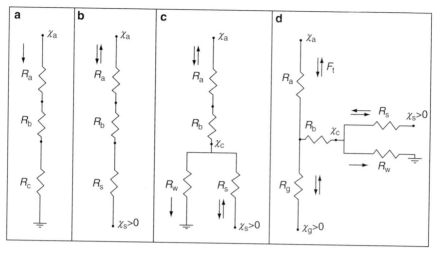

Fig. 15.4 Different types of resistance models for describing exchange of gases between the surface and the atmosphere, with increasing complexity: (**a**) deposition velocity model, (**b**) bi-directional exchange model, (**c**) bi-directional exchange model with a cuticular pathway, and (**d**) two-layer bi-directional exchange model with a cuticular pathway and an emission potential at the ground. Here $\chi_a = c_{air}$; $\chi_c = c_{canopy}$; $\chi_s = c_{stomata}$, χ_c is the concentration outside the leaf, $R_a(z_{ref})$, R_b, R_s and R_w are the aerodynamic, the leaf boundary layer, the stomatal and the wet surfaces resistances for NH_3, respectively (After Nemitz et al. 2001)

15.5.2 Resistance Analogue Models for Exchange of Gases Between the Surface and the Atmosphere

Models describing the exchange or gases between the atmosphere and the surface are usually based on the resistance analogy similar to that of Ohm's law for electricity (Dämmgen et al. 1997). This analogy relies on the assumption of constant flux layer, which is satisfied in a stationary surface layer with no local advection. Although this assumption would make these models not applicable to local dispersion, there always exists a layer near the ground where no advection occur and these models can be applied (Loubet et al. 2001). Figure 15.4 shows different models for the dry deposition of gases that can either be used as model at the canopy scale, in which case the resistances are per unit surface of ground (noted with an uppercase R), or at the leaf scale, in which case the resistances are given per unit surface of leaf (noted with a lowercase r).

15.5.2.1 Deposition Velocity Model

Figure 15.4a shows the simplest model of gaseous dry deposition, whereby the surface is assumed a perfect sink. The aerodynamic resistance $R_a(z_{ref})$ describes the transport by turbulent diffusion in the atmosphere from z to the laminar boundary layer. It is a function of the friction velocity and the atmospheric stability. The laminar boundary layer resistance R_b on molecular diffusion, while the surface resistance R_c (the canopy as a whole, or the water surface, etc.) depends on processes going on at the surface. In the model of Fig. 15.4a, the flux F can be expressed as a function of the dry deposition velocity $v_d(z_{ref})$ or the resistances $R_a(z_{ref})$, R_b and R_c:

$$F = -v_d(z_{ref})\, c_{air}(z_{ref}) = -\frac{c_{air}(z_{ref})}{R_a + R_b + R_c} \qquad (15.1)$$

where $c_{air}(z_{ref})$ is the concentration in the air at reference height z_{ref}. The aerodynamic resistance $R_a(z_{ref})$ is given by:

$$R_a(z_{ref}) = \frac{1}{\kappa\, u_*}\left[\ln\left(\frac{z_{ref}-d}{z_{0m}}\right) - \Psi_h\left(\frac{z_{ref}-d}{L}\right) + \Psi_h\left(\frac{z_{0m}}{L}\right)\right] \qquad (15.2)$$

where $\kappa = 0.4$ is the von Karman's constant, u_* is the friction velocity (m s^{-1}), which is a measure for the turbulence and increases with wind speed for the same atmospheric stability, d is the displacement height (m), which is about 0.7 times the height of the vegetation. If individual vegetation elements are packed closely together, then the top of the surface begins to act as a displaced surface, called the displacement height; z_{0m} the surface roughness length (m), which is a measure for the roughness of the surface, and is about 0.1 times the height of the vegetation. It is the height at which the wind speed is zero in an extrapolated logarithmic wind profile. Ψ_h is a correction function for atmospheric stability, L is the Monin-Obukhov length (m) and is a measure of the atmospheric stability. If L is very large (negative or positive) the atmosphere has a neutral stratification, whereas if L is relatively small and positive the atmosphere is stable, and if L is relatively small and negative the atmosphere is unstable. For neutral conditions: $\Psi_h = 0$. For stable conditions: $\Psi_h = -5(z_{ref} - d)/L$. For unstable conditions: $\Psi_h = 2\ln((1 + x^2)/2)$, with $x = (1 - 15(z_{ref} - d)/L)^{1/4}$.

The laminar boundary layer resistance R_b for gases transfer through vegetation is often parameterized as follows (Hicks et al. 1987):

$$R_b = \frac{2\left(\dfrac{Sc_g}{Pr}\right)^{\frac{2}{3}}}{\kappa\, u_*} \qquad For\ NH_3 : R_b = \frac{1.89}{\kappa\, u_*} \qquad (15.3)$$

where Pr is the Prandtl number (dimensionless; value 0.72), Sc is the Schmidt number for gases (for NH_3: about 0.662). Sc decreases with the diffusivity of the gas and does not depend much on temperature.

From Eqs. [15.2] and [15.3] it can be seen that $R_a(z_{ref})$ and R_b decreases with increasing u_*. Moreover, v_d cannot become higher than $V_{max} = 1/(R_a + R_b)$. The rather simple model of Fig. 15.4a gives a good description of NH_3 deposition to semi-natural and natural vegetation, which have a low nitrogen status.

15.5.2.2 Stomatal Compensation Point

NH_3 exchange with plants receiving large amount of nitrogen is however known to be bi-directional, due to the existence of a non-zero stomatal compensation point concentration C_s, which mainly varies with the nitrogen status and the temperature of the plant (Sutton et al. 1995a; Schjoerring 1997; Loubet et al. 2002). Indeed, agricultural crops have NH_3 compensation points that vary from 0.1 to 20 ppb (0.07–14 μg m^{-3}) (Schjoerring et al. 2000), and even temporarily up to 40 μg m^{-3} NH_3 (Fig. 15.5), which may lead to large emissions from crops (Schjoerring 1991; Husted and Schjoerring 1996; Holtan-Hartwig and Bøckman 1994). The equations for Fig. 15.4b hence should rather be used in such case. The flux between the surface and the atmosphere is then given by:

$$F = -\frac{c_{air}(z_{ref}) - c_s}{R_a + R_b + Rs} \tag{15.4}$$

where R_s is the stomatal resistance. The stomatal compensation point C_s results from the thermodynamic and chemical equilibrium between the ammonium concentration in the apoplast (NH_4^+), and the gaseous NH_3 concentration in the sub-stomatal cavity. This equilibrium is mainly dependent on temperature and pH of the apoplastic solution (Schjoerring 1997; Smith R.I. et al. 2000). Following Schjoerring (1997), and Nemitz et al. (2000b):

$$C_s = 4.79 \ 10^{-12} \ . \ \Gamma \ . \ \exp\left(10396 \frac{T_{leaf} - 25}{298 \times (T_{leaf} + 273)}\right) \tag{15.5}$$

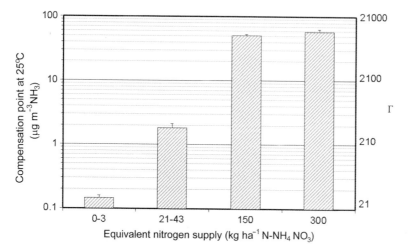

Fig. 15.5 Compensation point concentration at 25°C and Γ values in maize leaves during the vegetative growth (from five leaves to flowering), as a function of an equivalent nitrogen supply. Error-bars correspond to standard errors

where Γ is the ratio of the molar concentrations $[NH_4^+]/[H^+]$ in the apoplast (Nemitz et al. 2000b). Γ expresses a potential compensation point independent of temperature, which is representative of the plant and the ecosystem: small Γ (typically 20–100) for non-fertilised ecosystems and large Γ for fertilised ecosystems (typically 300–3,000). Even fertilised ecosystems such as grazed grassland show a great temporal variability in Γ (e.g. Schjoerring 1997; Loubet et al. 2002).

Getting into more details, the value of Γ for semi-natural grassland with short grass in Germany, which received fertilizer at a rate of 70 kg ha^{-1} a^{-1} N, was about 1,000. This is lower than observed for grassland in the UK, which received 300 kg ha^{-1} a^{-1} N where values for Γ were observed of 3,000 (long grass, 15 cm high) and 13,000 (short grass, 5 cm high) (Sutton et al. 1998b). For a Dutch heathland that was not fertilized, but is situated in an area with a high N deposition a value of Γ was observed of 1,200, which is large for semi-natural vegetation (Nemitz et al. 2004a). Husted et al. (2000) found for oilseed rape (*Brassica napus ssp. napus*) Γ values of 200–500. As an example, Fig. 15.5 shows the effect of Nitrogen supply on C_s for maize (Loubet et al., 1998, unpublished data):

The compensation point can either be determined from the NH_4^+ and H^+ concentration in the leaf apoplastic solution (e.g. Husted et al. 2000), or inferred from flux measurements both over vegetation surfaces (Fléchard et al. 1999) or within chambers (Hill et al. 2001). The compensation varies with plant species, nitrogen status, and growth stage as well as leaf age. However, on an annual basis a net emission is observed from agricultural crops that will depend on the plant species, meteorological conditions and stress due to drought, diseases or pests. An annual average emission is of the order of 1–5 kg ha^{-1} a^{-1} N during a growing season (Schjoerring and Mattsson 2001), which is consistent with micrometeorological measurements (e.g. Milford et al. 2001a).

In semi-natural areas like heathland and forest, the compensation point is generally so low that it does not play a role and only dry deposition occurs. In some cases, however, when air concentrations are very low, under very dry conditions or at high temperatures emission has been observed even from semi-natural ecosystems (e.g. Langford and Fehsenfeld 1992; Erisman et al. 1994a; Sutton et al. 1995c; Fléchard and Fowler 1998; Andersen et al. 1999).

15.5.2.3 Cuticular Pathway

As a highly soluble gas, NH_3 is readily adsorbed onto wet or humid surfaces, such as the cuticle of the leaves and the stems (Burkhardt and Eiden 1994; van Hove et al. 1989). The last process is often represented by a cuticular deposition pathway acting in parallel to the stomatal pathway (Sutton et al. 1995a, Fig. 15.4c, see Appendix 1 for detailed equations). The resistance to cuticular deposition of NH_3, R_w, is often modelled as a function of relative humidity RH, as follows (van Hove et al. 1989; Sutton et al. 1995a; Nemitz et al. 2000b):

$$R_w = R_{wmin} \exp\left(\frac{100 - RH}{\beta_{Rw}}\right) \tag{15.6}$$

where R_{wmin} ranges between 2 and 20 and β_{Rw} is of the order of 6–12. It should be noted however, that cuticular deposition to "wet surfaces" may also occur without apparent dew (Duyzer et al. 1992; Sutton et al. 1992; Wyers and Erisman 1998), which might be due to the presence of wet films at the leaf surface created by the conjunction of stomatal evaporation and the presence of hygroscopic aerosols deposited preferentially near the stomates (Burkhardt and Eiden 1994). The empirical parametrization of Eq. [15.6] incorporates the effect of hygroscopy, but does not distinguish any dependence of R_w on R_s. Smith R.I. et al. (2000) gives an alternative empirical expression of both temperature and relative humidity. Nemitz et al. (2001) included the dependence of R_w on the SO_2/NH_3 concentration ratio.

It should be recognized that the parametrization of R_w is, however, a steady state simplification of a dynamic bi-directional exchange of NH_3 with leaf cuticles. An initial dynamic model of this effect was provided by Sutton et al. (1995a, 1998a). A simplification of that model was that the leaf surface pH needed to be specified. Fléchard et al. (1999) advanced this substantially, by developing a model that simulated leaf surface pH in response to wet and dry deposition processes and derived bi-directional cuticular exchange using a dynamical model that takes into account the uptake of different soluble pollutants their chemistry in the water layer on the leaf. In these models, the steady state value of R_w is effectively replaced by a capacitance of the leaf surface, a capacitance charge, with exchange limited by an adsorption/desorption resistance (R_d).

Near sources, cuticular deposition is likely to be very high in humid climate due to large NH_3 concentrations. Throughfall measurements near intensive source which are reported in the literature may be indicative of cuticular deposition, although they might be subject to uncertainties due to dry deposition onto collectors, flooding under high rain events, or biochemical transformations of NH_x deposited onto leaves (Theobald et al. 2001; Dämmgen et al. 2005; Erisman et al. 2005). Theobald et al. (2001) report throughfall ranging 3% and 4% of the NH_3 emitted by a source releasing between 500 and 2,800 kg $N-NH_3$ year^{-1}. It should be noted that at very high concentrations the value of R_w tends to increase, due to a partial saturation of the leaf surface sink. The presence of SO_2 will increase the dry deposition of NH_3. For that reason, R_w is sometimes modeled as a function of the ratio NH_3/SO_2 (Nemitz et al. 2001), or by empirical approaches based on observational data (Pitcairn et al. 2004).

15.5.2.4 Two Layer Exchange Model

A two-layer resistance scheme (Fig. 15.4d, Nemitz et al. 2001, see Appendix 2 for detailed equations) takes into account the ground level source which can either be an emission from a fertilizer or from decomposing leaf litter (Denmead et al. 1976; Sutton et al. 1993c; Nemitz et al. 2000a). Moreover, the two-layer model allows reproducing the absorption by leaves of a fraction emitted from the ground (Nemitz et al. 2001). The soil or litter compensation point may be very large, especially in nitrogen rich litter (Husted et al. 2000; Nemitz et al. 2001; Sutton et al. 2001a), where Γ can reach more than 100,000 (Sutton et al. 2006). The two-layer models have shown to be very useful in modelling the effect of the canopy structure one NH_3 exchange with the atmosphere (Personne et al. 2008).

15.6 Deposition of Particulate Ammonium

The dry deposition of particles to the ground can be represented by the scheme of Fig. 15.4a, but with $R_c = 0$, and with a parallel pathway correspond to a resistance inverse to the settling velocity of the particle V_s. The computed deposition velocity for particles becomes simply (Slinn 1982; Zhang et al. 2001):

$$V_d = V_s + \frac{1}{R_a(z_{ref}) + R_{bpart}} \qquad (15.7)$$

where R_{bpart} is the boundary layer resistance for particle, which depends on Brownian diffusion. V_d depends strongly on the particle size, the characteristics of the surface (roughness) and u^*. The dry deposition velocity V_d of NH_4^+ containing particles for neutral atmospheric conditions was estimated by Erisman et al. (1994b) as:

$$V_d = \frac{u_*}{A} \qquad (15.8)$$

where V_d and u^* are in m s^{-1}, A = 500 (dimensionless) for low vegetation and A = 100 for forests. Measurements show that the dry deposition velocity of NH_4^+ containing particles to moorland or grass is of the order of 0.2 cm s^{-1}, with a large uncertainty justifying the rough parameterisation of Eq. [15.8] (Sutton et al. 1993c; Duyzer 1994). The dry deposition velocity to forests is higher than to moorland.

The dry deposition velocity of NH_3 is potentially relatively high and is about a factor of 10 higher than that of particulate NH_4^+. This means that NH_x after conversion from NH_3 to NH_4^+ is not dry deposited very well and is transported over long distances. The only efficient removal process for particulate NH_4^+ is wet deposition (Asman and Janssen 1987).

15.6.1 Measured Dry Deposition Velocities of NH₃ to (semi-) Natural Vegetation

Measurement of NH_3 deposition velocity to semi-natural vegetation is reported in many studies (Table 15.1). Table 15.1 shows higher V_d for forest than for moorland and grassland, which reflects the higher u_* over forest.

Table 15.1 Example dry deposition velocity and canopy resistance data for NH_3 for semi-natural ecosystems (moorland, unfertilised grassland and forests), as found in the literature. The deposition velocity or R_c approaches correspond to the simpler model of Fig. 15.4a

Ecosystem type	v_d (cm s^{-1})	R_c (s m^{-1})	References
Moorland	1–4	<10[a]	Sutton et al. (1992)
	1.9[b]	0–150	Duyzer (1994)
	1.2	Dry: 61	Fléchard and Fowler (1998)[c]
		Wet: 23	
		Snow: 56	
		Frozen: 50–100	
	1.5–2.0	3–6	Sutton et al. (1993c)
Unfertilised grassland	0.13–1.4	–	Hesterberg et al. (1996)[d]
	–	5–27	Sutton et al. (1997)[e]
Forest			
Sitka spruce, European larch, Lodgepole pine, Noble fir	6.6	6	Sutton et al. (1993c)[f]
Douglas fir	2.5	20–25	Duyzer et al. (1994)[f]
Douglas fir	3.2	–	Wyers et al. (1992)[f]
Norway spruce	0.88 (stable atmosphere) 1.8–4.0 (other conditions)	–	Andersen et al. (1999)[f]
Parameterisation			
Humid semi-natural ecosystems and forests		Daytime wet: 500	Erisman et al. (1994b)
		Night-time dry: 1,000	
		Night-time wet: 0	

[a] During frozen conditions $R_c = 50 – 200$ s m^{-1}.
[b] Measurements made mainly during daytime. Estimated annual average 24-h $V_d = 1.4$ cm s^{-1}.
[c] Emission observed during 6% of the time.
[d] A compensation point between 3 and 6 ppbv (2.1–4.2 ∝g m^{-3}) was observed.
[e] Sometimes emission occurs.
[f] Emission observed during some periods. It should be noted that, the number of experimental data as well as the experimental conditions may bias the interpretation (daytime over-represented, low wind speeds underrepresented). Moreover, since $V_d(z_{ref})$ depends on z_{ref}, the V_d may not be directly comparable. Evaporation of NH_3 from NH_4^+ containing aerosols deposited onto leaves may lead to slight underestimation of V_d (Nemitz et al. 2004a, b; Nemitz and Sutton 2004).

V_d is often larger than what can be accounted for by stomatal uptake (Duyzer 1994). This is especially clear from nighttime measurements. The dry deposition velocity increases with surface wetness (Wyers and Erisman 1998) and decreased with NH_3 concentration (Fléchard and Fowler 1998). Due to the large non-stomatal dry deposition, dry deposition velocity of NH_3 is often larger than for SO_2 or O_3. It should be noted, however, that there are many other studies where deposition velocity values were not actually reported, but instead net (bi-directional) fluxes together with parametrizations of Γ and R_w were given.

15.7 Wet Deposition

15.7.1 General

Cloud droplets and raindrops are usually acidic. If NH_3 is absorbed by cloud droplets or raindrops it reacts with H^+ to form NH_4^+. Both NH_3 and NH_4^+ containing particles can be removed by absorption in clouds droplets or scavenging below clouds. In general in-cloud scavenging of particulate NH_4^+ contributes most to the NH_4^+ concentration in rainwater. This holds, however, not for the contribution of a point source to the wet deposition close to this point source. There are two reasons for that:

- At short distances from the source the NH_3 plume has usually not reached the clouds and for that reason in-cloud scavenging of the NH_3 originating from the source will not occur.
- Most of the NH_3 will not yet have reacted with acid compounds (H_2SO_4 in aerosols or gaseous HNO_3 and HCl) so close to the source (see Section 15.4) and for that reason most of the NH_x will be in the form of NH_3.

For that reason it can be expected that much of the NH_x from a nearby source is removed by below-cloud scavenging of NH_x, which is not as efficient as in-cloud scavenging of NH_x. In the following below-cloud scavenging of NH_3 will be discussed in more detail. A short description of in-cloud scavenging of NH_4^+ is discussed in Appendix 3.

Wet deposition of NH_x originates partly from groups of sources within 50–100 km (wet deposition of scavenged NH_3) and partly from long-range transport (wet deposition of scavenged NH_4^+) (Asman and van Jaarsveld 1992). It has been shown from experimental results that wet deposition of NH_x is correlated with the emission density on scales of 20–100 km (Asman and van Jaarsveld 1992; Park and Lee 2002; Aneja et al. 2003). Within 0.5–1 km from a source the contribution of the source to wet deposition of NH_x is much less than the contribution to dry deposition. This is caused by the fact that the plume has not been mixed up at this distance and the NH_3 concentration at ground level is relatively high. Wet deposition is determined by the average concentration over the whole plume height and not by the much higher ground-level concentration. Due to its limited importance at the very local scale wet deposition is not taken into account in most local models: Danish OML-DEP (Olesen 1995), the UK LADD (Dragosits et al. 2002), French FIDES (Loubet et al. 2001) and MODDAAS (Loubet et al. 2006). Conversely, wet deposition is included in DEPO1 (Asman 1998).

Sometimes high NH_4^+ concentrations are measured with open precipitation collectors near sources. It is likely that a large fraction of these high concentrations are due to dry deposition of NH_3 to the (wet) surface of the rain collector.

15.7.1.1 Below-Cloud Scavenging of NH_3

Below clouds, NH_3 is taken up by falling raindrops (typical radius 0.1–1 mm). Falling raindrops have a relatively high speed (0.71–6.5 m s^{-1}). Moreover, they have a much smaller surface to volume ratio than cloud droplets. As a consequence, with the exception of small raindrops, it is unlikely that any droplet becomes NH_3-saturated in the plume. For convective conditions, a maximum scavenging coefficient can be derived following Asman (1995):

$$\lambda_b = 9.85 \times 10^{-5} \left(\frac{I}{I_{ref}} \right)^{0.616}$$

(15.9)

where λ_b is the below-cloud scavenging coefficient for NH_3 (s^{-1}). This gives the fraction of the plume concentration that is removed per second; I is the rainfall rate (mm h^{-1}, unit usually reported) and I_{ref} is the reference rainfall rate in the same unit as I. Equation [15.9] shows that the removal rate of NH_3 by below-cloud scavenging increases with the rainfall rate. The uncertainty on λ_b is about a factor of 2. If a wind speed of 2 m s^{-1} occurs during a rain event, about 5% and 20% of the NH_3 is removed from the plume at 1,000 m downwind of the source for rainfall rates of 1 and 10 mm h^{-1}, respectively. If this is combined with the frequency of rain events (5–10% of the time in humid regions), we estimate that an order of magnitude of between 0.25–2% of NH_3 scavenged by wet-deposition up to 1,000 m downwind from sources. This amount is much smaller than typical near source removal by dry deposition.

15.7.1.2 Below-Cloud Scavenging of Particulate NH_4^+

The efficiency of below-cloud scavenging of particles depends on the particle size (distribution) and the raindrop size distribution, which is a function of the rainfall rate. In the atmosphere NH_4^+ is predominately found in the fine particles (0.1 < diameter < 1 μm). For this size range the scavenging coefficient is very low: of the order of 1×10^{-7} s^{-1} for a rainfall rate of 0.5 mm h^{-1} and 1×10^{-6} s^{-1} for a rainfall rate of 25 mm h^{-1}. It is likely, however, that there are also larger particles (1 μm < diameter < 10 μm) emitted by housings that may have adsorbed NH_3 and therefore contain NH_4^+. These will be scavenged at a higher rate, but will also be dry deposited at a relatively high rate, compared to smaller NH_4^+ containing particles.

15.8 Short Description of Existing Models NH$_x$ Local Deposition

There are many different types of local atmospheric transport models for ammonia and ammonium. The reason for that is that they are made for different purposes. One of the things that can vary is e.g. the scale. Some models are only suited to model the deposition up to 1 km from one source. Other models are able to give local depositions for a whole country. Some models are especially developed to study the results of one NH_3 field experiment, whereas others are models that describe the transport of and deposition of air pollutants in general. Which processes are incorporated into which detail in a model depends on the purpose of the model. If e.g. the recapture of NH$_3$ emitted from the soil by overlaying vegetation should be described it is necessary to have a multi-layer model that has the soil and vegetation in different layers. Sometimes a cascade of models is used, where the output of long-range transport models forms the background deposition onto which the results of a local deposition model are added as is the case with the combination DDR/LRTAP (Asman and Maas 1986), DEPO1/TREND (Asman et al. 2004), FRAME/LADD (Theobald et al. 2004) and DEHM-REGINA in combination with OML-DEP (see Table 15.2). An overview of dispersion models, which may be suitable for modelling local dispersion of NH$_3$ and be coupled with deposition of NH$_3$, is given in Appendix 4.

15.8.1   DDR

DDR is a 3-D Gaussian surface depletion model (Asman and Maas 1986; Asman et al. 1989). The way dry deposition is treated is derived from Horst (1977), but in addition a pseudo first order reaction velocity (for transformation of NH$_3$ to NH$_4^+$) is added. Moreover, dry and wet deposition and the existence of a mixing height are built into the model. A surface depletion model was chosen, because the dry deposition velocity of NH$_3$ is so large that dry deposition cannot be described adequately by a source depletion model (where it is assumed that the concentration decreases equally throughout the plume due to dry deposition). The basic concept of a surface depletion model is that the deposition flux to the surface is represented by sources (or sinks) at the earth's surface with an equivalent negative source strength, The concentration distribution is then calculated as the sum of the non-depositing plume from the primary source plus the (negative plumes from all upwind surfaces that account for dry deposition. The dry deposition is then calculated as the near-surface concentration multiplied by the dry deposition velocity. This model is a statistical model in the sense that it uses frequency distributions of Pasquill stability classes and an average wind speed for each wind-direction for each class. The model has been used to model the NH$_x$ deposition close to farms (0–500 m), as well as the background depositions on a 5 × 5 km^2 scale in the Netherlands including foreign contributions that were partly calculated with a long-range transport model that covers whole Europe (Asman and Janssen 1987). The dry deposition was modelled using dry deposition velocities.

15.8.2   TREND/OPS Model

TREND and the derived short-range model OPS is a Gaussian diffusion model that uses meteorological statistics and can cover whole Europe. (Asman and van Jaarsveld 1992; van Jaarsveld 1995, 2004). It is used to calculate concentrations and depositions over longer time periods (month–10 years). A meteorological pre-processor creates the meteorological input for the model from e.g. hourly meteorological observations. The meteorological input of the model consists of

Table 15.2 A non-exhaustive list of local atmospheric transport and deposition models for NH_x

Model	Brief description	Parameters	Pluses	Minuses	Scale (km)	Big leaf/ Multi-layer	References
DDR	3-D statistical Gaussian surface depletion model, first order reaction, mixing height	Frequency distribution wind speed, Pasquill classes, precipitation statistics and rate for each wind direction	Can cover a very large area with great detail, both point and area sources, can cover long periods	Meteorology, z_0, v_d everywhere the same, no detailed chemistry	0.01–500	Dry deposition velocities	Asman and Maas 1986; Asman et al., 1989
TREND/OPS	3-D statistical Gaussian dispersion model, first order reaction depending on actual concentrations, mixing height	Representative u_*, L, precipitation statistics and rate for each stability class for each wind direction	Can cover a very large are with great detail, both point and area sources, some possibilities for variation in z_0 and R_c, can cover long periods	Meteorology the same everywhere, no detailed chemistry	0.01–2,000	Big leaf and two-layer model	Asman and van Jaarsveld, 1992, 1995, 2004
LADD	Hybrid model, with Lagrangian column following the mean flow and vertical diffusion law, coupled with surface exchange; one version has been extended to include compensation point	u_*, z_{0m}, d, L S_{srce}, source location R_s, R_w (mins and slopes) C	3D Fast Land use predefined C_s and R_s used	No inside canopy transfer	0.001–2,000	Big-leaf	Hill, 1998

(continued)

Table 15.2 (continued)

Model	Brief description	Parameters	Pluses	Minuses	Scale (km)	Big leaf/ Multi-layer	References
DEPO1	Gaussian-3D model coupled with canopy compensation point resistance model, Includes reaction, dry deposition of NH_4^+ and wet deposition (below-cloud/in-cloud) of NH_3 and NH_4^+, and the existence of a mixing height.	u_*, z_{0m}, d, L_{MO} S_{srce}, source location R_s, R_w (mins and slopes) C_s, rainfall rate, scavenging ratios, pseudo-first order reaction rate	Can treat compensation points, point sources, areas sources, can cover a large area and long periods	Meteorology and z_{0m} constant understand	0.01–500	Big-leaf	Asman 1998; Asman et al. 2004
FIDES	Analytical solution of the diffusion equation coupled with resistance analogue model at the canopy scale in 2D	z_{0m}, d, u_*, L S_{srce}, size source R_s, R_w C_s	Process based Fast Few parameters	Only in 2D z_{0m} constant no side flux	0.001–5,000	Big-leaf	Loubet et al. (2001)
MODDAAS	Lagrangian Stochastic dispersion model coupled with a leaf scale resistance analogue model in 2D	$U(z_{ref})$, $L(z_{ref})$ S_{srce}, source location R_s, R_w (mins and slopes) C_s lad(z) and h(z)	Process based Multi-sources Multi-canopies Inside canopy transfer	Only 2D Slow Large no. of parameters turbulence parameterised	0.001–2,000	Multi-layer	Loubet et al. (2006)
DAMOS	The DAMOS (Danish Ammonia Modelling System) is a combination of the Eulerian long-range transport model DEHM-REGINA and the local scale plume model OML-DEP	u_*, z_{0m}, L S_{srce}, source location pseudo-first order reaction rate, land use	High temporal and geographical resolution in emissions – good description of transport	High demand for computer time	0.4–1,000	Big-leaf	Hertel et al., 2006; Frohn et al., 2001; Tilmes et al, 2002; Olesen, 1995

information on the frequency of occurrence of different stability classes for different wind directions. For each class for each wind direction there is a representative value of u^* and L, information on the mixing height and information on the precipitation (precipitation probability, precipitation intensity, length of rainfall period). The model takes into account diurnal and seasonal variations of the meteorological parameters. The wind speed and diffusivity in the model are height-dependent. Close to the source the model is based on a detailed description of the plume, whereas further away where the plume has reached the mixing height the process descriptions are simplified. Dry deposition of NH_3 and NH_4^+ are modelled using the big-leaf model or a two-layer model and wet deposition of both components is described with below-cloud and in-cloud scavenging coefficients. The model is a source depletion model that however, mimics the results of a surface depletion model. The model uses a pseudo-first order reaction rate to describe the reaction from NH_3 to NH_4^+. The reaction rate can be based on the results of detailed chemistry models taking into account actual concentrations of NH_3, SO_2 and NO_x. The model can treat both point sources and area sources.

15.8.3   DEPO1 and Derived Models

DEPO1 (Asman 1998) is a steady state model, where vertical diffusion is treated with a K-model and cross-wind horizontal diffusion with friction and stability dependent. The model assumes the same surface roughness length everywhere. Wind speed and diffusivity in the model are height-dependent. The dry deposition velocities are modelled using the big leaf approach. The model can treat compensation points. Compensation points and surface resistances can vary over the model area. A pseudo-first order reaction rate of NH_3 to NH_4^+ is included in the model, but other components and more complicated reactions can be included as well. Below-cloud and in-cloud scavenging for NH_3 and NH_4^+ are calculated using scavenging coefficients. For below-cloud scavenging of NH_3 the scavenging coefficients of Asman (1995) are used. The model uses hourly meteorology (u^*, L, rainfall rate) as an input, but is also used to calculate deposition over long periods (10 years). DEPO1 was used to derive transfer matrices that were build into a GIS-system that was used to calculate the deposition on a $100 \times 100\,m^2$ scale for an area of about $80 \times 80\,km^2$ size and which gives the user the opportunity to study the effects of emission reductions or different spatial distributions in an easy way (Asman et al. 2004). Recently the model was improved and now includes a two-layer model including ground and leaf compensation points.

15.8.4   LADD

The LADD (Local Atmospheric Dispersion and Deposition) model is used to simulate atmospheric dispersion of NH_3 within domains of up to a few kilometres (Hill 1998; Dragosits et al. 2002). LADD is a Lagrangian model that simulates atmospheric dispersion and surface deposition by moving a vertical column of air along straight-line trajectories across a grid. This air column is divided into layers of increasing depth up to the height of the planetary boundary layer and moves across the grid at a rate equal to the mean wind speed for the trajectory direction. As the column moves across the grid, NH_3 is emitted into the layers containing sources and is mixed vertically within the column at a rate determined by the turbulent diffusion coefficient (K). This coefficient is calculated from the mean wind speed for the trajectory, the roughness length (z_0) assigned to the grid square, the height within the column, the boundary layer height and the atmospheric stability following the methods of Pasquill and Smith (1983) and Ayra (1988). Deposition from the lowest layer to the surface is calculated using z_0, the resistance of the surface to NH_3 deposition (R_c), atmospheric stability and mean wind speed. Within the model each grid square in the domain is assigned a value of z_0 and R_c, which are dependent on the land use. For each trajectory direction, parallel trajectories are modelled sequentially until the entire domain has been covered. The trajectory direction is then increased by a user-defined increment (e.g. 1°) and the process is repeated for all directions. Wind direction frequencies (for each 10° sector) are used to weight the contribution to NH_3 concentrations from each trajectory. The model input data are the emission strength and height for each grid square, the land cover for each grid square, the mean wind speed and the wind direction probability for each 10° sector, the height of the atmospheric boundary layer and the NH_3 concentrations at the domain boundaries. Once all trajectories have been modelled, the mean NH_3 deposition and the mean concentration (at various heights) for each grid square are output as well as the amount of NH_3 that is exported out of the domain.

15.8.5   FIDES-2D

The FIDES model (*Flux Interpretation by Dispersion and Exchange over Short Range*) is a steady state, two-dimensional model described in Loubet et al. (2001). No chemical reactions are considered in the atmosphere. The canopy height h and the roughness length z_0 are assumed constant, and wind speed and turbulence are assumed to be horizontally uniform. The big-leaf assumption is made, which considers the canopy as a unique layer at height z_s (larger than z_0), for heat, momentum, evaporation and NH_3. To assess the concentration $C(x,z)$ and the fluxes, the model (as does the MODDAAS-2D model) is based on the general superposition principle (Thomson 1987), which relates the concentration at a location (x,z), $C_a(x,z)$, to the source strength at another location (x_s,z_s), $S(x_s,z_s)$, with the use of a dispersion function $D(x,z/x_s,z_s)$ (in s m³), which can also be seen as a source-receptor matrix. A negative source strength (in $\propto$g NH_3 m⁻¹ s⁻¹) denotes a sink. The dispersion model, derived from Huang (1979), describes the diffusion from a line source, based on the assumption that wind speed and diffusivity are power functions of height. The deposition model is the two-pathway resistance model of Fig. 15.4c.

15.8.6 *MODDAAS-2D*

The MODAAS-2D model results from the coupling of a Lagrangian Stochastic Dispersion (LS) model with a leaf scale ammonia exchange model. MODDAAS-2D is similar to FIDES-2D in that it is a steady state and two-dimensional model, with no chemistry, which is also based on the superposition principle. However, the model is different in that it is a multi-layer model which considers explicitly the transfer through the canopy. The dispersion matrix approach (which is also used in FIDES) is applied: the concentration in each grid cell is the sum of the contribution of all the sources weighed by the dispersion matrix element which has a unit of $m\ s^{-1}$. The LS model gives the dispersion matrix, using a parameterised or a given flow, following the approach of (Thomson 1987). The source terms S_i depending on the concentration in the grid cell C_i are resolved by solving the linear system corresponding to the superposition principle. The radiation inside the canopy is estimated with an exponential attenuation. The exchange model in each grid cell is that of Fig. 15.4c. The emission and deposition terms of the exchange flux are split as in FIDES in order to get a linear system for the concentration in each grid cell of the canopy. Details of the MODDAAS-2D can be found in (Loubet et al. 2006).

15.8.7 *DAMOS*

The *Danish Ammonia Modelling System* (DAMOS) is a combination of the Eulerian long-range transport model DEHM (Christensen 1997; Frohn et al. 2002) and the Gaussian local scale transport-deposition model OML-DEP (Olesen 1995) for the dry deposition making use of the surface depletion method from (Horst 1977) with a pseudo first order reaction velocity (NH_3 to NH_4^+, Asman et al. 1989). The DEHM calculations are performed for the entire Northern Hemisphere with two-way nesting; the outer domain using a $150 \times 150\,km$ resolution, for Europe a $50 \times 50\,km$ resolution is applied, and for Denmark and nearby areas using a $16.67 \times 16.67\,km$ resolution. These calculations are based on meteorological data generated by MM5 (Grell et al. 1994). The local scale model OML-DEP is applied for a $16 \times 16\,km$ domain that covers the nature area for which detailed deposition mapping is needed. DEHM background concentrations of ammonia and sulphur dioxide are obtained for each hour by interpolation between up to three grid cells upwind from the OML-DEP domain. Meteorological data are from the MM5. OML-DEP calculations are performed for 40×40 receptor points evenly distributed over the domain each representing a $400 \times 400\,m$ area. The dry deposition velocities are in both DEHM and OML-DEP performed with the same module which is based on the methodology in the EMEP model (Simpson et al. 2003). The ammonia emissions are computed using the parameterisations with high spatial and temporal resolution (Gyldenkaerne et al. 2005a; Skjoth et al. 2004). The high resolution in the inventories has shown to be very important for the model performance (see the discussion in Hertel et al. 2006).

15.8.8   Other Models

An interesting approach has been used in Switzerland where mountains make modelling quite challenging. Spatially very detailed emission inventories were made for NH_3 ($200 \times 200 \, m^2$ or less), that formed the input for the calculations. In stead of using complicated atmospheric transport models they used a function of deposition vs. distance that was developed for the Netherlands (average for 10 years, averaged over all wind directions) (Rihm and Kurz 2001). Although this should not be done in principle as the Swiss climate differs from the Dutch climate, a good correlation was obtained between modelled and measured values for 17 sites. Later Thöni et al. (2004) refined the method adjusting the function distance vs. so that an optimum correlation was obtained for this function that then should be more representative of the Swiss situation. one of the reasons why the method is a bit questionable is that the dry deposition in Western Europe can show differences of a factors 2–5 between different wind directions, mainly because the difference in frequency. Nevertheless this type of approach, or a spatially very detailed model such as DEPO1 could then be used to redistribute the dry deposition of NH_3 calculated for $50 \times 50 \, km^2$ in such a way within the grid element that a reasonable input to nature areas can be estimated.

15.8.9   Available Datasets for Validating Local Scale Deposition Models

Validation of the NH_3 recapture model is a crucial point. Although NH_3 deposition should be used for validation, it is often not measured. Most datasets only contain concentration measurements at several distances downwind from a source, which is either controlled or measured. Table 15.3 gives an overview of the known datasets that could be used for validating local recapture models.

15.9   Methods for Effect Assessment and Air Monitoring in NH_3 Hotspots

15.9.1   Key Issues for Effects Assessment and NH_3 Monitoring in Hot-Spots

Interest in the effects of NH_3 concentrations and deposition in hot-spot areas often arises when a semi-natural site (e.g. nature reserve) is affected by NH_x deposition from nearby sources. Frequently in these cases, national- or regional-scale dispersion and deposition modelling is not sufficiently detailed to assess whether vulnerable semi-natural sites are at risk or not. For example, national-scale modelling with

Table 15.3 Known datasets that could be used to validate short-range NH_x deposition models

Dataset name	Year	Brief description	References
Burrington Moor (Grassland)	1998	Slurry spread on a U shape 20 m width × 600 m long band.	Sutton et al. (1997)
		NH_3 emission measured with a 5 heights, denuder, mass balance.	Loubet et al. (2006)
		Concentration measured at 1 height down to 300 m downwind	
Bretagne (Grassland)	1999	Slurry spread on a 30 m width × 90–150 m long band.	Loubet et al. (2006)
		NH_3 emission measured with a 3 heights, bubbler, mass balance.	
		Concentration measured at 1 height down to 180 m downwind	
Davron (maize)	1997	Controlled NH_3 line source of 200 m long.	Loubet et al. (2003)
		NH_3 emission controlled with a flow-meter and measured with a 7 heights Ferm denuder mass balance.	
		Concentration measured at 7 heights within the maize canopy and above down to 160 m downwind	
Woodland	2001	Controlled NH_3 line source of 30 m long.	Theobald et al. (2001)
		NH_3 emission controlled with a flow-meter	
		Concentration measured at several heights within the forest down to 60 m downwind	
		Turbulence measured	
		Throughfall measurements	

a spatial resolution of 5 km can provide an estimate of average concentration and deposition for the grid-square containing the site but due to the spatial variability of the emission density within the grid element, the concentration and deposition at the site of interest maydiffer from the grid average. Local dispersion and deposition modelling can provide a better estimate of concentration or deposition at the site and can also take into account the location of sources relative to the site and detailed land cover and land use information. Modelled concentration and deposition data can be used to assess the potential impact on the site (e.g. by comparison with effect thresholds) and these analyses can be used by governments, local authorities or conservation organisations to determine the impact on the site in relation to conservation, human health and planning issues. The local-scale modelling is often done in conjunction with field measurements or long-term monitoring to provide an alternative method of impact assessment as well as data that can be used to verify the outputs of the modelling. The following paragraphs give an overview of existing measurement and monitoring techniques, the application of local-scale modelling and how these can be combined to estimate effects on semi-natural sites.

15.9.2   Review of Methods for Monitoring NH₃ Concentrations and Deposition in Hot Spots

NH_3 concentrations vary in time and for that reason one should at least monitor for a full year in order to get an impression of the seasonal variation. If the measurements have to be used to compare with model results, the temporal resolution should preferably be of the order of 1–2 h in order to capture the, often large, diurnal variation. Seasonal variation is to a large extent determined by agricultural activities (spreading of manure), but also by variations in temperature and wind speed. Diurnal variations are rather caused by variations in wind speed (friction velocity, atmospheric stability) and temperature. These factors affect the emission, but also the dispersion as well as the dry deposition. In this way, all processes that lead to dry deposition are co-dependent on the same meteorological factors. As the emission rate shows spatial variations, the associated concentrations and depositions do so, too. Measurements should be placed at the locations that the effects need to be determined (e.g. edge of nature reserves closest to source or an ecosystem type of interest) as well as locations that would be useful for verifying any dispersion modelling (e.g. logarithmically spaced downwind of a source). The requirement for long-term monitoring at multiple locations means that it is normally only possible to use low-cost monitoring techniques.

Two complementary sampling strategies may be used for measuring ammonia in local hot-spot areas. Firstly, low temporal resolution monitoring (e.g. monthly resolution) provides information on spatial patterns and long-term trends. By using low-cost methods for such sampling, many sites can be compared, allowing assessments of local horizontal gradients and spatial representativity. The disadvantage of such methods is that they do not allow detailed diurnal interactions to be easily assessed. Therefore, a second, complementary approach is to conduct detailed time resolved (e.g. half hourly) continuous measurements of NH_3 concentration at one or two sites and relate the measured concentrations to wind direction, local meteorology and emissions. The second approach requires expensive continuous devices that generally require mains electricity supply, which can limit the flexibility where such measurements can be made.

Among the simpler methods for measuring NH_3 concentrations are passive diffusion samplers (e.g. Thijsse et al. 1998; Tang et al. 2001), acid-coated denuders (Ferm 1979) and filter packs (Appel et al. 1988; Sutton et al. 1993a, b). Diffusion tubes rely on the process of NH_3 diffusion in a laminar flow towards the inner-coated surface of the coated denuder that is much faster than that of the relevant particles containing the same analyte (NH_4). The particles containing NH_4 will pass the tube and will precipitate on the coated filter. The ammonium can then be extracted in the lab for analysis and the mean atmospheric NH_3 concentration over the exposure period can be calculated from the mass of ammonium present (Ferm 1979). Acid-coated denuders are usually glass tubes coated on the inside with an acid. Atmospheric air is drawn through the denuder by a pump and the NH_3 in the air reacts with the acid to form an ammonium salt. This technique allows the selective removal of NH_3 on the denuders and subsequent collection of particulate NH_4^+ on a downstream filter pack (e.g. Ferm 1979; Sutton et al. 2001b; Tang et al. 2003). In a similar way to the diffusion tubes, after exposure, the ammonium is extracted and the mean atmospheric NH_3 concentration can be calculated from the mass of ammonium present and the volume of air pumped during that period. The Ferm denuder method is tuned for hourly to daily measurement periods (e.g. Ferm 1979). More recently, a modified system architecture has been used to allow simple denuders to measure monthly time-integrated samples, making them more suited to low cost, long term monitoring (Sutton et al. 2001a).

Filter packs work on a similar principle to diffusion tubes except atmospheric air is drawn through the acid-coated filter using a pump, with a pre-filter used to collect particulate compounds (Appel et al. 1988; Sutton et al. 1993a). After exposure, the filters are analysed in a similar way to those in diffusion tubes. However, phase interactions in the collected samples give rise to uncertainty to the partitioning between gaseous NH_3 and aerosol NH_4^+. Therefore, the data are typically reported as the sum of the nitrogen species (TIA, Total Inorganic Ammonium). A developing method of monitoring NH_3 concentrations is the use of bio-monitors such as lichens (Van Dobben and Ter Braak 1998; Wolseley et al. 2006). Physiological measurements on the lichens (e.g. nitrogen content) can be linked to atmospheric concentrations of NH_3 and therefore provide a method of estimating long-term atmospheric concentrations of NH_3 (see posters, this workshop). Concentration measurements can be compared with threshold concentrations (such as critical levels) to estimate the effects of toxicity to ecosystems and humans. It is also important to assess effects by estimating the deposition (wet and dry) to the site in question. It is possible to estimate dry-deposition rates from concentration measurements but the estimates are often uncertain. Methods to do this will be explored later in this section.

It is possible to measure rates of dry deposition directly using a range of techniques. The most accurate methods of dry deposition measurement are eddy-covariance (Famulari et al. 2004), continuous gradient techniques (Sutton et al. 2001b) and relaxed eddy accumulation (e.g. Fowler et al. 2001; Nemitz et al. 2001). However, these techniques tend to be the most expensive and, for the reasons given above about the need for multiple measurements over long time periods, will not be covered in this document. It is however possible to carry out low-cost gradient techniques. The COTAG (COnditional Time-Averaged Gradient) technique uses the low-cost denuders (Sutton et al. 2001b) to measure atmospheric NH_3 concentrations at several heights to derive time-integrated exchange rates (e.g. Fowler et al. 2001). The equations needed to calculate a deposition (or emission) flux are only valid for a certain range of atmospheric stability conditions and to allow use of these equations, air is only sampled during periods with these atmospheric conditions. Provided this sampling time represents the vast majority of the measurement period, the sampled atmospheric concentrations can be used to calculate the flux (Fowler et al. 2001). Throughfall measurements can also be used to estimate deposition to a plant canopy (e.g. Draiijers et al. 1994; Cape et al. 1995). Throughfall is the precipitation that is collected on the ground below the canopy and represents the sum of dry and wet deposition to the leaves minus the NH_x that has entered the leaves either via the stomatal or cuticular routes or has flowed down the plant stems/trunks (stemflow). In the AMBER experiment (Theobald et al. 2004), throughfall measurements were made beneath a tree canopy to estimate the amount of dry deposition to the trees from an artificial NH_3 line source. It should be noted that throughfall measurements will underestimate deposition of NH_x since the component that has entered the leaves and the stemflow are not included. Dry deposition to leaves and branches can be estimated by washing the branches or leaves with deionised water after a measured period following the last precipitation event. In a similar way to the throughfall measurements, the NH_x washed off the leaves will represent the dry deposition to the plant minus the NH_x that has entered the plant either via the stomatal or cuticular routes and stemflow since the last precipitation event.

Total deposition (wet plus dry) to short vegetation can also be estimated using the ammonium concentrations of pots of vegetation before and after being placed at the site of interest (Sommer and Jensen 1991; Leith et al. 2004). It is unadvisable to use bulk wet deposition measurements (i.e. wet and dry deposition to an open collector) as an estimate of total NH_x deposition, since the deposition characteristics of the collector will be different to that of vegetation the deposition rate will not be representative for the ecosystem in question.

As stated above, it is possible to relate concentration measurements to deposition estimates. Exchange of NH_3 with surface vegetation is a complex process. In some situations the exchange is bi-directional resulting in emissions from the vegetation to the atmosphere as well as deposition from the atmosphere to the vegetation. Models such as those detailed before can be used to calculate a deposition flux from a measured concentration (Nemitz et al. 2000b). However, these models require a lot of site specific input data (e.g. plant and soil physiological data, micro-meteorological data) and are often, therefore, too detailed to use for effects assessments. Deposition is the dominant exchange process close to sources of NH_3 since the atmospheric concentrations are generally higher than the compensation points of the vegetation. The deposition process can be simplified substantially by reducing the parameters down to just one; the deposition velocity. This avoids the complexity of having to parameterise many site-specific factors. In this simple scheme the deposition rate is calculated by multiplying the NH_3 concentration at the ground (or just above) by a dry deposition velocity specific to the ecosystem (Erisman et al. 1994b; Neirynck et al. 2005). This approach needs to be used with knowledge of the uncertainties since it is not a very accurate method but is useful for screening purposes to calculate an approximate deposition rate.

15.9.3 Dispersion Modelling Techniques for Effects Assessments

Dispersion modelling is often used in a effects assessments either on its own or in conjunction with monitoring. There is no substitute for accurate measurements for effects assessments but the use of modelling instead provides a faster and often cheaper way for these assessments. When done in conjunction with monitoring, the measurement data provide a method of model verification and also can be used for model improvement and development. It is important that the correct type of model is used for the task. Factors governing the type of model necessary depend on factors such as the process that you are trying to estimate (e.g. dry deposition, atmospheric concentration, advection, recapture), the input data available, the spatial/temporal scale necessary and the resources (time, equipment, money etc.) available. Once a suitable model has been selected, the input data necessary for the model must be collected (e.g. meteorological, soil, land cover, source data). The model is then run using these input data and the required output (concentrations, deposition etc.) is obtained. If monitoring data are available they can then be used to verify the model output. The output data are the used to estimate the potential impacts to the any vulnerable sites within the modelling domain. In addition to simulating real impacts, models can be used to investigate scenarios by varying the input data to look at the effects of source strengths, source locations, land cover changes, climate change etc.

15.9.4 Review of Effects Assessment of NH_3 in Hot Spots and Experiments

After the completion of the modelling and/or monitoring the data can to be used to estimate the potential effects of the NH_3 concentrations and deposition on the receptor of interest: semi-natural ecosystems or humans. Most assessments take the form of comparing concentration and deposition estimates (measured or modelled) with relevant thresholds. For effects on semi-natural ecosystems ammonia concentrations are usually compared with 'critical levels', which are defined as "the concentration of a pollutant in the atmosphere, below which vegetation is unlikely to be damaged according to present knowledge" (Posthumus 1988). It should be noted that the critical level value for ammonia ($8 \mu g$ NH_3 m^{-3}) is now considered by many scientists to be out-of-date and not sufficiently precautionary. In particular, it can be shown that exceedance of the critical loads occurs at much lower concentrations than the critical levels. (i.e. ammonia concentrations of $>1–2 \mu g$ m^{-3} are sufficient in most cases to lead to exceedance of the critical load). This indicates that the issue of most concern from ammonia is the indirect impact of nitrogen from ammonia deposition, rather than the direct toxic effect of ammonia concentrations. Deposition rates are usually compared with 'critical loads' which are defined as "*the amount of pollutant deposited below which significant harmful effects on specified elements of the environment do not occur, according to present knowledge*" (Nilsson and Grennfelt 1988). Therefore if the deposition rate is less than the critical load (i.e. it is not exceeded), then it is usually assumed that significant harm to the ecosystem will not occur. Much recent work has gone into defining critical loads for different ecosystem types. The most recent culmination of this work was the Expert Workshop of 2002 in Switzerland (Bobbink et al. 2003). This meeting agreed on many of the currently used empirical critical loads for nitrogen deposition for a range of ecosystem types. Many of these critical loads were based on data from experimental field studies.

Following the effects assessment, information on potential exceedances of critical loads and levels can be used by government departments, planning committees or conservation bodies to conclude whether an impact on a semi-natural site will probably occur as a result of the process (ammonia source) being studied.

15.10   Main Results from Modelling and Monitoring Deposition near Hot Spots

15.10.1   Example of NH₃ Deposition Modelling in a Tree-Belt Downwind from a Controlled Source

The dispersion of NH_3 was modelled through a tree belt of 10 m height, located 15 m downwind of a line source placed at 2 m above ground. Figure 15.6 shows that the highest deposition and concentration occurs with the smallest wind-speed

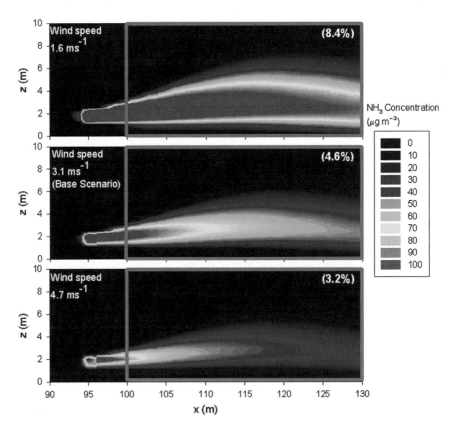

Fig. 15.6 NH_3 concentration field in a tree-belt of 30 m long and 10 m height, 5 m downwind from an intensive line source placed at 2 m above grassland for different wind speeds, as modelled with the MODDAAS-2D model

because with increasing wind speed the NH_3 is more diluted into air and also it passes under the crown of the trees, which lies roughly above 4 m. This illustrates the singularity of local deposition which depends on dilution/dispersion and deposition, as compared to deposition over homogeneous surface where deposition basically increases with decreasing aerodynamic resistance and hence increasing wind speed (see e.g., Seinfeld and Pandis 1998).

15.10.2   Sensitivity of Local Deposition of NH_x to Environmental Conditions

The modelled developed for NH_3 recapture near hot spots, have been used to assess the influence of environmental conditions on local dispersion. Asman (1998) has shown that NH_3 recapture at 2,000 m downwind from a farm can represent up to 60% of the emissions, and that it increases with (i) increasing source height, (ii) increasing atmospheric stability, (iii) decreasing wind-speed, (iv) increasing surface roughness, and (v) decreasing compensation point of the surface. Asman (1998) has also shown, based on hourly meteorological data spanning over 2 years, that most frequently NH_3 recapture varies between 10% and 40% of the emission. Loubet et al. (2006) have confirmed, based on the within canopy transfer mode MODDAAS, that the predominant factors controlling short-range deposition are turbulent mixing at the source height, which is influenced by wind-speed, atmospheric stability, surface roughness and source height, as well as the stomatal compensation point Cs, the cuticular resistance R_w, and the stomatal resistance. In these studies information was given on the accumulated deposition as a function of the distance to the source, but not on the deposition as a function of the distance to the source.

Figures 15.7, 15.8 and 15.9 illustrate the sensibility of local recapture of NH_x within 200 m downwind from hot spots to environmental conditions, based on simulations using the FIDES-2D model (Loubet et al. 2001). The environmental factors looked at are those affecting the turbulent mixing (source height, wind speed (U), roughness length (z_0), stability of the surface boundary layer), as well as those affecting the transfer rate to the surface C_s, R_w and R_s. Particle deposition is also considered to discuss the issue of NH_4^+ from farm buildings, as either recently created aerosols or dust transporting a fraction of nitrogen. The base scenario considers:

1. An area source emitting 500 μg NH_3 m^{-1} s^{-1} (roughly corresponding to 800 kg year^{-1} for a 50 × 5 m wide building) infinite in the crosswind direction (along-wind dispersion) of 5 m width, at $h_s = 2$ m height
2. A wind speed of 3 m s^{-1} at the source height
3. A roughness length (z_0) of 10 mm
4. A zero displacement height (d)
5. Thermal neutrality of the boundary layer
6. A stomatal resistance $R_s = 80$ s m^{-1}
7. A cuticular resistance $R_w = 20$ s m^{-1}, and
8. A compensation point $C_s = 0$ μg m^{-3} NH_3

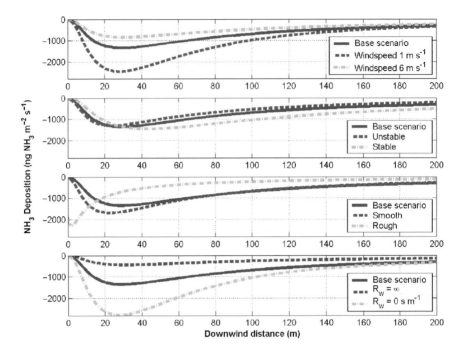

Fig. 15.7 NH_3 deposition downwind from a $500 \propto g\ NH_3\ m^{-1}\ s^{-1}$ source of 5 m width at 2 m height, as a function of downwind distance, as estimated with the FIDES-2D model. The base scenario is described in the text. Several effects are considered: wind speed at the source height (1 or 6 m s^{-1}); boundary layer thermal stratification (unstable $L = -20$ m, stable $L = 20$ m, where L is the Monin and Obukov length); surface roughness (Smooth $z_0 = 1$ mm, Rough $z_0 = 1,000$ mm); surface resistance (only stomatal absorption $R_w = \infty$, surface completely absorbing $R_c = R_b$, where R_b is the boundary layer resistance of the canopy)

The model used for particles deposition is adapted from the FIDES model: the same dispersion model is assumed (which is a strong hypothesis for particles larger than 10–20 μm), while the surface exchange scheme considers a resistance estimated following Seinfeld and Pandis (1998).

Figure 15.7, shows the qualitative effect of environmental conditions on local deposition:

- Increasing wind speed leads to decreased local deposition of NH_3, due to more dilution of the emitted NH_3, at the source and downwind.
- Surface layer stability favours local deposition of NH_3 due to decreased turbulence, and hence dilution, whereas instability decreases local deposition, due to increased turbulence. Note also that the distance of peak deposition increases in the order unstable < neutral < stable, due to the plume width increasing faster under unstable conditions.
- The rougher the surface is the larger deposition is very close toe the source, but also the faster it decreases with distance, hence in the end, rougher surface leads to smaller cumulated deposition (see Fig. 15.9). This is due to rougher surfaces

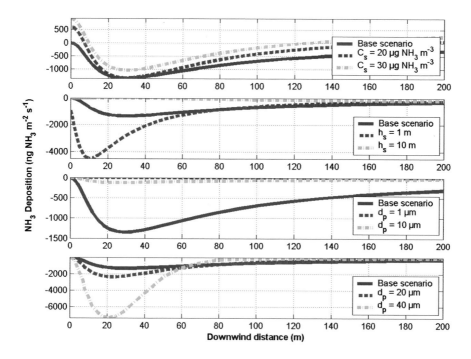

Fig. 15.8 NH$_3$ deposition downwind from a 500 µg m^{-1} s^{-1} NH$_3$ source of 5 m width at 2 m height, as a function of downwind distance, as estimated with the FIDES-2D model. The base scenario is described in the text. Several effects are considered: canopy compensation point (canopy receiving large amount of nitrogen C_s = 20 µg m^{-3} NH$_3$, canopy saturated with nitrogen C_s = 30 µg m^{-3} NH$_3$); source height (source very close to the ground hs = 1 m, source at the roof of a building hs = 10 m); particles of 1 or 10 µm and dust of 20 and 40 µm

inducing larger turbulence rate and hence faster dilution of the plume (which also reaches more quickly the ground).

- Perfectly absorbing surfaces (which is represented by R_w = 0 s m^{-1}) leads to larger local deposition than surfaces with stomatal and cuticular absorption (base scenario), which itself is lower than stomatal absorption only ($R_w = \infty$). It should be noted however local NH$_3$ deposition is not linearly related to the "deposition velocity", because increasing deposition rate also leads to a faster depletion of the plume and hence decreases deposition further downwind.

Similarly Fig. 15.8 shows that:

- An increase of the stomatal compensation point C_s decreases local deposition of NH$_3$ and can even lead to emissions of NH$_3$ at distances larger than 100 m downwind from hot spots.
- Increase in source height decreases local deposition of NH$_3$ and inversely. However, similarly to effects of increased surface deposition, increased deposition with smaller height also decreases deposition at further distance, due to increased depletion of NH$_3$ from the plume. The peak distance also increases

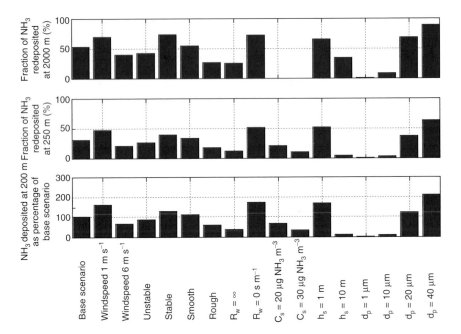

Fig. 15.9 Fraction of NH_3 re-deposited at 2,000 and 200 m downwind from a 500 μg NH_3 m⁻¹ s⁻¹ source of 5 m width at 2 m height, as a function of downwind distance, as estimated with the FIDES-2D model. The base scenario is described in the text. Several effects are considered: wind speed at the source height (1 or 6 m s⁻¹); boundary layer thermal stratification (unstable $L = -20$ m, stable $L = 20$ m, where L is the Monin and Obukov length); surface roughness (Smooth $z_0 = 1$ mm, Rough $z_0 = 1,000$ mm); surface resistance (only stomatal absorption $R_w = \infty$, surface completely absorbing $R_c = R_b$, where R_b is the boundary layer resistance of the canopy); canopy compensation point (canopy receiving large amount of nitrogen $C_s = 20$ μg m⁻³ NH_3, canopy saturated with nitrogen $C_p = 30$ μg m⁻³ NH_3); source height (source very close to the ground hs = 1 m, source at the roof of a building hs = 10 m); particles of 1 or 10 μm and dust of 20 and 40 μm. At 2,000 m, the fraction re-deposited for the scenarios $C_s = 20$ and 30 μg m⁻³ NH_3 are not shown for clarity: the cumulated emission at 2,000 m downwind amount 65% and 125% of the emission from the farm, respectively

with increasing source height, as it then takes more distance for the plume to reach the ground.

- Particles of 1 μm show almost no deposition, while 10 μm particles show deposit of about a tenth of the base scenario, whereas particles of 20 and 40 μm show larger deposit than the base scenario. It should however be noted that the FIDES-2D model for particles does not properly describes the dispersion of particles of 20–30 μm and that hence these results should be taken as qualitative.

Figure 15.9 synthesises the results of Figs. 15.7 and 15.8, by displaying the fraction of NH_3 emitted that is recaptured at 200 m from the source for each scenario. The deposition at 2,000 m is also shown to allow comparison with the work of Asman (1998). This allows to quantitatively compare the effect of each scenario. Figure 15.9 shows that up to 30% of NH_3 is recaptured at 200 m downwind from the source in the base scenario and that it varies between 12% and 55% depending on the environmental conditions (not considering particles). Apart from stability and wind speed, which are influential but can not be easily modified by abatement techniques (though wind-speed at the source height might be), the most influential parameters are (1) the source height, (2) the surface roughness, (3) the surface resistance, and (4) the stomatal compensation points. Hence, abatement techniques should preferentially focus on one of those parameters. In a simplified view, abatement techniques should try to maximise deposition very close to the source. Results of Fig. 15.9 suggests that this could be achieved by shading the emission from wind (both by setting the source at the lowest possible height, or having trees around), and by ensuring an effective sink around it (well watered canopy, with well evaporating vegetation on the ground). One effective way might be to have tall trees with small LAI and an under-storey well watered arable crop.

15.10.3 Comparison of Three Approaches to Model Recapture of NH_3 to Tree Belt Near Hotspot

Figure 15.10 shows the recapture of NH_3 to a tree belt surrounding an intensive source of $11.9 \, kg \, a^{-1}$ N for several scenarios (base scenario given in Table 15.4). The tree canopy presents a crown located in the upper part of the canopy, which allows a flow below the crown to take place. Three models are compared: the FIDES model (Loubet et al. 2001), the LADD model (Hill 1998), both models not taking account of within canopy horizontal transfer, and the MODDAAS model (Loubet et al. 2006), which take account of within canopy transfer. Since the three models do not used similar inputs, the FIDES and the LADD models were adapted so as to give the same conditions as the MODDAAS model.

- **LADD**: In LADD, the canopy height is not used and therefore a surrogate parameter was needed. The roughness length (z_0) was used since this scales approximately linearly with canopy height. Leaf area index (LAI) is also not used by LADD and instead, an adjustment to the canopy resistance was made, assuming that it is inversely proportional to LAI.
- **FIDES**: The canopy is adjacent to the source making it difficult to simulate the effect of sources at different upwind distances. The best approximation was to spread the source further upwind to simulate moving the source in that direction e.g. to simulate a source 10 m upwind of the canopy, the source region covered 0–10 m upwind of the canopy edge and the emission strength reduced accordingly. In a similar way to LADD, z_0 was used as a surrogate for canopy height and the canopy resistances (stomatal and cuticular) were adjusted to simulate different LAIs.

In Fig. 15.10 the units of the y-axes are the deposition to the canopy per unit width in the crosswind direction. The source used in the scenarios is assumed to be a line source running in the crosswind direction and therefore the deposition in these plots represents the modelled deposition from each metre of source. For example, the

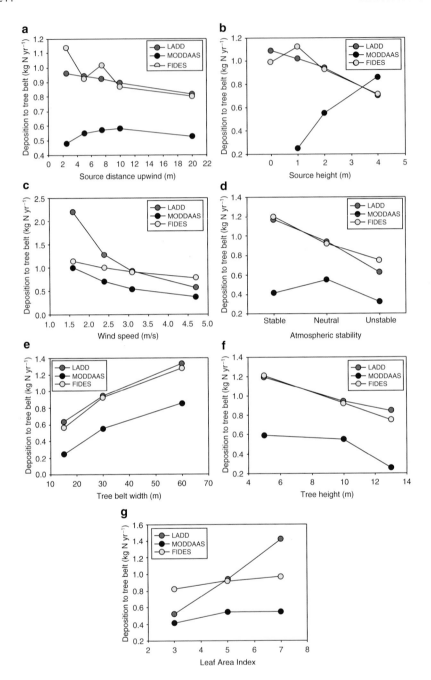

Fig. 15.10 Ammonia dry deposition to a modelled tree belt (per crosswind unit length) for seven scenarios using the FIDES, the LADD and the MODDAAS models. The units of the y-axes are deposition per m in the direction perpendicular to the wind, therefore (for a source strength of 11.9 kg a^{-1} N per unit source length) a deposition of 2 kg a^{-1} N represents a recapture of 16.8% of the emitted NH$_3$

Table 15.4 Parameters used in the base scenario of Fig. 15.10 for modelling recapture of NH_3 emitted from a farm to a surrounding tree belt

	Source height	Distance from source	Canopy length	Canopy height	Leaf Area Index	Wind speed at 2 m	Atmos. stability
	(m)	(m)	(m)	(m)		(m s^{-1})	
Base scenario	2	5	30	10	5	3.1	Neutral

NH_3 release rate is 11.9 kg a^{-1} N per unit crosswind distance. Therefore a deposition in the plot of 2 kg a^{-1} N represents a recapture of 16.8% of the emitted NH_3. In general, FIDES gives a similar deposition rate to LADD, which is greater than that from MODDAAS.

The effect of moving the source upwind (Fig. 15.10a) is to decrease the deposition in FIDES and LADD since the NH_3 is given more time to mix vertically before reaching the canopy. For MODDAAS, the effect is different with a peak in deposition when the source is 10 m upwind of the canopy. This is due to the canopy structure, since most of the recapture surfaces (leaves and branches) are above a height of about 4 m. If the source is too close to the canopy the NH_3 will pass underneath the densest part of the canopy in the MODDAAS model: an effect that cannot be simulated by the FIDES and the LADD models.

Figure 15.10b shows the effect of changing the source height. LADD gives greater recapture for lower sources and so does FIDES apart from when the source is at ground level. This is due to both a larger time for the plume to reach the ground while it is mixed, and a larger mixing efficiency of the plume due to increasing turbulent diffusivity with canopy height (and hence roughness). A ground level source was not modelled by MODDAAS but, for the heights modelled, the recapture increased with source height. This is also because of the canopy structure; the lower the source, the more NH_3 passes under the region of high canopy density in the MODDAAS model.

Increasing the wind speed (Fig. 15.10c) decreases the recapture for all three models with LADD giving the largest gradient response. This is probably because the advection wind speed used in LADD is constant with height whereas FIDES and MODDAAS use a wind speed profile that changes with height.

Both LADD and FIDES display a decrease in recapture as the atmospheric stability changes from stable through neutral to unstable (Fig. 15.10d). This is due to the increased mixing of the atmosphere dispersing the NH_3 more vertically. The response of the MODDAAS model peaks at neutral stability. The low recapture during stable conditions is probably due to the NH_3 plume staying close to the ground and therefore not reaching the densest part of the canopy.

Increasing the tree belt width (i.e. the downwind extent of the canopy), increases the recapture in all three models in a very similar way (Fig. 15.10e), whilst increasing the tree height has the effect of reducing the recapture (Fig. 15.10f). This last effect may appear unexpected and requires interpretation of the different models. A taller tree canopy increases vertical mixing over the tree belt leading to more rapid dilution of NH_3 concentrations in the LADD and FIDES models, which has the effect of reducing the deposition. In the case of the MODDAAS model, increased tree height has the effect of 'moving' the densest part of the canopy upwards and away from the NH_3 plume, which for the base run was released at 2 m height, so that the plume passed under the model trees. Caution is therefore needed interpreting the results of this sensitivity test. Using MODDAAS as the most detailed of the models, it could also be shown (a) that design of the woodland structure could increase the ammonia recapture of ammonia passing under the main canopy and (b) that recapture of a plume the height of which was matched to the height of maximum canopy density, would increase with tree height.

Changing the LAI increases the deposition modelled by FIDES only slightly whereas for the LADD model the increase is proportional to the LAI (Fig. 15.10g). This is because in FIDES it is the stomatal and cuticular resistances that are changed but in LADD it is the canopy resistance. For the MODDAAS model, increasing the LAI from 3 to 5 increases the recapture because there is a higher density of recapture surfaces but as the LAI is increased to 7, the canopy is getting too dense and the NH_3 plume is funnelled below the region of greatest density and therefore there is little change in the recapture.

The example simulations in Fig. 15.10 shows that detailed models gives better insight than simplified models into the processes involved in local recapture by tree belt. The FIDES simulations in Fig. 15.9 suggested that maximum recapture should be achieved by shading the emission from wind and ensuring an effective sink around it. Figure 15.10 confirms this statement but also gives precisions in what it means in terms of source location: the source should be located upwind of the densest part of the canopy in order for the plume not to pass below the crown of the canopy.

15.10.4   Simulation of NH_x Deposition at Landscape Scales

The question of integrating local recapture of NH_x at regional and continental scale is a major concern in order to evaluate the weight of local to diffuse deposition. However, in order to integrate from the local scale to larger scales, there is the need to consider the intermediate "landscape scale", at which the deposition to each individual hot spot can be aggregated to provide estimates of averaged NH_x deposition at larger scales. Other information on NH_x deposition can also be estimated at this scale, such as the percentage area, which exceeds nitrogen critical load. This intermediate "landscape scale" is also the scale at which the abatement techniques can be reasoned as a whole. Asman et al. (2004) have modelled both the background deposition of NH_x and NO_y with the TREND model on a $5 \times 5\,km^2$ grid (van Jaarsveld 1995), and the local NH_x deposition using the DEPO1 model on a much finer scale ($100 \times 100\,m^2$). Figure 15.11 shows the modelled sum of NH_x

and NO_y, wet and dry deposition in this area. It can be seen that the background deposition is very large in this area (larger than $12.5\,kg\ ha^{-1}\ a^{-1}$ N), but also that deposition around hotspots can reach values larger than $50\,kg\ ha^{-1}\ a^{-1}$ N over areas as large as 2 km, and even values larger than $100\,kg\ ha^{-1}\ a^{-1}$ N over smaller areas. Figure 15.11 also shows how patchy the NH_x deposition is, which illustrates the difficulty of measuring the concentration/deposition in such an area from only a few measurements.

In the UK, the LANAS (Landscape Analysis of Nitrogen and Abatement Scenarios) project carried out under the Global Nitrogen Enrichment (GaNE) programme of the NERC linked together four N flow models (farmyard, arable field, grazed field and atmospheric dispersion) to simulate the N flows and interactions

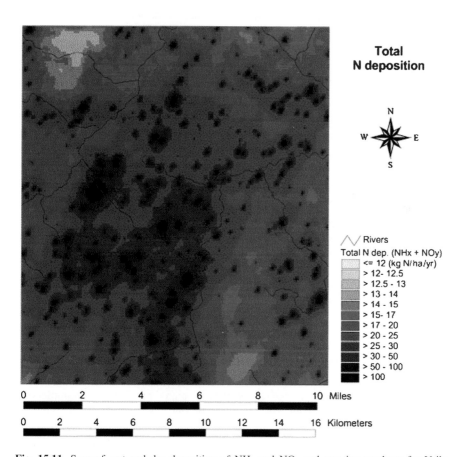

Fig. 15.11 Sum of wet and dry deposition of NH_3 and NO_x and reaction products for Vejle County (domain 270 km²) on a $100 \times 100\,m^2$ scale (kg ha⁻¹ a⁻¹ N), as modelled using the TREND model for background NH_x and NO_y deposition (all European sources) and the DEPO1 model for local NH_x deposition (From Asman et al. 2004)

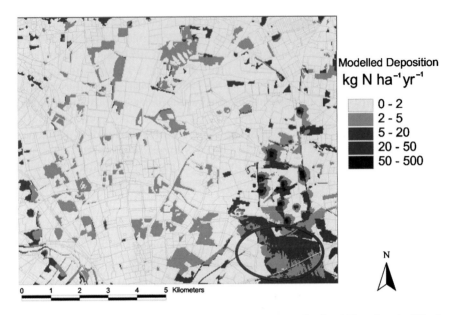

Fig. 15.12 Map of NH_3 dry deposition for an area of East Anglia (8 × 12 km, domain 96 km²), England calculated by the LANAS model at a 25 m grid resolution (From Theobald et al. 2004)

within a study area in East Anglia (Theobald et al. 2004). The study area consisted of arable land, rough grazing and intensive pig and poultry rearing. Figure 15.12 shows a map of the total N deposition flux for the study area highlighting the large predicted deposition fluxes near to the intensive pig and poultry rearing activities (in the east). These deposition rates were estimated by the LADD model. Since this model calculates NH_3 dry deposition velocities based on the land cover type (using the parameters z_0 and R_c), the deposition flux is influenced strongly by the land cover type. This is apparent in Fig. 15.12 for a heathland (highlighted by the red oval) where the deposition flux is higher than that to the surrounding land cover types (arable and grassland).

15.10.5   Main Monitoring Results

Duyzer et al. (2001) conducted a study with many passive NH_3 samplers in four agricultural areas in the Netherlands. For this type of sampler continuous comparison with standard concentration measurement methods are needed. The standard deviation for the annually averaged concentration is 1–5% at a typical concentration of 20 μg m⁻³ NH_3. The concentrations vary typically between 10 and 40 μg m⁻³ NH_3 within a few kilometres within these areas. It was shown that when eight evenly distributed samplers were placed in a 5 × 5 km² area the accuracy in the average concentration for the whole area was better than 30%. This means that it is impossible to get an idea of the average NH_3 concentration within a country from measurements at a few sites only.

Velders et al. (2002) report on a project where the NH_3 concentration was measured during 1 year with passive samplers, where one sampler was placed in each of the $15 \times 15 \, km^2$ grid elements that were covering the whole Netherlands (159 sites). This project was started to obtain better information on the concentration patterns over the country. Kriging was used as a technique to interpolate the concentration between the sites. The average annual concentration over the country was 6.6 μg m^{-3}, whereas the average concentration of the stations that were continuously measuring with AMOR monitors was 7.8 μg m^{-3} for the same period. The error in duplo measurements at one site is about 1 μg m^{-3}. The error in the interpolation for the Netherlands as a whole is about 2.4–2.8 μg m^{-3}. This means that the relative error in the concentrations is about 50–80% in the areas with low concentrations and 15% in the areas with the highest concentrations (15 μg m^{-3}). The average NH_3 concentration in the Netherlands varied from 4.0 μg m^{-3} in autumn 2000 to 9.6 μg m^{-3} in May 2001. Locally there can be large influences of point sources leading to concentrations of between 20 and 70 μg m^{-3}.

Smits et al. (2005) conducted a study in an agricultural area in the Netherlands to investigate potential discrepancies between modelled and measured NH_3 concentrations and possible uncertainties in the emissions. In this study very detailed information farms and on activities on the farms as a function of time was collected during 1 year in a $3 \times 3 \, km^2$ are, and some less detailed information for a larger surrounding area. With this information and meteorological measurements in this area a very detailed emission inventory (place, time) was made, which was used as an input to the OPS atmospheric transport and deposition model. At the same time the NH_3 was measured continuously at two sites in this area using an AMOR ammonia monitor. During the same period the average NH_3 concentration was measured during 14-day period at 50 sites within the area. Figure 15.13 shows how the measured and modelled average NH_3 concentrations for the whole area compared with the modelled concentrations. One

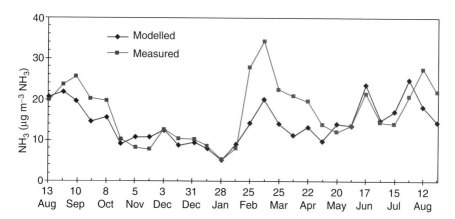

Fig. 15.13 Modelled vs. measured area averaged NH_3 concentrations for a $3 \times 3 \, km^2$ area in the Netherlands, for the period July 2002–September 2003 using the OPS model (Smits et al. 2005)

of the conclusions is that there is a large discrepancy in spring which probably could be caused by saturation of the vegetation with NH_3 (a high compensation point) and maybe also by an underestimation of emission during application of manure in periods with high solar radiation. Although the emissions are not measured and hence there is uncertainty in the emission rates it could be worthwhile for modellers to have access to this data set. This is maybe possible in the future if the farmers would agree to that, which would mean that the data would become available in an anonymous form.

15.11   Conclusions and Recommendations

Hot spots may be defined as areas with high NH_3 emission density, which are mainly farms and their surroundings, where grazing and manure application take place, as well as slurry spread fields. The emissions from hot spots depend on the number of animals in the farm, the manure handling systems, but also depend upon environmental conditions, especially temperature and wind-speed, which are important parameters to consider for modelling deposition near hot spots. Other nitrogen components may be emitted by farms, such as NO_x, N_2O, N_2, VON, or primary aerosols. Other reactive species (DMS, DMDS, H_2S), also emitted by farms may interact with NH_x deposition near hot spots.

Deposition of NH_x within 1 km from hot spots ranges from 2% to 60% of emitted NH_x, and is mainly due to dry deposition of NH_3, since wet deposition, in a temperate climate, is evaluated as less than 5% recapture of the emitted NH_x. Moreover, photochemical reactions and chemical reactions with gaseous acids are unlikely to greatly affect local dispersion and deposition of NH_x near hot spots.

Dry deposition of NH_3 near hot spots results from a combination of turbulent dispersion, stomatal absorption, and non-stomatal deposition:

- Turbulent dispersion depends upon the topography of the site, the shape of the farm buildings, the structures of the surrounding canopies, the height at which NH_x is released into the atmosphere, the wind speed and atmospheric stability. Dry deposition is sensitive to all of these parameters. The height of release, or more generally the wind speed at the source height, appears to be one of the most important parameter. However, little has been studied on the detailed turbulence surrounding the source and hence no definitive conclusions can be drawn onto which parameter is the most influential on dispersion in the case of farm buildings. In the case of emission from fields, there is less unknowns.
- Deposition due to stomatal absorption is large, and hence the "green" LAI of vegetation surrounding hot spots, as well as the vegetation water stress are major parameters influencing NH_x deposition.
- Deposition of NH_3 to external surfaces of plants is also a major influential process, which is also dependent upon the leaf area and the leaf wetness. Non-stomatal deposition can increase local NH_x deposition by a factor of 2 or 3 as compared to stomatal uptake.

A range of models of local deposition of NH_x exist, either based on Lagrangian or Eulerian dispersion models of different complexity. The NH_3 deposition schemes range from deposition velocities to multi-layer approaches, big-leaf being the most common scheme. Some of the models take account of wet and aerosol deposition, as well as chemical reactions in the gas phase. Their spatial scales range from 1 m downwind from the source to several kilometres, with models being more adapted for very short range and others more adapted to mesoscale NH_x deposition. Some of the models have been intensively validated against datasets that are available to the community. These models have proven their usefulness for studying the sensitivity of NH_x deposition to environmental conditions as well as the efficiency of abatement techniques. Some example of coupling of local scale NH_x deposition models with larger scale chemistry models and GIS emission databases also exist.

A range of methods have been developed and proved adapted for long term monitoring of NH_x in the vicinity of hot spots. These methods are advantageous if used in conjunctions with dispersion models as they can be used both for estimating deposition of NH_x near hot spots and NH_x emissions from the farm with inverse modelling. There is however no available routine method to estimate NH_x dry deposition under local advection conditions.

It should be noted that there a lack of systematic studies of NH_x deposition near hot spots at the European scale, which could tell the magnitude of NH_x recaptured near hot spots at such a scale. The scale is important in that it means a range of farm managements, ecosystems, and climates that spans over the whole Europe.

From existing studies, it appears that the main uncertainties on estimating NH_x recapture near hot spots are linked with:

- Good estimates of hotspots sources and their seasonal and daily variability
- Measurements of NH_x dry deposition near hot spots to validate the local deposition models, which were, until now, mostly validated against monitored NH_3 concentrations
- Modelling correctly nighttime dispersion and deposition of NH_x near hot spots
- Modelling the effects of farm buildings and tree belts of different shapes on the turbulent dispersion and the deposition of NH_x
- Examining the potential saturation of the cuticular sink of NH_3 under strong NH_3 load and its subsequent de-saturation after rain events
- Parametrizing the stomatal compensation point of the vegetation over the season, and its possible evolution due to deposited NH_x (and other compounds emitted from the farm)
- Getting more information on the relative emissions of other N compounds from hot spots and the potential impacts of other reactive species emissions on NH_x deposition near hot spots (sulphur compounds and acid compounds)

Acknowledgements The work described here was supported by funding from the Department for Environment, Food and Rural Affairs of the UK, the European research project NitroEurope IP, COST Action 729 and the European Science Foundation NinE programme.

References

Adams P.J., Seinfeld J.H., Koch D., Micley L., Jacob D. (2001) General circulation assessment of direct radiative forcing by the sulfate-nitrate-ammonium-water inorganic aerosol system. Journal of Geophysical Research, 106(D1), 1097–1111.

Andersen H.V., Hovmand M.F. Hummelshøj P., Jensen N.O. (1999) Measurements of ammonia concentrations, fluxes and dry deposition velocities to a spruce forest 1991–1995. Atmospheric Environment, 33, 1367–1383.

Anderson N., Strader R., Davidson C. (2003) Airborne reduced nitrogen: ammonia emissions from agriculture and other sources. Environment International, 29, 277–286.

Aneja V.P., Nelson D.R., Roelle P.A., Walker J.T., Battye W. (2003) Agricultural ammonia emissions and ammonium concentrations associated with aerosols and precipitation in the southeast United States. Journal of Geophysical Research-Atmospheres, 108(D4), 4152, doi:10.1029/2002JD002271.

Asman W.A.H. (1992) Ammonia emission in Europe: updated emission and emission variations. Report 228471008, National Institute of Public Health and Environmental Protection, Bilthoven, The Netherlands, 88 pp.

Asman W.A.H. (1995) Parameterization of below-cloud scavenging of highly soluble gases under convective conditions. Atmospheric Environment, 29, 1359–1368.

Asman W.A.H. (1998) Factors influencing local dry deposition of gases with special reference to ammonia. Atmospheric Environment, 32, 415–421.

Asman W.A.H. (2002) Die Modellierung lokaler Ammoniak-Depositionen im Umfeld von Stallgebäuden (Modelling local ammonia deposition near livestock buildings, in German). In: Emissionen der Tierhaltung. Grundlagen, Wirkungen, Minderungsmassnahmen. KTBL-Schrift 406, KTBL-Schriften-Vertrieb im Landwirtschaftsverlag GmbH, Münster, Germany, pp. 295–319.

Asman W.A.H., Janssen A.J. (1987) A long-range transport model for ammonia and ammonium for Europe. Atmospheric Environment, 21, 2099–2119.

Asman W.A.H., Maas J.F.M. (1986) Schatting van de depositie van ammoniak en ammonium in Nederland t.b.v. het beleid in het kader van de hinderwet. (Estimation of the deposition of ammonia and ammonium for environmental impact assessment in the Netherlands. In Dutch). Report R 86 8, Institute for Meteorology and Oceanography, State University Utrecht, The Netherlands, 107 pp.

Asman W.A.H., van Jaarsveld J.A. (1992) A variable-resolution transport model applied for NH_x in Europe. Atmospheric Environment, 26A, 445–464.

Asman W.A.H., Pinksterboer E.F., Maas H.F.M., Erisman J.W., Waijers-Ypelaan A., Slanina J., Horst T.W. (1989) Gradients of the ammonia concentration in a nature reserve - model results and measurements. Atmospheric Environment, 23, 2259–2265.

Asman W.A.H., Hutchings N.J., Sommer S.G., Andersen J., Münier B., Génermont S., Cellier P. (2004) Emissions of ammonia. In: Friedrich R., Reis S. (eds.) Emissions of air pollutants. Springer, Berlin, Germany, 111–143.

Aylor D.E., Yansen W., Miller D.R. (1993) Intermittent wind close to the ground within a grass canopy. Boundary-Layer Meteorology, 66, 427–448.

Ayra S.P. (1988) Introduction to micrometeorology. Academic, San Diego, CA, 307 pp.

Baldocchi D. (1988) A multi-layer model for estimating sulfur dioxide deposition to a deciduous oak forest canopy. Atmospheric Environment, 22, 869–884.

Baldocchi D.B. (1992) A Lagrangian Randomn-walk model for simulating water vapor, CO2 and sensible heat flux densities and scalar profiles over and within a soybean canopy. Boundary-Layer Meteorology, 61, 113–144.

Battye W., Aneja V.P., Roelle P.A. (2003) Evaluation and improvement of ammonia emissions inventories. Atmospheric Environment, 37, 3873–3883.

Bjerg B., Kai P., Morsing S., Takai H. (2004) CFD analysis to predict close range spreading of ventilation air from livestock buildings. Agricultural Engineering International. Manuscript BC 03 014. Vol VI, August 2004.

Bobbink R., Ashmore M., Braun S., Flückiger W., Van den Wyngaert I.J.J. (2003) empirical nitrogen critical loads for natural and semi-natural ecosystems: 2002 update. Empirical Critical Loads for Nitrogen – Expert workshop. SAEFL, Berne.

Bouvet T., Wilson J.D., Tuzet A. (2006) Observation and modelling of heavy particle dispersion in a windbreak flow. Journal of Applied Meteorology and Climatology, 45, 1332–1349.

Bouwman A.F., Lee D.S., Asman W.A.H., Dentener F.J., van der Hoek K.W. (1997) A global high-resolution emission inventory for ammonia. Global Biogeochemical Cycle, 11, 561–587.

Burkhardt J., Eiden R. (1994) Thin water films on coniferous needles. Atmospheric Environment, 28, 2001–2017.

Bussink D.W., Oenema O. (1998) Ammonia volatilization from dairy farming systems in temperate areas: a review. Nutrient Cycling in Agroecosystems, 51, 1352–2310.

Cape J.N., Sheppard L.J. et al. (1995) Throughfall deposition of ammonium and sulphate during ammonia fumigation of a Scots pine forest. Water Air and Soil Pollution, 85(4), 2247–2252.

Chavez C., Coufal C.D., Carey J.B., Lacey R.E., Beier R.C., Zahn J.A. (2004) The impact of supplemental dietary methionine sources on volatile compound concentrations in broiler excreta. Poultry Science, 83, 901–910.

Christensen J.H. (1997) The Danish Eulerian hemispheric model - A three-dimensional air pollution model used for the Arctic. Atmospheric Environment, 31, 4169–4191.

Chrysikopoulos C.V., Lynn M.H., Roberts P.V. (1992) A three dimensional steady-state atmospheric dispersion-deposition model for emissions from a ground-level area source, Atmospheric Environment, 26, 747–757.

Dämmgen U., Erisman J.W. (2005) Emission, transmission, deposition and environmental effects of ammonia from agricultural sources. In: Kuczynski T., Dämmgen U., Webb J., Myczko A. (eds.) Emissions from European agriculture. Wageningen Academic Publishers, Wageningen, The Netherlands, pp. 97–112.

Dämmgen U., Hutchings N.J. (2008) Emissions of gaseous nitrogen species from manure management - a new approach. Environmental Pollution, 154, 488–497.

Dämmgen U., Grünhage L., Jäger H.-J. (1997) The description, assessment and meaning of vertical fluxes of matter within ecotopes - a systematic consideration. Environmental Pollution, 96, 249–260.

Dämmgen U., Erisman J.W., Cape J.N., Grünhage L., Fowler D. (2005) Practical considerations for addressing uncertainties in monitoring bulk deposition. Environmental Pollution, 134, 535–548.

De Jong J.J.M., Klaassen W. (1997) Simulated dry deposition of nitric acid near forest edges. Atmospheric Environment, 31, 3681–3691.

Denmead O.T., Freney J.R., Simpson J.R. (1976) A closed ammonia cycle within a plant canopy. Soil Biology & Biochemistry 8, 161–164.

Dentener F.J., Crutzen P.J. (1994) A three-dimensional model of the global ammonia cycle. Journal of Atmospheric Chemistry, 19, 331–369.

Diau E.W.G., Tso T.L., Lee Y.P (1990) Kinetics of the reaction OH + NH$_3$ in the range 273–433 K. Journal of Physical Chemistry, 94, 5261–5265.

Dlugi R., Kins L., Kohler D., Kohler E., Reusswig K., Reuder J., Roider G., Ruoss K., Schween J., Zelger M. (1997) Studies on deposition, emission and chemical transformation above and within tall vegetation. In Slanina P.D.S (ed.) Biosphere-atmosphere exchange of pollutants and trace substances: experimental and theoretical studies of biogenic emissions and of pollutant deposition. Springer, Berlin, pp. 433–437.

Döhler H., Dämmgen U., Berg, W., Bergschmidt A., Brunsch R., Eurich-Menden B., Lüttich M., Osterburg B. (2002) Adaptation of the German emission calculation methodology to international guidelines, determination and forecasting of ammonia emissions from German agriculture, and scenarios for reducing them by 2010 (in German, summary in English), Texte 05/02 Umweltbundesamt (Berlin).

Draaijers G.P.J., Van Ek R., Bleuten W. (1994) Atmospheric deposition in complex forest landscapes. Boundary-Layer Meteorology, 69, 343–366.

Dragosits U., Sutton M.A., Place C.J., Bayley A.A. (1998) Modelling the spatial distribution of agricultural ammonia emissions in the UK. Environmental Pollution Nitrogen, the Confer-N-s First International Nitrogen Conference 1998, 102, 195–203.

Dragosits U., Theobald M.R., Place C.J., Lord E., Webb J., Hill J., ApSimon H.M., Sutton M.A. (2002) Ammonia emission, deposition and impact assessment at the field scale: a case study of sub-grid spatial variability. Environmental Pollution, 117, 147–158.

Durenkamp M., De Kok L.J. (2002) The impact of atmospheric H$_2$S on growth and sulfur metabolism of *Allium cepa* L. Phyton (Horn), 42(3): 55–63.

Duyzer J.H. (1994) Dry deposition of ammonia and ammonium aerosols over heathland. Journal of Geophysical Research, 99, 18757–1873.

Duyzer J.H., Verhagen H.L.M., Weststrate J.H. (1992) Measurement of the dry deposition flux of NH$_3$ on to coniferous forest. Environmental Pollution, 75(1), 3–13.

Duyzer J.H., Verhagen H.L.M., Weststrate J.H., Bosveld F.C., Vermetten A.W.M. (1994) The dry deposition of ammonia onto a Douglas fir forest in the Netherlands. Atmospheric Environment, 28, 1241–1253.

Duyzer J.H., Nijenhuis B., Westrate H. (2001) Monitoring and modelling of ammonia concentrations and deposition in agricultural areas of the Netherlands. Water Air and Soil Pollution: Focus, 1, 131–144.

EEA (2003) Europe's environment - the third assessment. The third assessment report. European Environment Agency, Copenhagen, Denmark, 341 p.

Ellermann T., Andersen H.V., Bossi R., Brandt J., Christensen J., Frohn L.M., Geels C., Kemp K., Løfstrøm P., Mogensen B., Monies C. (2006) Atmospheric deposition. NOVANA (In Danish: Atmosfærisk deposition. NOVANA). National Environmental Research Institute. Technical Report, 595, pp. 1–66.

Eriksson E. (1952) Composition of atmospheric precipitation. A. Nitrogen compounds. Tellus, 4, 215–232; 296–303.

Erisman J.W., Wyers P. (1993) Continuous measurements of surface exchange of SO$_2$ and NH$_3$: Inplications for their possible interactions process. Atmospheric Environment, 27, 1937–1949.

Erisman J.W., Vermetten A.W.M., Asman W.A.H., Waijers-Ypelaan A., Slanina J. (1988) Vertical distribution of gases and aerosols: the behaviour of ammonia and related components in the lower atmosphere. Atmospheric Environment, 22, 1153–1161.

Erisman J.W., van Elzakker B.G., Mennen M.G., Hogenkamp J., Zwart E., van den Beld L., Römer F.G., Bobbink R., Heil G., Raessen M., Duyzer J.H., Verhage H., Wyers G.P., Otjes R.P., Möls J.J. (1994a) The Elspeetsche Veld experiment on surface exchange of trace gases: summary of results. Atmospheric Environment, 28, 487–496.

Erisman J.W., van Pul A., Wyers P. (1994b) Parameterization of surface resistance for the quantification of atmospheric deposition of acidifying pollutants and ozone. Atmospheric Environment, 28, 2595–2607.

Erisman J.W., Vermeulen A., Hensen A., Fléchard C., Dämmgen U., Fowler D., Sutton M., Grunhage L., Tuovinen J.P. (2005) Monitoring and modelling of biosphere/atmosphere exchange of gases and aerosols in Europe. Environmental Pollution, 133, 403–413.

Famulari D., Fowler D., Hargreaves K., Milford C., Nemitz E., Sutton M.A., Weston K. (2004) Measuring eddy covariance fluxes of ammonia using tunable diode laser absorption spectroscopy. Water Air and Soil Pollution: Focus, 4, 151–158.

Fangmeier A., Hadwiger-Fangmeier A., van der Eerden L., Jäger H.-J. (1994) Effects of atmospheric ammonia on vegetation – a review. Environmental Pollution, 86, 43–82.

Farquhar G.D., Firth P.M., Wetselaar R., Weir B. (1980) On the gaseous exchange of ammonia between leaves and the environment: determination of the ammonia compensation point. Plant Physiology, 66, 710–714.

Ferm M. (1979) Method for determination of atmospheric ammonia. Atmospheric Environment, 13, 1385–1393.

Finlayson-Pitts B.J., Pitts J.N. (1986) Atmospheric chemistry. Fundamentals and experimental techniques. Wiley, New York, 1098 pp.

Fléchard C.R., Fowler D. (1998) Atmospheric ammonia at a moorland site. II: Long-term surface-atmosphere micrometeorological flux measurements. Quarterly Journal of the Royal Meteorological Society, 124, 759–791.

Fléchard C.R., Fowler D., Sutton M.A., Cape J.N. (1999) A dynamic chemical model of bi-directional ammonia exchange between semi-natural vegetation and the atmosphere. Quarterly Journal of the Royal Meteorological Society, 125, 2611–2641.

Flesch T.K., Wilson J.D., Harper L.A., Crenna B.P. (2005) Estimating gas emissions from a farm with an inverse-dispersion technique. Atmospheric Environment, 39, 4863–4874.

Fournier N., Pais V.A., Sutton M.A., Weston K.J., Dragosits U., Tang Y.S., Aherne J. (2002) Parallelisation and application of a multi-layer atmospheric transport model to quantify dispersion and deposition of ammonia over the British Isles. Environmental Pollution, 116, 95–107.

Fowler D., Pitcairn C.E.R., Sutton M.A., Fléchard C., Loubet B., Coyle M., Munro R. C. (1998) The mass budget of atmospheric ammonia in woodland within 1 km of livestock buildings. Environmental Pollution, 102.

Fowler D., Coyle M., Fléchard C., Hargreaves K.J., Nemitz E., Storeton-West R., Sutton M.A., Erisman J.W. (2001) Advances in micrometeorological methods for the measurement and interpretation of gas and particle nitrogen fluxes. Plant and Soil, 228(1), 117–129.

Frohn L.M., Christensen J.H., Brandt J. (2002) Development of a high-resolution nested air pollution model - the numerical approach. Journal of Computational Physics, 179, 68–94.

Fukumoto Y., Osada T., Hanajima D., Haga K. (2003) Patterns and quantities of NH_3, N_2O and CH_4 emissions during swine manure composting without forced aeration - effect of compost pile scale. Bioresource Technology, 89, 109–114.

Gash J.H.C. (1985) A note on estimating the effect of a limited fetch on micrometeorological evaporation measurements, Boundary-Layer Meteorology, 35, 409–413.

Génermont S., Cellier P. (1997) A mechanistic model for estimating ammonia volatilization from slurry applied to bare soil. Agricultural and Forest Meteorology, 88, 145–167.

Gilliland A.B., Dennis R.L., Roselle S.J., Pierce T.E. (2003) Seasonal NH_3 emission estimates for the eastern United States based on ammonium wet concentrations and an inverse modeling method. Journal of Geophysical Research-Atmospheres, 108(D15), Art. No. 4477 AUG 13 2003.

Grell G.A., Dudhia J., Stauffer D.R.A. (1994) Description of the fifth-generation Penn State/NCAR mesoscale model (MM5). [NCAR/TN-398+STR], −117. NCAR Technical Note.

Gyldenkærne S., Skjøth C.A., Christensen J., Ellermann T., Frohn L.M., Brandt J., Hertel O. (2005a) A high resolution ammonia emission inventory for regional scale air pollution models. Journal of Geophysical Research, 110, D07108, doi:10.1029/2004JD005459.

Gyldenkaerne S., Skjoth C.A., Hertel O., Ellermann T. (2005b) A dynamical ammonia emission parameterization for use in air pollution models. Journal of Geophysical Research-Atmospheres, 110(D7), Art. No. D07108 APR 13 2005.

Harper L.A., Denmead O.T., Sharpe R.R. (2000) Identifying sources and sinks of scalars in a corn canopy with inverse lagrangian dispersion analysis. II. Ammonia. Agricultural and Forest Meteorology, 104, 75–83.

Harrison R.M., Kitto A.-M.N. (1992) Estimation of the rate constant for the reaction of acid sulphate aerosol with NH_3 gas from atmospheric measurements. Journal of Atmospheric Chemistry, 15, 133–143.

Hensen A., Loubet B., Mosquera J., van den Bulk W.C.M., Erisman J.W., Daemmgen U., Milford C., Löpmeier F.J., Cellier P., Sutton M.A. (2008a) Estimation of NH_3 emissions from a naturally ventilated livestock farm using local scale atmospheric dispersion modelling. Biogeoscience Discussions (under review).

Hensen A., Nemitz E., Flynn M., Blatter A., Jones S., Sørensen L.L., Hensen B., Pryor S., Jensen B., Otjes R.P., Cobussen J., Loubet B., Erisman J.W., Gallagher M.W., Neftel A., Sutton M.A. (2008b) Inter-comparison of ammonia fluxes obtained using the Relaxed Eddy Accumulation technique. Biogeoscience Discussions (under review).

Hertel O. (1995) Transformation and deposition of sulphur and nitrogen compounds in the marine boundary layer. D.Sc.. thesis, University of Bergen, 1-10-1995. Roskilde, Denmark, National Environmental Research Institute, Ph.D. thesis reports, 215 pp.

Hertel O., Skjøth C.A., Lofstrøm P., Geels C., Frohn L.M., Ellermann T., Madsen P. V. (2006) Modelling nitrogen deposition on a local scale - a review of the current state of the art. Environmental Chemistry, 3, 317–337.

Hesterberg R., Blatter A., Fahrni M., Rosset M., Neftel A., Eugster W., Wanner H. (1996) Deposition of nitrogen-containing compounds to an extensively mangaged grassland in central Switzerland. Environmental Pollution, 91, 21–34.

Hill J. (1998) Applications of computational modelling to ammonia dispersion from agricultural sources. Ph.D. thesis, Imperial College, Centre for Environmental Technology, University of London.

Hill P.W., Raven J.A. Loubet B., Fowler D., Sutton M.A. (2001) Comparison of gas exchange and bioassay determinations of the ammonia compensation point in Luzula sylvatica (Huds.) Gaud. Plant Physiology, 125, 476–487.

Holtan-Hartwig L., Bøckman O.C. (1994) Ammonia exchange between crops and air. Norwegian Agricultural Science (Suppl. 14).

Horst T.W. (1977) Surface depletion model for deposition from a gaussian plume. Atmospheric Environment, 11, 41–46.

Horst T.W., Slinn W.G.N. (1984) Estimates of pollution profiles above finite area sources. Atmospheric Environment, 18, 1339–1346.

Houghton J.T., Ding Y., Griggs D.J., Noguer M., van der Linden P.J., Dai X., Maskell K., Johnson C.A. (eds.) (2001) Climate change 2001: the scientific basis. Contribution of Working Group I to the Third Assessment Report of the Intergovernmental Panel on Climate Change (IPCC). Cambridge University Press.

Hov O., Hjollo B.A., Eliassen A. (1994) Transport distance of ammonia and ammonium in Northern Europe.1. Model description. Journal of Geophysical Research-Atmospheres, 99, 18735–18748.

Huang C.H. (1979) A theory of dispersion in turbulent shear flow. Atmospheric Environment, 13, 453–463.

Huijsmans J.F.M., Hol J.M.G., Vermeulen G.D. (2003) Effect of application method, manure characteristics, weather and field conditions on ammonia volatilization from manure applied to arable land. Atmospheric Environment, 37, 3669–3680.

Husted S., Schjoerring J.K. (1996) Ammonia fluxes between oilseed rape plants and the atmosphere in response to changes in leaf temperature, light intensity and relative air humidity. Interactions with stomatal conductance and apoplastic NH_4^+ and H^+ concentrations. Plant Physiology, 112, 67–74.

Husted S., Schjoerring J.K., Nielsen K.H., Nemitz E., Sutton M.A. (2000) Stomatal compensation points for ammonia in oilseed rape plants under field conditions. Agricultural and Forest Meteorology, 105, 371–383.

Hutchings N.J., Sommer S.G., Andersen J.M., Asman W.A.H. (2001a) A detailed ammonia emission inventory for Denmark. Atmospheric Environment, 35, 1959–1968.

Hutchings N.J., Sommer S.G., Andersen J.M., Asman W.A.H. (2001b) Modelling the Danish ammonia emission. Atmospheric Environment, 35, 1959–1968.

Irvine M.R., Gardiner B.A., Hill M.K. (1997) The evolution of turbulence across a forest edge. Boundary-Layer Metheorology, 84, 491–496.

Jarvis S.C., Pain B.F. (1990) Ammonia volatilisation from agricultural land. Proceedings of the Fertiliser Society 298, 35. The Fertiliser Society, Peterborough.

Ketzel M., Louka P., Sahm P., Guilloteau E., Sini J.F., Moussiopoulos N. (2002) Intercomparison of numerical urban dispersion models; part II: Street Canyon in Hannover, Germany. Water Air and Soil Pollution: Focus, 2, 603–613.

Klaassen W. (1991) Average fluxes from heterogeneous vegetated regions. Boundary-Layer Meteorology, 58, 329–354.

Krupa S.V. (2003) Effects of atmospheric ammonia (NH_3) on terrestrial vegetation: a review. Environmental Pollution, 124, 179–221.

Langford A.O., Fehsenfeld F.C. (1992) Natural vegetation as a source or sink for atmopsheric ammonia: a case study. Science, 255, 581–583.

Leith I.D., van Dijk N., Pitcairn C.E.R., Sheppard L.J., Sutton M.A. (2004) Bioindicator methods for nitrogen based on transplantation: standardised model plants. Chapter 10, In: Sutton M.A., Pitcairn C.E.R., Whitfield C.P. (eds.) Bioindicator and biomonitoring methods for assessing the effects of atmospheric nitrogen on statutory nature conservation sites. Final report to JNCC for contract F90-01-535. CEH, Edinburgh, pp. 107–112.

Lim T.T., Heber A.J., Ni J.Q., Sutton A.L., Shao P. (2003) Odor and gas release from anaerobic treatment lagoons for swine manure. Journal of Environmental Quality, 32, 406–416.

Lin J-.S., Hildemann L.M. (1997) A generalized mathematical scheme to analytically solve the atmospheric diffusion equation with dry deposition. Atmospheric Environment, 31, 59–71.

Loubet B., Cellier P. (2001) Experimental assessment of atmospheric ammonia dispersion and short range dry deposition in a maize canopy. Water Air and Soil Pollution: Focus, 1, 157–166.

Loubet B., Milford C., Sutton M.A., Cellier P. (2001) Investigation of the interaction between sources and sinks of atmospheric ammonia in an upland landscape using a simplified dispersion-exchange model. Journal of Geophysical Research-Atmospheres, 106, 24183–24195.

Loubet B., Milford C., Hill P.W. Tang Y.S., Cellier P., Sutton M.A. (2002) Seasonal variability of apoplastic NH_4^+ and pH in an intensively managed grassland. Plant and Soil, 238, 97–110.

Loubet B., Cellier P., Génermont S., Laville P., Flura D. (2003) Measurement of short-range dispersion and deposition of ammonia over a maize canopy. Agricultural and Forest Meteorology, 114, 175–196.

Loubet B., Cellier P., Milford C., Sutton M.A. (2006) A coupled dispersion and exchange model for short-range dry deposition of atmospheric ammonia. Quarterly Journal of the Royal Meteorological Society, 132, 1733–1763.

McCulloch R.B., Stephen F.G., Murray J., George C., Aneja V.P. (1998) Analysis of ammonia, ammonium aerosols and acid gases in the atmosphere at a commercial hog farm in eastern North Carolina, USA. Environmental Pollution, 102(S1), 263–268.

Melillo J.M., Steudler P.A., Aber J.D., Bowden R.D. (1989) Atmospheric deposition and nutrient cycling. In: Andreae M.O., Schimel D.S. (eds.) Exchange of trace gases between terrestrial ecosystems and the atmosphere. Wiley-Interscience, New York, pp. 263–280.

Milford C., Theobald M.R., Nemitz E., Sutton M.A. (2001a) Dynamics of ammonia exchange in response to cutting and fertilising in an intensively-managed grassland. Water Air and Soil Pollution: Focus, 1(5–6), 167–176.

Milford C., Hargreaves K.J., Sutton M.A, Loubet B., Cellier P. (2001b) Fluxes of NH_3 and CO_2 over upland moorland in the vicinity of agricultural land. Journal of Geophysical Research-Atmospheres, 106, 24169–24181.

Möller D. (2003) Luft. Chemie, Physik, Biologie, Reinhaltung, Recht. de Gruyter, Berlin, 750 pp.

Monteith J.L., Unsworth M.H. (1990) Principles of environmental physics, 2nd edition. Arnold, London, 291 pp.

Neirynck J., Kowalski A.S. et al. (2005) Driving forces for ammonia fluxes over mixed forest subjected to high deposition loads. Atmospheric Environment, 39(28), 5013–5024.

Nemitz E., Sutton M.A. (2004) Gas-particle interactions above a Dutch heathland: III. Modelling the influence of the NH_3-HNO_3-NH_4BO_3 equilibrium on size-segregated particle fluxes. Atmospheric Chemistry and Physics, 4, 1025–1045.

Nemitz E., Sutton M.A., Gut A., San José R., Husted S., Schjoerring J.K. (2000a) Sources and sinks of ammonia within an oilseed rape canopy. Agricultural and Forest Meteorology (Ammonia Exchange Special Issue), 105(4), 385–404.

Nemitz E., Sutton M.A., Schjoerring J.K., Husted S., Wyers G.P. (2000b) Resistance modelling of ammonia exchange over oilseed rape. Agricultural and Forest Meteorology (Ammonia Exchange Special Issue), 105(4), 405–425.

Nemitz E., Milford C., Sutton M.A. (2001) A two-layer canopy compensation point model for describing bi-directional biosphere-atmosphere exchange of ammonia. Quarterly Journal of the Royal Meteorological Society, 127, 815–833.

Nemitz E., Gallagher M.W., Duyzer J.H., Folwer D. (2002) Micrometeorological measurements of particle deposition velocities to moorland vegetation. Quarterly Journal of the Royal Meteorological Society, 128(585), 2281–2300.

Nemitz E., Sutton M.A., Wyers G.P., Jongejan P.A.C. (2004a) Gas-particle conversions above a Dutch heathland: I. Surface exchange fluxes of NH_3, SO_2, HNO_3 and HCl. Atmospheric Chemistry and Physics, 4, 989–1005.

Nemitz E., Sutton M.A., Wyers G.P., Mennen M.G., van Putten E.M., Gallagher M.W. (2004b) Gas-particle conversions above a Dutch heathland: II. Concentrations and surface exchange fluxes of atmospheric particles. Atmospheric Chemistry and Physics, 4, 1007–1024.

Nenes A., Pandis S.N., Pilinis C. (1998) ISORROPIA: a new thermodynamic equilibrium model for multiphase multicomponent inorganic aerosols. Aquatic Geochemistry, 4, 123–152.

Olendrzynski K., Jonson J.E., Bartnicki J., Jakobsen H.A., Berge E. (2000) EMEP Eulerian model for acid deposition over Europe. International Journal of Environment and Pollution, 14, 391–399.

Olesen H.R. (1995) Regulatory dispersion modeling in Denmark. International Journal of Environment and Pollution, 5, 412–417.

Olesen J.E., Sommer S.G. (1993) Modelling effects of wind-speed and surface cover on ammonia volatilization from stored pig slurry. Atmospheric Environment, 27, 2567–2574.

Oudendag D.A., Luesink H.H. (1998) The manure model: manure, minerals (N, P and K), ammonia emission, heavy metals and the use of fertiliser in Dutch agriculture. Environmental Pollution, 102(S1), 241–246.

Pain B.F., van der Weerden T.J., Chambers B.J., Phillips V.R., Jarvis S.C. (1998) A new inventory for ammonia emissions from UK agriculture. Atmospheric Environment, 32, 309–313.

Park S.U., Lee Y.H. (2002) Spatial distribution of wet deposition of nitrogen in South Korea. Atmospheric Environment, 36, 619–628.

Pasquill F., Smith F.B. (1983) Atmospheric diffusion, 3rd edition. ISBN 0-85312-426-4 and ISBN 0-470-27404-2.

Personne E., Loubet B., Cellier C., Sutton M.A. et al. (2008) A two-layer resistance model for the exchange of heat and ammonia over a grassland: evaluation during a cut and a fertilisation. Biogeoscience Discussions (under review).

Philip J.R. (1959) The theory of local advection: 1. Journal of Meteorology, 16, 535–547.

Pinder R.W., Pekney N.J., Davidson C.I., Adams P.J. (2004) A process-based model of ammonia emissions from dairy cows: improved temporal and spatial resolution. Atmospheric Environment, 38, 1357–1365.

Pinder R.W., Adams P.J., Pandis S.N., Gilliland A.B. (2006) Temporally resolved ammonia emission inventories: current estimates, evaluation tools, and measurement needs. Journal of Geophysical Research-Atmospheres, 111, Art. No. D16310 AUG 25 2006.

Pitcairn C.E.R., Leith I.D., Sheppard L.J., Sutton M.A., Fowler D., Munro R.C., Tang Y.S., Wilson D. (1998) The relationship between nitrogen deposition, species composition and foliar nitrogen concentrations in woodland flora in the vicinity of livestock farms. Environmental Pollution, 102(S1), 41–48.

Pitcairn C.E.R., Skiba U.M., Sutton M.A., Fowler D., Munro R., Kennedy V. (2002) Defining the spatial impacts of poultry farm ammonia emissions on species composition of adjacent woodland groundflora using Ellenberg Nitrogen Index, nitrous oxide and nitric oxide emissions and foliar nitrogen as marker variables. Environmental Pollution, 119.

Pitcairn C.E.R, Leith I.D., Shepperd L.J., van Dijk N., Tang Y.S., Wolseley P., James P., Sutton M.A. (2004) Field inter-comparison of different bio-indicator methods to assess the impacts of nitrogen deposition. Annex 1, In: Sutton M.A., Pitcairn C.E.R., Whitfield C.P. (eds.) Bioindicator and biomonitoring methods for assessing the effects of atmospheric nitrogen on statutory nature conservation sites. Final report to JNCC for contract F90-01-535. CEH, Edinburgh, pp. 142–180.

Poggi D., Katul G.G., Albertson J.D. (2004) Momentum transfer and turbulent kinetic energy budgets within a dense model canopy. Boundary-Layer Meteorology, 111, 589–614.

Posthumus A.C. (1988) Critical levels for effects of ammonia and ammonium. In UNECE Proceedings of the Bad Harzburg Workshop. UBA, Berlin, pp. 117–127.

Raupach M.R. (1989) Stand overstorey processes. Philosophical Transactions of the Royal Society of London Series B, 324, 175–190.

Riedo M., Milford C., Schmidt M., Sutton M.A. (2002) Coupling soil-plant-atmosphere exchange of ammonia with ecosystem functioning in grasslands. Ecological Modelling, 158, 83–110.

Rihm B., Kurz D. (2001) Deposition and critical loads of nitrogen in Switzerland. Water Air and Soil Pollution, 130, 1223–1228.

Rodean H.C. (1996) Stochastic lagrangian models of turbulent diffusion. Meterological monographs, Vol. 26, 83. American Meteorological Society, Boston, MA.

Roelofs J.G.M., Kempers A.J., Houdijk A.L.F.M., Jansen J. (1985) The effect of air-borne ammonium on Pinus nigra var. maritima in the Netherlands. Plant Soil, 84, 45–56.

Rosnoblet J., Theobald M., Génermont S., Gabrielle B., Cellier P. (2007) How the use of a mechanistic model of ammonia volatilisation in the field may improve national ammonia volatilisation inventories. In: Monteney G.J. (ed.) Ammonia emissions in agriculture. Wageningen Academic Publishers, Wageningen, The Netherlands.

Saltelli, A. and Hjorth, J., 1995. Uncertainty and Sensitivity Analyses of Oh-Initiated Dimenthyl Sulfide (Dms) Oxidation-Kinetics. Journal of Atmospheric Chemistry, 21(3), 187–221.

Sawford B.L., Guest F.M. (1991) Lagrangian statistical simulation of the turbulent motion of heavy particles. Boundary-Layer Meteorology, 54, 147–166.

Schade G.W., Crutzen P.J. (1995) Emission of aliphatic amines from animal husbandry and their reactions: potential source of N_2O and HCN. Journal of Atmospheric Chemistry, 22, 319–346.

Schjoerring J.K. (1991) Ammonia emission from the foliage of growing plants. In: Sharkey T.D., Holland E.A., Mooney H.A. (eds.) Trace gas emissions by plants, pp. 267–292.

Schjoerring J.K. (1997) Plant-atmosphere ammonia exchange. Quantification, Physiollogy regulation and interaction with environmental factors. D.Sc. thesis, Royal Veterinary and Agricultural University, Copenhagen, Denmark, 55 pp.

Schjoerring J.K., Mattsson M. (2001) Quantification of ammonia exchange between agricultural cropland and the atmosphere: measurements over two complete growth cycles of oilseed rape, wheat, barley and pea. Plant Soil, 228, 105–115.

Schjoerring J.K., Husted S., Mattsson M. (1998) Physiological parameters controlling plant-atmosophere ammonia exchange. Atmospheric Environment (Ammonia Special Issue), 32(3), 491–498.

Schjoerring J.K., Husted S., Mäck G., Nielsen K.H., Finnemann J., Mattsson M. (2000) Physiological regulation of plant-atmosphere ammonia exchange. Plant Soil, 221, 95–102.

Seedorf J., Hartung J., Schroder M., Linkert K.H., Pedersen S., Takai H., Johnsen J.O., Metz J.H.M., Groot Koerkamp P.W.G., Uenk G.H. (1998a) A survey of ventilation rates in livestock buildings in Northern Europe. Journal of Agricultural Engineering Research, 70, 39–47.

Seedorf J., Hartung J., Schroder M., Linkert K.H., Pedersen S., Takai H., Johnsen J.O., Metz J.H.M., Groot Koerkamp P.W.G., Uenk G.H. (1998b) Temperature and moisture conditions in livestock buildings in Northern Europe. Journal of Agricultural Engineering Research, 70, 49–57.

Seinfeld J.H., Pandis S.N. (1998) Atmospheric chemistry and physics. From air pollution to climate change. Wiley, New York.

Shon Z.-H, Kim K.-H., Swan H., Lee G., Kim Y.-K. (2005) DMS photochemistry during the Asian dust-storm period in the Spring 2001: model simulations vs. field observations. Chemosphere, 58, 149–161.

Shuttleworth W.J., Wallace J.S. (1985) Evaporation from sparse crops – an energy combination theory. Quarterly Journal of the Royal Meteorological Society, 111, 839–855.

Singles R.J., Sutton M.A., Weston K.J. (1998) A multi-layer model to describe the atmospheric transport and deposition of ammonia in Great Britain. Atmospheric Environment, 32(3), 393–399.

Skjøth C.A., Hertel O., Gyldenkaerne S., Ellermann T. (2004) Implementing a dynamical ammonia emission parameterization in the large-scale air pollution model ACDEP. Journal of Geophysical Research-Atmospheres, 109(D6), Art. No. D06306 MAR 23 2004.

Slinn W.G.N. (1982) Prediction for particle deposition to vegetative surfaces. Atmospheric Environent, 16, 1785–1794.

Smith D., Spanel P., Jones J.B. (2000) Analysis of volatile emissions from porcine faeces and urine using selected ion flow tube mass spectrometry. Bioresource Technology, 75, 27–33.

Smith F.B. (1957) The diffusion of smoke from a continuous elevated point source into a turbulent atmosphere. Journal of Fluid Mechanics, 2, 49–76.

Smith R.I., Fowler D., Sutton M.A., Fléchard C., Coyle M. (2000) Regional estimation of pollutant gas dry deposition in the UK; model description, sensitivity analyses and outputs. Atmospheric Environment, 34, 3757–3777.

Smits M.C.J., van Jaarsveld J.A., Mokveld L.J., Vellinga O., Stolk A., van der Hoek K.W., van Pul W.A.J. (2005) Het 'VELD' project: een gedetailleerde inventarisatie van de ammoniakemissies en – concentraties in een agrarisch gebied. (The 'VELD' project: a detailed inventory of ammonia emissions and concentrations in an agricultural area, in Dutch). Report 429, Agrotechnology and Food Innovations, Wageningen, The Netherlands, p. 183.

Søgaard H.T., Sommer S.G., Hutchings N.J., Huijsmans J.F.M., Bussink D.W, Nicholson F. (2002) Ammonia volatilization from field-applied slurry – the ALFAM model. Atmospheric Environment, 36, 3309–3319.

Sommer S.G., Jensen E.S. (1991) Foliar absorption of atmospheric ammonia by ryegrass in the field. Journal of Environmental Quality, 20(1), 153–156.

Sommer S.G., Bannink A., Chadwick D., Misselbrook T., Harrison R., Hutchings N.J., Menzi H., Monteny G.J., Ni J.Q., Oenema O., Webb J. (2006) Algorithms determining ammonia emission from buildings housing cattle and pigs and from manure stores. Advances in Agronomy, 89, 261–335.

Sørensen S., Falbe-Hansen H., Mangoni M., Hjorth J., Jensen N.R. (1996) Observation of DMSO and $CH_3S(O)OH$ from the gas phase reaction between DMS and OH. Journal of Atmospheric Chemistry, 24, 299–315.

Stelson A.W., Seinfeld J.H. (1982) Relative humidity and temperature dependence of the ammonium nitrate dissociation constant. Atmospheric Environment, 21, 983–992.

Su H.B., Shaw R.H., Paw K.T., Moeng C.H., Sullivan P.P. (1998) Turbulent statistics of neutrally stratified flow within and above spars forest from large-eddy simulation and field observations. Boundary-Layer Meteorology, 88, 363–397.

Sutton M.A., Moncrieff J.B., Fowler D. (1992) Deposition of atmospheric ammonia to moorlands. Environmental Pollution, 75, 15–24.

Sutton M.A., Fowler D., Moncreiff J.B., Storeton-West R.L. (1993a) The exchange of atmospheric ammonia with vegetated surfaces. I. Fertilized vegetation. Quarterly Journal of the Royal Meteorological Society, 119, 1047–1070.

Sutton M.A., Fowler D., Moncreiff J.B. (1993b) The exchange of atmospheric ammonia with vegetated surfaces. I: Unfertilized vegetation. Quarterly Journal of the Royal Meteorological Society, 119, 1023–1045.

Sutton M.A., Asman W.A.H., Schjoerring J.K. (1993c) Dry deposition of reduced nitrogen. In: Lövblad G., Erisman J.W., Fowler D. (eds.) Models and methods for the quantification of atmospheric input to ecosystems. (Report of the UNECE Göteborg Workshop, 11/1992), Nordic Council of Ministers, Copenhagen, pp. 125–143. [Revised version published in Tellus 46B, 255–273].

Sutton M.A., Schjorring J.K., Wyers G.P. (1995a) Plant-atmosphere exchange of ammonia. Philosophical Transactions of the Royal Society London Series A, 351, 261–278.

Sutton M.A., Burkhardt J.K., Guerin D., Fowler D. (1995b) Measurement and modelling of ammonia exchange over arable croplands. In: Heij G.J., Erisman J.W. (eds.) Acid rain research: do we have enough answers? Elsevier, Amsterdam, The Netherlands, pp. 71–80.

Sutton M.A., Fowler D., Burkhardt J.K., Milford C. (1995c) Vegetation atmosphere exchange of ammonia: canopy cycling and the impacts of elevated nitrogen inputs. Water Soil and Air Pollution, 85, 2057–2063.

Sutton M.A., Milford C., Dragosits U., Singles R., Fowler D., Ross C., Hill R., Jarvis S. C., Pain B.P., Harrison R., Moss D., Webb J., Espenhahn S.E., Halliwell C., Lee D. S., Wyers G.P., Hill J., ApSimon H.M. (1997) Gradients of atmospheric ammonia concentrations and deposition downwind of ammonia emissions: first results of the ADEPT Burrington Moor experiment. In: Pain B.P., Jarvis S.C. (eds.) Gaseous exchange with grassland systems. CAB International, Wallingford, UK.

Sutton M.A., Milford C., Dragosits U., Place C.J., Singles R.J., Smith R.I., Pitcairn C.E.R., Fowler D., Hill J., ApSimon H.M., Ross C., Hill R., Jarvis S.C., Pain B.F., Phillips V.C., Harrison R., Moss D., Webb J., Espenhahn S.E., Lee D.S., Hornung M., Ullyett J., Bull K.R., Emmett B.A., Lowe J., Wyers G.P. (1998a) Dispersion, deposition and impacts of atmospheric ammonia: quantifying local budgets and spatial variability. Environmental Pollution, 102, 349–361.

Sutton M.A., Burkhardt J.K., Guerin D., Nemitz E., Fowler D. (1998b) Development of resistance models to describe measurements of bi-directional ammonia surface-atmosphere exchange. Atmospheric Environment, 32, 473–480.

Sutton M.A., Milford C., Nemitz E., Theobald M.R., Hill P.W., Fowler D., Schjoerring J.K., Mattsson M.E., Nielsen K.H., Husted S., Erisman J.W., Otjes R., Hensen A., Mosquera J., Cellier P., Loubet B., David M., Genermont S., Neftel A., Blatter A., Herrmann B., Jones S.K., Horvath L., Fuhrer E.C., Mantzanas K., Koukoura Z. (2001a) Biosphere-atmosphere interactions of ammonia with grasslands: experimental strategy and results from a new European initiative. Plant and Soil, 228.

Sutton M.A., Tang Y.S., Miners B., Fowler D. (2001b) A new diffusion denuder system for long-term, regional monitoring of atmospheric ammonia and ammonium. Water Air and Soil Pollution: Focus, 1, 145–156.

Sutton M.A., Nemitz E., Schjoerring J.K., Mattsson M., Hensen A., Cellier P., Loubet B., Roche R., Neftel A., Horvath L., Weidinger T., Meszaros R., Rajkai K., Theobald M.R., Gallagher M.W., Burkhardt J., Dammgen U. (2008) Dynamics of ammonia exchange with cut grassland: objectives and measurement strategy of the GRAMINAE integrated experiment. Biogeoscience Discussions 5, 3347–3407.

Tang Y.S., Cape J.N., Sutton M.A. (2001) Development and types of passive samplers for NH_3 and NO_x. The Scientific World, 1, 513–529.

Tang Y.S., Sutton M.A., Love L., Hasler S., Sansom L., Hayman G. (2003) Monitoring of nitric acid, particulate nitrate and other species in the UK. CEH, Edinburgh, 30 pp.

Theobald M.R., Milford C., Hargreaves K.J., Sheppard L.J., Nemitz E., Tang Y.S., Phillips V.R., Sneath R., McCartney L., Harvey F.J., Leith I.D., Cape J.N., Fowler D., Sutton M.A. (2001) Potential for ammonia recapture by farm woodlands: design and application of a new experimental facility. The Scientific World, 1(S2), 791–801.

Theobald M.R., Dragosits U., Place C.J., Smith J.U., Sozanska M., Brown L., Scholefield D., Del prado A., Webb J., Whitehead P.G., Angus A., Hodge I.D., Fowler D., Sutton M.A. (2004) Modelling nitrogen fluxes at the landscape scale Water Air and Soil Pollution: Focus, 4(6), 135–142.

Theobald M.R. et al. (2005) An assessment of how process modelling can be used to estimate agricultural ammonia emissions and the efficacy of abatement techniques. DEFRA project 1M0130 report, 27 pp. + 28 appendixes. CEH, Edinburgh.

Thijsse Th.R., Duyzer J.H., Verhagen H.L.M., Wyers G.P., Wayers A., Möls J.J. (1998) Measurement of ambient ammonia with diffusions tube samplers. Atmospheric Environment, 32, 333–337.

Thomson D.J. (1987) Criteria for the selection of stochastic models of particle trajectories in turbulent flows. Journal of Fluid Mechanics, 180, 529–556.

Thöni L., Brang P., Braun S., Seitler E., Rihm B. (2004) Ammonia monitoring in Switzerland with passive samplers: patterns, determinants and comparison with modelled concentrations. Environmental Monitoring and Assessment, 98, 93–107.

Van Breemen N., van Dijk H.F.G. (1988) Ecosystems effects of atmospheric deposition of nitrogen in the Netherlands. Environmental Pollution, 54, 249–274.

Van Dobben H.F., Ter Braak C.J.F. (1998) Effects of atmospheric NH_3 on epiphytic lichens in the Netherlands: the pitfalls of biological monitoring. Atmospheric Environment, 32, 551–557.

van Hove L.W.A., Adema E.H., Vredenberg W.J., Pieters G.A. (1989) A study of the adsorption of NH_3 and SO_2 on leaf surfaces. Atmospheric Environment, 23, 1479–1486.

Van Jaarsveld J.A. (1995) Modelling the long-term atmospheric behaviour of pollutants on various spatial scales. Ph.D. thesis, University of Utrecht, The Netherlands.

Van Jaarsveld J.A. (2004) The operational priority substances model. Report 500045001/2004, National Institute for Public Health and the Environment (RIVM), Bilthoven, The Netherlands, 156 pp.

Van Jaarsveld J.A., Bleeker A., Hoogervorst N.J.P. (2000) Evaluatie ammoniak emissiesredukties met behulp van metingen en modelberekeningen (evaluation of ammonia emission reductions using measurements and model calculations). Report 722108025, National Institute of Public Health and the Environment, Bilthoven, The Netherlands, 62 pp.

Velders G.J.M., van der Meulen A., van Jaarsveld J.A., van Pul W.A.J., Dekkers A.L.M. (2002) Ruimtelijke verdeling van ammoniakkoncentraties in Nederland gemeten met passieve samplers (Spatial distribution of ammonia concentrations in the Netherlands measured with passive samplers, in Dutch). Report 722601006/2002, National Institute of Public Health and the Environment (RIVM), Bilthoven, The Netherlands, 34 pp.

Vestreng V., Støren E. (2000) Analysis of the UNECE/EMEP emission data. MSC-W status report 2000. Norwegian Meteorological Institute, Blindern, Oslo, Norway.

Wathes C.M., Phillips V.R., Holden M.R., Sneath R.W., Short J.L., White R.P.P., Hartung J., Seedorf J., Schroder M., Linkert K.H. (1998) Emissions of aerial pollutants in livestock buildings in Northern Europe: overview of a multinational project. Journal of Agricultural Engineering Research, 70, 3–9.

Wilson J.D. (1982) An approximate analytical solution to the diffusion equation for short-range dispersion from a continuous ground-level source. Boundary-Layer Meteorology, 23, 85–103.

Wilson J.D. (2000) Trajectory models for heavy particles in atmospheric turbulence: comparison with observations. Journal of Applied Meteorology, 39, 1–49.

Wilson J.D., Sawford B.L. (1996) Review of lagrangian stochastic models for trajectories in the turbulent atmosphere. Boundary-Layer Meteorology, 78, 191–210.

Wolseley P., James P.W., Theobald M.R., Sutton M.A. (2006) Detecting changes in epiphytic lichen communities at sites affected by atmospheric ammonia from agricultural sources. Lichenologist, 38(2), 161–176.

Wyers G.P., Erisman J.W. (1998) Ammonia exchange over coniferous forest. Atmospheric Environment, 32, 441–451.

Wyers G.P., Vermeulen A.T., Slanina J. (1992) Measurement of dry deposition of ammonia on a forest. Environmental Pollution, 75, 25–28.

Zeng P.T., Takahashi H. (2000) A first-order closure model for the wind flow within and above vegetation canopies. Agricultural and Forest Meteorology, 103, 301–313.

Zhang L., Gong S., Padro J., Barrie L. (2001) A size-segregatted particle dry deposition scheme for an atmospheric aerosol module. Atmospheric Environment, 35, 540–560.

Appendix 1: A Single-Layer Model for Exchange with Stomata and Cuticule

This appendix gives the set of equations that describes the bi-directional exchange model of Fig. 15.4c. The equation for the flux for model in Fig. 15.14 can be derived easily by assuming that a canopy compensation point exists (c_{canopy} in e.g. kg m^{-3}), which is linked to the stomatal resistance and the stomatal compensation point on the one hand and to the resistance for deposition to water layers r_w (m s^{-1}) on the other hand. The flux from the atmosphere to the canopy and including the soil is then:

$$F\left(z_{ref}\right) = -\frac{\left(c_{air}\left(z_{ref}\right) - c_{canopy}\right)}{r_a\left(z_{ref}\right) + r_b} \tag{15.10}$$

The flux can then be split up in a flux to the stomata and a flux to the canopy (including the soil). The flux to the stomata is:

$$F_s = -\frac{\left(c_{canopy} - c_{stomata}\right)}{r_s} \tag{15.11}$$

The flux to the canopy (including the soil is):

$$F_w = -\frac{c_{canopy}}{r_w} \tag{15.12}$$

Mass conservation implicates that:

$$F(z_{ref}) = F_s + F_w. \tag{15.13}$$

In these equations c_{canopy} can now be eliminated:

$$F(z_{ref}) = -\frac{c_{air}\left(z_{ref}\right)}{r_a\left(z_{ref}\right) + r_b} + \frac{\dfrac{c_{air}\left(z_{ref}\right)}{\left(r_a\left(z_{ref}\right) + r_b\right)} + \left(\dfrac{c_{stomata}}{r_s}\right)}{1 + \left(r_a\left(z_{ref}\right) + r_b\right)\left(\dfrac{1}{r_s} + \dfrac{1}{r_w}\right)} \tag{15.14}$$

Note that this is the equation of Smith R.I. et al. (2000) corrected for a typing error.

Appendix 2: Two-Layer Model for Exchange with Stomata, Cuticule and the Ground

This appendix gives the set of equations that describes the two-layer bi-directional model of Fig. 15.4d. As with the single-layer model with stomatal and cuticular resistances the canopy compensation point c_{canopy} has to be eliminated by expressing

it in compensation points and resistances (Nemitz et al. 2001). In this model also a concentration $c_{air}(z_{0m})$ is needed which is the concentration in the air just above the canopy (kg m^{-3}). The canopy compensation point can be found from (Nemitz et al. 2001):
    The total flux F_t (kg m^{-2} s^{-1}) is:

$$F_t = -\frac{(c_{air} - c_{air}(z_{0m}))}{r_a} \quad (15.15)$$

The flux F_s from the canopy to the stomata is:

$$F_s = -\frac{(c_{canopy} - c_{stomata})}{r_s} \quad (15.16)$$

The flux F_w from the canopy to the cuticula is:

$$F_w = -\frac{c_{canopy}}{r_w} \quad (15.17)$$

The flux F_f from the air just above the canopy to the canopy is:

$$F_f = -\frac{(c_{air}(z_{0m}) - c_{canopy})}{r_b} \quad (15.18)$$

The flux F_g from the air just above the canopy and the ground is:

$$F_g = -\frac{(c_{air}(z_{0m}) - c_{ground})}{r_{ac} + r_{bg}} \quad (15.19)$$

In this equation c_{ground} is the compensation point at ground level (kg m^{-3}), r_{ac} is the in-canopy aerodynamic resistance (s m^{-1}) and r_{bg} is the ground boundary layer resistance (s^{-1}).
    Mass conservation implicates that:

$$F_f = F_s + F_w \quad (15.20)$$

and

$$F_t = F_f + F_g \quad (15.21)$$

Which after manipulation gives

$$c_{air}(z_{0m}) = \frac{\dfrac{c_{air}}{r_a} + \dfrac{c_{ground}}{r_g} + \dfrac{c_{canopy}}{r}}{\dfrac{1}{r_a} + \dfrac{1}{r_b} + \dfrac{1}{r_g}} \quad (15.22)$$

In this equation $r_g = r_{ac} + r_{bg}$.

An alternative expression for $c_{air}(z_{0m})$ can be obtained:

$$c_{air}(z_{0m}) = c_{canopy}\left(1 + \frac{r_b}{r_s} + \frac{r_b}{r_w}\right) - \frac{c_{stomata}r_b}{r_s} \tag{15.23}$$

An equation for c_{canopy} can be simplified to yield:

$$c_{canopy} = \left[\frac{c_{air}}{r_a r_b} + c_{stomata}\left(\frac{1}{r_a r_s} + \frac{1}{r_b r_s} + \frac{1}{r_g r_s}\right) + \frac{c_{ground}}{r_b r_g}\right] \times$$

$$\left[\frac{1}{r_a r_b} + \frac{1}{r_a r_s} + \frac{1}{r_a r_w} + \frac{1}{r_b r_g} + \frac{1}{r_b r_s} + \frac{1}{r_b r_w} + \frac{1}{r_g r_s} + \frac{1}{r_g r_w}\right] \tag{15.24}$$

Once c_{canopy} is known F_s and F_w can be found and then also the sum of them: F_f. From F_f then $c_{air}(z_{0m})$ can be found and then F_t and F_g can be calculated. The in-canopy resistance r_{ac} (s m^{-1}) at height ($d + z_{0m}$) (sum of the roughness length and the displacement height) is (Nemitz et al. 2000b):

$$r_{ac}(d + z_{0m}) = \int_0^{d+z0m} \frac{1}{K_H(z)} dz \tag{15.25}$$

where KH is the eddy diffusivity for heat (m^2 s^{-1}), which is also often taken to describe exchange of gases and particles.

$$r_{ac} = \frac{\alpha(d + z_{0m})}{u_*} \tag{15.26}$$

where α (z) is a dimensionless height dependent constant (Nemitz et al. 2000b).

For neutral conditions, Nemitz et al. (2000b) mention that Shuttleworth and Wallace (1985) provide the following equation, which only holds for neutral atmospheric conditions:

$$\alpha(d + z_{0m}) = \frac{h_c}{kn(h_c - d)}\left[\exp(n) - \exp\left(n\left(1 - \frac{d + z_{0m}}{h_c}\right)\right)\right] \tag{15.27}$$

where h_c is the height of the canopy, k is the von Karman's constant (dimensionless; value: 0.4) and n an exponential-decay constant (see e.g. Monteith and Unsworth 1990). No information is apparently available on a parameterization for stable or unstable conditions. Nemitz et al. (2001) note that for unstable conditions (free convection) it is likely that scaling with w* would be more appropriate.

The boundary layer resistance at ground surface r_{bg} (s m^{-1}) can be parameterized as (Shuepp 1977):

$$r_{bg} = \frac{Sc - \ln\left(\frac{\delta_0}{z_1}\right)}{ku_{*g}} \tag{15.28}$$

where Sc is the Schmidt number (dimensionless), δ_0 is the distance above ground (m) were the molecular diffusivity equals the eddy diffusivity, and z_1 is the upper height of the logarithmic wind profile that forms above the ground of which u_g^*/k is the slope. Nemitz et al. (2001) note that there are no comprehensive datasets for hc, $\alpha(z)$, u_g^*, δ_0 and z_1 for a wide range of plant species and vegetation stages.

Appendix 3:   In-Cloud Scavenging of NH_3 and NH_4^+

Although in-cloud scavenging of NH_3 and NH_4^+ is probably not a process that might be highly important for deposition of NH_x near hot spots, a short description of its magnitude is given in this appendix.

Due to the high solubility of NH_3 in the acidic cloud droplets and the large surface to volume ratio, most NH_3 in clouds will be taken up into cloud droplets within a few seconds. Aerosols of the size that contains most NH_4^+ act as condensation nuclei, so most NH_4^+ in clouds will also be found in cloud droplets. For these compounds the scavenging rate is determined by the removal rate of cloud water by precipitation. The in-cloud scavenging coefficient $\lambda_{i,NH3,NH4}$ for NH_3 and NH_4^+ is:

$$\lambda_{i,NH_3,NH_4} = 3.5 \times 10^{-4} I_{mm}^{0.78} \text{ (for clouds with a temperature} \geq 0°C) \quad (15.29)$$

$$\lambda_{i,NH_3,NH_4} = 2.4 \times 10^{-4} I_{mm} \text{ (for clouds with a temperature} < 0°C) \quad (15.30)$$

The rainfall rate decreases with the cloud height. From the equations it can be seen that both NH_3 and NH_4^+ are removed at a somewhat higher rate by in-cloud scavenging in the lower part of the cloud than by below-cloud scavenging.

APPENDIX 4:   Short Description of Dispersion Models that Are Used in Local Deposition Models for NH_3

Two classes of dispersion models are often considered: Eulerian models, which solve the time variation of a concentration field, and Lagrangian models, which follows the plume or the air column in space. Both Eulerian and Lagrangian approaches lead to a range of models from very simple (Gaussian-like models) to fairly complex models (CFD, Lagrangian Stochastic models). As NH_3 deposition near hot spots is large and evolves quickly with distance, it may be necessary to consider both dispersion within the canopy and the flow distortion due to the farm buildings and elements of vegetation, which may be treated with specific CFD codes. This section gives an overview of the existing dispersion models used or applicable to NH_3 dispersion and deposition near hot spots.

Gaussian-like models: Gaussian models (Gash 1985), are a solution of the Advection-Diffusion Equation (ADE) with the assumption of constant wind speed (u) and diffusivity (K_z). Analytical solutions of the ADE that include variation of u and K_z with height are numerous e.g. (Smith 1957; Philip 1959; Huang 1979; Wilson 1982). There are existing analytical models including a deposition velocity (Chrysikopoulos et al. 1992; Lin and Hildemann 1997), though these models are not well adapted for NH_3 exchange, which is bi-directional. Other analytical models based on Lagrangian similarity, such as the (Horst and Slinn 1984) model, are also well suited for modelling NH_3 deposition over the short-range (Asman 1998). However, some ill-defined parameters that need to be adjusted for different stability class make them less attractive. The approach proposed by (Raupach 1989) is very interesting for within-canopy dispersion, but its use is less justified for above-canopy transfer. Other approaches, which gives similar results but are more flexible, consist in solving numerically the above-canopy advection-diffusion equation (Asman 1998). The main limitation of the previous models is that they only consider above-canopy dispersion and hence cannot deal with side fluxes (De Jong and Klaassen 1997),

Air column models: Air column models are hybrid models that follow the air column as Lagrangian models, and solve the diffusion vertically in the column in an Eulerian framework. They are also based on constant wind speed (the column velocity), but consider a diffusivity that evolves with height. The LADD model (Hill 1998), the FRAME model (Singles et al. 1998), the TREND model (Asman 1992) are all air column models. The main limitations of these models are as Gaussian-like models that they cannot deal with within side fluxes, but also that they poorly deal with lateral dispersion.

Within canopy dispersion models and CFDs: There are a number of dispersion models that could potentially be used for modelling dispersion within the canopy and henceforth dispersion in a woodland near a farm building. These are first order Eulerian models e.g. (Zeng and Takahashi 2000) which limits are that they are based on flux-gradient relationships, which may be broken due to intermittency of the turbulence within the canopy e.g. (Aylor et al. 1993). Second or higher order Eulerian models are used to overcome this problem (e.g. Poggi et al. 2004), though these introduce new parameters which are uneasy to set for within canopy transfer. CFD models are among these models, and they can be used to model the flow distortion around buildings. Large Eddy Simulation models are probably the most adapted for within canopy transfer modelling, though they also need a parameterisation of the sub-grid-scale variances (e.g. Su et al. 1998). One main interest of Eulerian dispersion models is that they can predict both turbulence and concentration, as affected by canopy structure.

The Lagrangian Stochastic (LS) are another class of models (Rodean 1996) that can be used for modelling within canopy dispersion. They are well adapted to simulate non-diffusive fluxes such as encountered in a canopy (Wilson and Sawford 1996). Their main weakness is that they need the turbulence field as input; hence a turbulence model is needed as input of LS dispersion models to simulate dispersion.

Chapter 16
Standardised Grasses as Biomonitors of Ammonia Pollution Around Agricultural Point Sources

Ian D. Leith, Netty van Dijk, Carole E.R. Pitcairn, Lucy J. Sheppard, and Mark A. Sutton

16.1   Introduction

The use of standardised plant species as biomonitors of air pollution has been a recognised technique for a number of years (De Temmerman et al. 2004). Grass species have been used extensively in a number of biomonitoring studies (Sant' Anna et al. 2004). These standardised grasses can be used as passive biomonitors, but are more effectively used as active bioaccumulators of a range of gaseous and particulate pollutants (Scholl 1971). The advantage of these standardised plants as bioaccumulators is their ability to accumulate a variety of inorganic and organic air pollutants without showing symptoms of visible injury. Standardised plants can also provide an indication of air pollution impacts over a short time period and are independent of the habitat substrate. Although the use of standardised biomonitors is extensively reported in the literature for the bioaccumulation of sulphurous compounds, heavy metals and trace elements (Klumpp et al. 2004) they have not been widely used as active bioaccumulators of N pollution.

Sommer (1988) exposed barley (*Hordeum vulgare var. Harry*) plants as a bioindicator of ammonia (NH_3) deposition along a 0–300 m transect from a dairy farm for 1 month. The tissue N content of the barley plants was increased closer to the farm reflecting the increased N deposition from ammonia. In a subsequent study, Sommer and Jensen (1991) exposed *Lolium multiflorum Lam.* along 0–130 m transects away from a dairy farm. Tissue N concentration was found to increase closer to the farm. Dry matter was also increased in close proximity to the farm but did not decrease significantly with distance from the farm. Hicks et al. (2000) used two graminiod species *Deschampsia flexuosa* (L.) and *Nardus stricta* L. as potential biomonitors of wet N deposition in upland areas.

Sutton et al. (2004) carried out a comprehensive review of existing biomonitoring methods for determining the impacts of N deposition on plant species and habitats. Part of this review was a field comparison study, which assessed a range of bioindicator

Ian D. Leith
Centre for Ecology & Hydrology, Bush Estate, Penicuik, Midlothian, EH26 0QB, United Kingdom

M. Sutton, S. Reis and S.M.H. Baker (eds), *Atmospheric Ammonia,*
© Springer Science + Business Media B.V. 2009

techniques, including standardised grasses. These inter-comparison methods were used in parallel, to assess the impacts of NH_3-N deposition at a poultry farm in the Scottish Borders. In a subsequent study (Leith et al. 2005) a short–list of robust biomonitoring methods (tissue %N and soluble ammonium concentration in bryophytes, Ellenberg N Index and epiphytic lichen diversity and standardised grasses) identified by Sutton et al. (2004) was assessed at a contrasting range of habitats, atmospheric N deposition and different N forms.

The aim of these two field studies (Sutton et al. 2004; Leith et al. 2005) was to identify robust biomonitoring methods, which could be used to assess the effects of atmospheric N on statutory nature conservation sites.

This paper reports the results from the above two transect studies using standardised grasses carried out at two poultry farms (Earlston, Scottish Borders; Sutton et al. 2004 and Piddles Wood, Dorset; Leith et al. 2005) with contrasting woodland habitats and two different grass species. The aim was to determine if grass transplants could be used as bioindicators to detect if N was having an impact on the selected habitats.

16.2   Methods

Trays of standardised grasses were exposed along a known NH_3 concentration gradient at two poultry farms over two different time periods Earlston (2002) and Piddles Wood Dorset (2004). *Lolium perenne* L. was used in the first study at Earlston (Sutton et al. 2004). However, in the second study (Leith et al. 2005) one of the aims was to establish if grass biomonitors could be used at sites which were dominated by wet N deposition. Therefore, a slower-growing grass species (*Deschampsia flexuosa*) was used at all sites in this study, including Piddles Wood.

16.2.1   Study Sites

16.2.1.1   Earlston, Scottish Borders

A transect of sampling sites was set up at a 120,000 bird poultry farm in the Scottish Borders on 1 October 2002. The site is a mixture of a 30-year-old *Pinus sylvestris* L. plantation with *Betula pubescens* Ehrh. L. and mature *Fagus sylvatica* L. The farm has been operational for approximately 20 years and has an emission rate of 4,800 kg N year^{-1}. Six positions were located in the managed mixed-coniferous woodland at 16, 30, 46, 76, 126 and 276 m to the north of the poultry farm. At each of the positions, one tray of *L. perenne* (six plants per tray) was placed out. The grass transplants were exposed for 38 days (1 October–7 November 2002).

16.2.1.2   Piddles Wood, Dorset

Piddles Wood is a designated *Site of Special Scientific Interest* (SSSI), notified in 1985 for its broadleaved, mixed lowland woodland (dominated by oak with coppiced hazel understorey) in north Dorset (area 63 ha). A large single poultry unit (approximately 100,000 birds) established in the mid-1990s is situated within 5 m of the woodland at its south-west edge. The grass biomonitoring system was established at Piddles Wood on 26 April 2004 with a single tray of six *Deschampsia flexuosa* pots placed at each of four distances (5, 20, 40 and 100 m) along a 250 m transect running in a NE direction away from the poultry farm. The standardised grass plants were exposed for 86 days (26 April–21 July 2004).

16.2.2   Plant Material

Lolium perenne (L.) is sown as a crop for fodder and grazing. It is a native perennial, common throughout the UK and predominantly a species of improved lowland pasture and hay meadows. It can also be found in a range of other habitats including inundated grasslands, waste-ground and roadsides (Preston et al. 2002).

Deschampsia flexuosa (L.) Trin. is a slow growing, evergreen, tufted or mat-forming polycarpic perennial grass. It is the most successful calcifuge grass species in the UK and is found in a wide range of unproductive habitats. It is very abundant in moorland, acidic heaths, open-woodland and plantations, but normally in drier areas of these habitats (Preston et al. 2002).

Standardised grasses for both studies were propagated at CEH Edinburgh; *L. perenne* was sown at a rate of 1.22 g per pot into round black pots (volume 3.6 l) using a standardised compost mix with no added fertiliser (peat:loam:grit in the ratio 4:1:1). Four wicks were inserted into each of the pots as part of the irrigation system (see irrigation section). Two weeks prior to exposure, all tillers were cut to a height of 2–3 mm above the rim of the pot.

Tillers from mature plant stock of *D. flexuosa* were propagated at CEH during the summer/autumn of 2003. The *D. flexuosa* seed had been originally supplied by Herbiseed, The Nurseries, Billingbear Park, Wokingham, Berkshire, RG 40 5RY. The rooted young plants were transferred to 1.1 l square black pots containing a peat:loam:grit compost (ratio 4:1:1) with no added fertiliser once they had rooted and had begun to develop. All the individual pots had a wicking system to provide irrigation on demand. All plants were over-wintered in an unheated glasshouse, and then hardened off in early spring 2004, when they were transferred outdoors prior to use at the five intensive sites. Full details of the propagation methods can be found in Sutton et al. (2004 Appendix I) and Leith et al. (2005; Section 4).

16.2.3   Irrigation System

As the sites could not be visited regularly (7–10 days between visits) an irrigation system was designed, which would provide the plants with sufficient rainwater between visits. The wicking system with a reservoir of water was based on the EuroBioNet system (EuroBionet instruction manual: www.eurobionet.com). At Earlston the pots were placed directly into a reservoir of rainwater. At Piddles Wood, the six plants were placed into a polystyrene moulding, which supported the plants above the $600 \times 400 \times 145$ mm HDPE rainwater reservoir tray (volume 28 l) thus allowing each plant's wick to absorb the rainwater that was used to fill the reservoir. All trays were protected from wildlife by garden mesh supported by 60 cm wooden posts.

16.2.4   Destructive Harvests

The plants from both sites were destructively harvested by cutting and collecting all the new tiller growth 1–2 cm above the level rim of the pot at the end of the set exposure period. The tillers were separated into green and brown coloured shoots (brown shoots comprised less than 1% of the total biomass at Earlston and Piddles Wood). The green tillers were used for %N, soluble NH_4-N concentration (Piddles Wood) and biomass determination, whereas the brown were only used for the biomass determination.

16.2.5   Nitrogen Accumulation Methods

16.2.5.1   Tissue N (% Dry Weight)

Green tiller samples were cleaned, dried at 70 °C for 48 h, ground to a fine powder using a ball mill and placed in 3 ml glass vials. All samples were analysed for %N using a Vario-EL elemental Analyser.

16.2.5.2   Soluble Ammonium Concentration

Soluble ammonium (NH_4-N) was also measured in the tillers of the standardised grasses. This method used fresh material and had been adapted and refined from a method by Loubet and Sutton (1999); Loubet et al. 2002. Frozen samples (1 g per bottle) were placed into 20 ml bottles. Ten millilitres of de-ionised water was added to each bottle and then left to extract for 4 h at room temperature (20 °C). The solution was filtered and then analysed for soluble ammonium using AMFIA (Ammonia Flow Injection Analysis) (see Appendix I; Leith et al. 2005).

16.2.5.3   Biomass Determination and Total N Inventory

The biomass tillers were dried at 70 °C for 48 h and then weighed. The dry weight of the green and brown tillers was added together to give a total biomass. The accumulated biomass is expressed on a dry weight basis. Total N per pot was derived from the tissue N and the biomass of above ground tillers. A full description of biomasss determinations are given in Section 5 (Leith et al. 2005).

16.2.6   Monitoring Ammonia Concentrations

16.2.6.1   Earlston

Ammonia concentrations were monitored at 1.5 m above ground level at each site using either passive ammonia diffusion samplers (site 1) or ALPHA samplers (Adapted low cost passive high absorption samplers: sites 2–6). There were three replicate ALPHA samplers at each of the six sites. The diffusion and ALPHA samplers were exposed for 38 days from 1 October 2002 until 7 November 2002, with two sampling periods (1–24 October 2002 and 24 October–7 November 2002) (Table 16.1).

16.2.6.2   Piddles Wood

Ammonia concentrations were measured at six locations along the transect at a height of 1.5 m above the ground for 3 months (April–August 2004) using ALPHA samplers (Tang et al. 2001). The ALPHA samplers were exposed for two periods (27 April–8 June and 8 June–21 July 2004) (Table 16.2). These samplers provide a time integrated concentration for the exposure period.

Table 16.1   Mean NH_3 concentrations along 300 m transect from Earlston poultry farm

Site no.	1	2	3	4	5	6
Distance from Poultry farm (m)	30	16	46	76	126	276
Mean NH_3 Concentration ($\mu g\ m^{-3}$)	69.6	29.1	12.8	8.0	3.4	0.58

Site 1 is east of the poultry farm. The other five sites are north of the farm in woodland.

Table 16.2   Ammonia air concentrations ($\mu g\ m^{-3}$) at Piddles Wood from 27/04/2004 until 21/07/2004 at different distances from the poultry unit

	Distance (m) from poultry farm	NH_3 ($\mu g\ m^{-3}$)		
		27/04–08/06	08/06–21/07	Average
Site 1	5	123	79.4	101
Site 2	20	27.9	34.7	31.4
Site 3	40	14.7	19.5	17.1
Site 4	100	1.94	2.0	2.0

16.2.7   Determination of Site-Specific NH₃ Effects Using Tissue N and Soluble NH₄-N Concentration

Cape et al. (2009, this volume) discuss in detail the concept of measurable difference for establishing a NH_3 threshold concentration for a specific site where the measured parameter i.e. tissue N concentration increases with NH_3 concentration. This method has been used in this study to determine the threshold NH_3 concentration for effects for the measured parameters from the two field studies. The method is illustrated on Fig. 16.2a. The best-fit equation of the line was determined by least-squares analysis and the 95% confidence limits calculated using Sigmaplot; Version 10. The upper 95% confidence limit directly above the lowest measured parameter data point (Fig. 16.2a: point A) is an estimate of the largest value which could be expected to fall within the local background range at the lowest measured NH_3 concentration. Extending the dotted line along to the point where the line intersects the fitted regression line (Point B) then down to the where it intersects the X-axis (Point C) indicates the lowest concentration that would be predicted to yield a measurement value above background.

16.3   Results

16.3.1   Ammonia Concentrations

Monitoring NH_3 concentrations using ALPHA samplers showed that the NH_3 concentrations declined exponentially along the NH_3 concentration gradient from both poultry farms. The highest NH_3 concentration ($100 \mu g \ m^{-3}$) was found at Piddles Wood. However, this site was located only 5 m from the poultry farm ventilation ducts. The background NH_3 concentration was higher at Piddles Wood ($1.5 \mu g \ m^{-3}$ at 250 m from the poultry farm) compared to the Earlston site at $0.58 \mu g \ m^{-3}$ (Tables 16.1, 16.2). The NH_3 concentrations measured along both transects were similar for approximately comparable distances.

16.3.2   Tissue N (% Dry Weight) and Soluble NH₄-N Concentration

16.3.2.1   Earlston

The %N and soluble NH_4-N concentrations in *L. perenne* increased with proximity to the poultry farm. Both biochemical assays showed log-linear relationships with NH_3 concentration (Fig. 16.1a, b). However, there was a much stronger relationship with soluble ammonium ($R^2 = 0.976$) than for tissue %N ($R^2 = 0.544$).

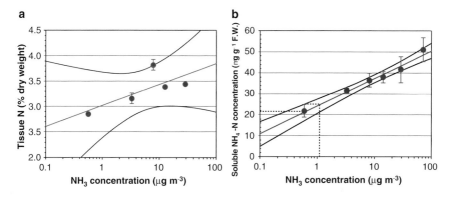

Fig. 16.1 (a) Tissue N (% dry weight) of *L. perenne* along a NH₃ transect from a poultry farm at Earlston in the Scottish Borders. The R² = 0.544 and the 95% confidence limits are shown. The errors are ±1 standard error. (**b**) Soluble ammonium concentration (μg g⁻¹) fresh weight for *L. perenne* along the NH₃ transect at Earlston. The R² = 0.976 and the 95% confidence limits are shown. The errors are ±1 standard error. The line intercepting the x-axis at 1.2 μg m⁻³ is the threshold NH₃ concentration limit

The soluble NH₄-N concentration more than doubled from 21 μg g⁻¹ fresh weight at a background NH₃ concentration of 0.6 μ g m⁻³ to 51 μg g⁻¹ fresh weight at 30 m from the farm (70 μg m⁻³).

16.3.2.2 Piddles Wood

Both the %N and soluble NH₄-N concentrations in *D. flexuosa* showed strong log-linear relationships with NH₃ concentration (R² = 0.953 and 0.999 respectively) along the NH₃ concentration gradient. However, comparison of the 95% confidence limits for both assays (Fig. 16.2a, b) show that the soluble NH₄-N concentration was a more precise indicator of response to enhanced N inputs i.e. the spread of the 95% limits is much greater in the %N (Fig. 16.2a). The tiller %N concentration doubled from 1.4% to 2.8% along the NH₃ gradient (2 μg m⁻³ background concentration to 100 μg m⁻³ at 5 m from the poultry farm). The soluble NH₄-N concentrations more than trebled along the transect from the poultry unit (3.1 μg g⁻¹ at 2 μg m⁻³ to 10.2 μg g⁻¹ at 100 μg m⁻³).

16.3.3 Biomass Determination and Total N Inventory

16.3.3.1 Earlston

There was active biomass accumulation in *L. perenne* in direct response to enhanced NH₃ along the concentration gradient from the poultry farm over the 38 day exposure period (Fig. 16.3a).

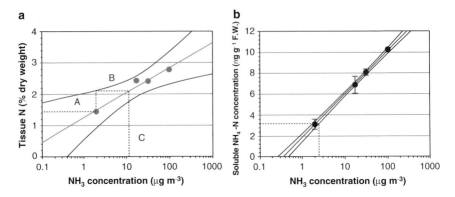

Fig. 16.2 (a) Tissue N (% dry weight) of *D. flexuosa* along a NH₃ transect from a poultry farm at a SSSI site at Piddles Wood in Dorset. The $R^2 = 0.953$ and the 95% confidence limits are shown. The errors shown are ±1 standard error. (**b**) Soluble ammonia concentration (μg g⁻¹) for *D. flexuosa* along the NH₃ transect at Piddles Wood, Dorset. The $R^2 = 0.953$ and the 95% confidence limits are shown. The errors shown are ±1 standard error. The line intercepting the x-axis at 2.5 μg m⁻³ is the threshold NH₃ concentration limit

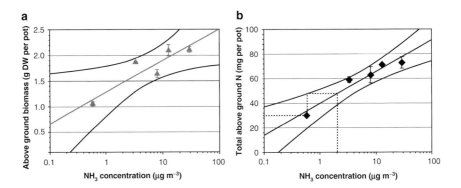

Fig. 16.3 (**a**) The effect on above ground biomass of *L. perenne* exposed along a NH₃ concentration gradient at a poultry farm at Earlston for 38 days. The $R^2 = 0.823$ and the 95% confidence limits are shown. (**b**) The relationship between the total N inventory (biomass x N concentration) along the NH₃ transect. The errors shown are ±1 standard error. The interception point of the dotted line on the x-axis at 2.0 μg m⁻³ is the threshold NH₃ concentration limit

The biomass accumulation in the grasses adjacent to the farm was approximately twice that of those grasses at the background site (1.1 g per pot at 0.59 μg m⁻³ NH₃ compared to 2.11 g per pot at 70 μg m⁻³ NH₃).

Although there was a strong log-linear relationship in biomass accumulation with NH₃ concentration ($R^2 = 0.822$), there is scatter in the data because the transect was through a mixed woodland which had variable light levels (Fig. 16.3a).

However, the influence of shading is largely cancelled out when the total N inventory is calculated. The determination of total N expressed as mg N per pot gives a substantially improved correlation with NH_3 exposure (Fig. 16.3b). The total N per pot increased log-linearly from 29 to 73 mg per pot with increasing NH_3 concentration ($R^2 = 0.935$).

There was also strong log-linear relationships for total biomass and Total N inventory with NH_3 concentration at Piddles Wood ($R^2 = 0.790$ and 0.990 respectively) Data not shown; see Leith et al. 2005. As with the Earlston data (Fig. 16.1b), the total N inventory was a more precise measure of N impact.

16.3.4   Determination of Site-Specific NH_3 Effects Using Tissue N and Soluble NH_4-N Concentration

From the log-linear responses for the measured parameters (soluble NH_4-N concentration, %N and total N inventory; Figs. 16.1b, 16.2a, b, 16.3b) a threshold NH_3 concentration has been estimated. The threshold NH_3 concentration for effects on *L. perenne* at the Earlston site was $1.1\,\mu g\ m^{-3}$ for soluble ammonium concentration (Fig. 16.1b). The threshold value for soluble NH_4-N concentration was slightly higher at Piddles Wood for *D. flexuosa* at $2.3\ NH_3\ \mu g\ m^{-3}$. This difference is probably due to the lower background NH_3 at the Earlston site than at Piddles Wood. The NH_3 threshold was found to be much higher at Piddles Wood for tissue N ($10.1\,\mu g\ m^{-3}$). This is probably due to the greater variability in %N resulting in wider 95% confidence limits.

16.4   Conclusions

- Both grass species (*L. perenne* and *D. flexuosa*) are considered suitable bioindicators of ammonia for sites adjacent to intensive livestock point sources.
- The standardised grasses of both species were found to be active accumulators of N in their foliage. They provide a snap-shot in time of N effects to a habitat and also give a quick response to exposure (typically of 30–90 days of NH_3 exposure).
- The biochemical methods (tissue N concentration and soluble NH_4-N concentration) were able to detect the increased NH_3 concentrations in both grass species at the Earlston and Piddles Wood poultry farms.
- The tissue N concentration in *L. perenne* increased from 2.8% to 3.4% along the ammonia concentration gradient, an increase of 17% whereas the soluble ammonium concentration more than double from 21 to $51\,\mu g\ g^{-1}$.
- The percentage increase in tissue N in *D. flexuosa* resulting from exposure to enhanced NH_3 concentrations at Piddles Wood (14%) was similar to that found for *L. perenne* (17%). However, the effect on soluble ammonium concentration

in *D. flexuosa* was much greater than for *L. perenne* with a fivefold increase from 2 to $10 \mu g \ g^{-1}$.

- The tissue N (% dry weight) signal in *D. flexuosa* was approximately 30% smaller at Piddles Wood than that found for *L. perenne* at Earlston poultry farm (with smaller NH_3 concentrations). However, this could be due to species difference in partitioning nitrogen and growth dilution.
- In both *L. perenne* and *D. flexuosa* the soluble NH_4-N concentration was the indicator most closely coupled with the ammonia concentration.
- Linear increases in accumulated above-ground biomass and total N inventory were found for both *L. perenne* and *D. flexuosa* with increasing NH_3 concentration. Total N inventory was found to be a more precise measure of ammonia effects than accumulated biomass or tissue N content.
- The significant NH_3 threshold concentration found for the *L. perenne* and *D. flexuosa* is very similar to that set by the expert panel at this Workshop of $2–3 \mu g \ m^{-3}$ for vascular plants.
- The grass biomonitors studies indicate that the increased NH_3 concentrations at both poultry farms (Earlston and Piddles Wood) are directly impacting the woodland groundflora, epiphytes and trees.
- This increased N from ammonia could have important impacts on the long-term integrity of the Piddles Wood SSSI, including an increasingly larger area surrounding the poultry unit, which could be under direct threat from accumulated N.

Acknowledgements We would like to thank the Joint Nature Conservation Committee (JNCC), the Scotland and Northern Ireland Forum For Environmental Research (SNIFFER), Scottish Natural Heritage (SNH), Natural England and the Centre for Ecology & Hydrology (CEH) for supporting this work.

References

Cape J.N., van der Eerden L.J., Sheppard L.J., Leith I.D. and Sutton M.A. (2009) Reassessment of critical levels for atmospheric ammonia. In: Sutton M.A., Reis S., Baker S.M.H.: Atmospheric Ammonia – Detecting emission changes and environmental impacts, 15–40, Springer publishers

De Temmerman L., Bell J.N.B., Garrec J.P., Klumpp A., Krause G.H.M., Tonneijk A.E.G. (2004) Biomonitoring of air pollutants with plants-considerations for the future. In: A. Klumpp, W. Ansel and G. Klumpp (eds.) Urban air pollution, bioindication and environmental awareness, 337–375. Goettingen: Cuvillier Verlag.

EuroBionet (2003) www.uni-hohenheim.de/eurobionet

Hicks W.K, Leith I.D., Woodin S.J., Fowler D. (2000) Can the foliar concentration of upland vegetation be used for predicting atmospheric nitrogen deposition? Evidence from field surveys. Environmental Pollution 107, 367–376.

Klumpp A., Klumpp G., Ansel W. (2004) Urban air quality in Europe- results of three years of standardised biomonitoring studies. In: A. Klumpp, W. Ansel and G. Klumpp (eds.) Urban air pollution, bioindication and environmental awareness, 337–375. Goettingen: Cuvillier Verlag.

Leith I.D., van Dijk N, Pitcairn C.E.R., Wolseley P.A., Whitfield C.P., Sutton M.A. (2005) Biomonitoring methods for assessing the impacts of nitrogen pollution: refinement and testing. JNCC Report 386, Peterborough.

Loubet B., Sutton M.A. (1999) Edinburgh protocol for measuring bulk NH4+ of grass leaves in GRAMINAE. In: M. Sutton and C. Milford (eds.) integrating measurements protocol. Centre for Ecology and Hydrology, Edinburgh, UK.

Loubet B., Milford C., Hill P.W., Tang Y.S., Cellier P., Sutton M. A. (2002) Seasonal variability of apoplastic NH_4^+ and pH in an intensively managed grassland. Plant and Soil 238, 97–110.

Preston C.D., Pearman D.A., Dines T.D. (2002) New atlas of the British & Irish flora. Oxford University Press, Oxford.

Sant' Anna S.M.R., Rinaldi M.C.P., Domingos M. (2004) Biomonitoring of air pollution in Sao Paulo city (Brasil) with *Lolium multiflorum* ssp. *Italicum* "Lena". In: A. Klumpp, W. Ansel and G. Klumpp (eds.) Urban air pollution, bioindication and environmental awareness, 303–308, Goettingen, Cuvillier Verlag.

Scholl G. (1971) Die Immissionsrate von Fluor als Maßstab für eine Immissionbegrenzung. VDI-Berichte 164, 39–45.

Sommer S.G. (1988) A simple biomonitor for measuring ammonia deposition in rural areas. Biology and Fertility of Soils 6, 61–64.

Sommer S.G., Jenson E.S. (1991) Foliar absorption of atmospheric ammonia by ryegrass in the field. Journal of Environmental Quality 20, 153–156.

Sutton M.A., Pitcairn C.E.R., Leith I.D., Sheppard L.J., van Dijk N., Tang Y.S., Skiba U., Smart S., Mitchell R., Wolseley P., James P., Purvis W., Fowler D. (2004) Bioindicator and biomonitoring methods for assessing the effects of atmospheric nitrogen on statutory nature conservation sites. In: M.A. Sutton, C.E.R. Pitcairn and C.P. Whitfield (eds.) JNCC Report No. 356, Peterborough.

Tang Y.S., Cape J.N., Sutton M.A. (2001) Development and types of passive samplers for monitoring atmospheric NO_2 and NH_3 concentrations. TheScientificWorld 1, 513–519.

Chapter 17
Soluble Ammonium in Plants as a Bioindicator for Atmospheric Nitrogen Deposition: Refinement and Testing of a Practical Method

Netty van Dijk, Ian D. Leith, Carole E.R. Pitcairn, and Mark A. Sutton

17.1 Introduction

Substrate nitrogen and soluble ammonium represent newly recognized parameters in the context of the bioindication of atmospheric nitrogen concentrations and deposition. The interest in these parameters originates from the analysis of plant carbon and nitrogen dynamics in ecosystem models in relation to the potential for plants to absorb or emit ammonia from the atmosphere, as regulated by the ammonia 'compensation point' (Sutton et al. 2004). The ammonia compensation point is a function of both agricultural and atmospheric N inputs. Enhanced nitrogen deposition raises the ammonia 'compensation point' and provides a limitation to further nitrogen deposition. In this context, the rates of atmospheric ammonia deposition, and hence total N deposition, depend partly on the extent to which the ecosystem has already responded to N deposition. As a consequence of these interactions, the level of substrate N or soluble ammonium (NH_4-N) may be a convenient indicator of accumulated nitrogen deposition and ecosystem response (Sutton et al. 2004).

Previous studies (Sutton et al. 2004; Leith et al. 2005) suggest that the relationship to atmospheric N is more precise for soluble ammonium (NH_4-N) than for total tissue N in plants. The steeper slope of the line in Fig. 17.1 for soluble NH_4-N (= free ammonium) indicates that NH_4-N is more sensitive to changes in ammonia air concentration than total tissue N (%N) and total soluble nitrogen (substrate nitrogen).

At a study carried out at Earlston poultry farm, Scottish Borders (Pitcairn et al. 1998) a good relationship was found between NH_3 air concentration and soluble NH_4-N in the pleurocarpous moss *Rhytidiadelphus triquetrus* (Fig. 17.2).

The standard extraction method for soluble NH_4-N, adapted from Loubet et al. (2002), requires plant material to be ground in liquid nitrogen. Although this method effectively breaks down the cell structure, it is time consuming. The aim was therefore to develop and validate a quicker, simpler and cheaper (= a more practical) extraction method without the use of liquid nitrogen.

N. van Dijk
Centre for Ecology & Hydrology, Bush Estate, Penicuik, Midlothian, EH26 0QB, United Kingdom

M. Sutton, S. Reis and S.M.H. Baker (eds), *Atmospheric Ammonia,*
© Springer Science+Business Media B.V. 2009

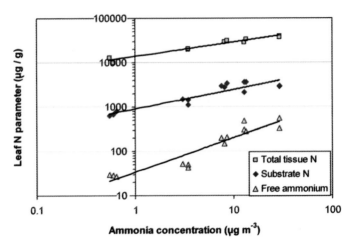

Fig. 17.1 Comparison of three bioindicator methods for nitrogen for the moss *Rhytidiadelphus triquetrus*: total tissue nitrogen, substrate N and soluble NH_4-N (= free ammonium). The largest relative response to NH_3 is shown for soluble NH_4-N

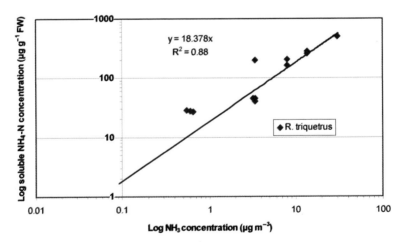

Fig. 17.2 Soluble NH_4-N in *Rhytidiadelphus triquetrus* in relation to NH_3 air concentration measured at Earlston poultry farm in the Scottish Borders

17.2 Development of a Practical Method for the Measurement of Foliar Soluble Nitrogen

Several extraction methods were tested along the standard method with the use of liquid nitrogen for comparison. Full details of the tested methods can be found in Leith et al. (2005, Appendix I).

Using frozen samples ($-20°C$) the vegetation with the added de-ionised water were placed in 20 ml vials and subjected the following extraction methods (Fig. 17.3):

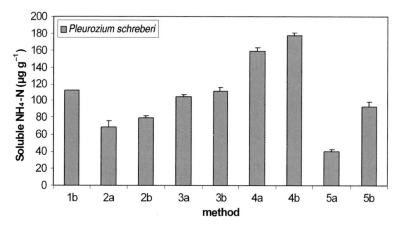

Fig. 17.3 Soluble NH$_4$-N concentrations in *Pleurozium schreberi* with different extraction methods. (**a**) Samples were analysed immediately after extraction and (**b**) samples had a 24 h 'leaking time'. Error bars are standard deviations

1. In water
2. In a sonic bath for 2 h
3. Autoclaving in water (110°C)
4. Autoclaving in sulphuric acid (110°C)
5. Standard method: grinding in liquid nitrogen

Two variants were examined: (a) samples were analysed immediately after extraction and (b) the samples had a 24 h 'leaking' time. Solutions were filtered (0.45 μm polypropylene disposable filter) and frozen (−20°C) until analysed for NH$_4^+$ by AMFIA (Ammonia Flow Injection Analysis).

Surprisingly the 'standard' method (5a, liquid N) gave the lowest concentrations of soluble NH$_4^+$ -N (Fig. 17.3), although the concentration is much higher after a 24 h 'leaking time'. Extraction in water gave similar results as the other tested methods while there is no need for the use of chemicals or expensive equipment. The same results were found when using another moss *Hypnum jutlandicum* or the grass *Lolium multiflorum*. The extraction method in water was found as the most simple and practical method.

Further tests of the extraction in water method showed that even after 48 h the equilibration process is not yet complete (Fig. 17.4). This supports the possibility that, in addition to free soluble NH$_4$ leakage, biological processes are responsible for more NH$_4$-N in the extraction solution. It is clearly preferable to measure only the free soluble NH$_4$-N at concentrations large enough to show differences between samples. For that reason an extraction or leaking time of 4 h was selected as the NH$_4$-N are high enough to be measured, while the measurements of effects of biological activity are kept small.

The new simplified extraction method for soluble NH$_4$-N is:

1. Weigh 1 g FW of frozen, cleaned moss/grass material and put it in a 20 ml bottle.
2. Add 10 ml de-ionised water.

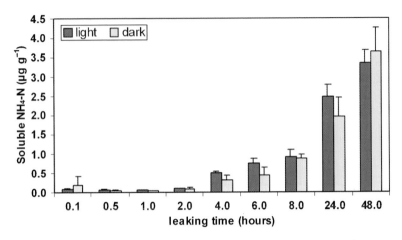

Fig. 17.4 The influence of light during and the extraction time and the lengths of the extraction time on NH$_4$-N concentration in *P. schreberi*. Extraction took place at 20°C. Error bars are standard deviations

3. Shake for 5 s and leave to extract for 4 h at lab temperature (20°C).
4. Shake again for 5 s.
5. Filter the solution through 0.45 μm pore size syringe filter.
6. Analyse the solution for NH$_4^+$ (by AMFIA) or store the solution in a freezer at −20°C and calculate the NH$_4$-N concentration expressed as μg g^{-1} FW (Fresh Weight).

17.3 Applying the Method at Sites with Different Ammonia Sources

The practical soluble NH$_4$-N method was applied in field studies at Piddles Wood (Dorset) and Whim Moss (Scottish borders). Both sites have different N point sources. At Piddles Wood the ammonia point source is a poultry unit (approximately 100,000 birds) which is adjacent to the edge of the woodland. Whim Moss is a manipulated field facility where moorland vegetation is exposed to either dry or wet N deposition along a transect. Dry N deposition is applied as gaseous NH$_3$ and wet N deposition is applied as NO$_3$ or NH$_4$. The NH$_3$ air concentrations at both field studies were measured with passive ALPHA samplers (Tang et al. 2001) along a gradient transect from the NH$_3$ source.

At Piddles Wood the moss species *E. praelongum* and *E. striatum* were collected as well as the grass *Deschampsia flexuosa* (see Chapter 16, Leith et al. 2009, this volume) along the transect and analysed for NH$_4$-N.

There is a strong relationship for both moss and grass (Fig. 17.5) between soluble NH$_4$-N and NH$_3$ air concentration. The steeper slope of the line in Fig. 17.5b for soluble NH$_4$-N shows again that NH$_4$-N is more sensitive to changes in ammonia air concentration than total tissue N (%N).

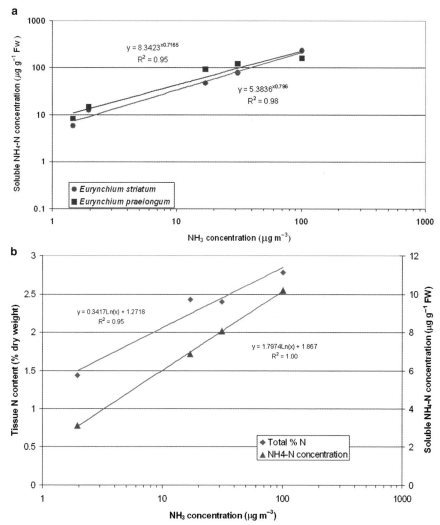

Fig. 17.5 Piddles Wood: the relationship between atmospheric NH_3 concentration and (**a**) soluble NH_4-N concentration in moss tissue in two *Eurhynchium* species and (**b**) N content and the soluble NH_4-N concentration in *Deschampsia flexuosa* (after 12 weeks of exposure) as transplants

At Whim Moss *Hypnum jutlandicum* samples were collected along a transect and analysed for NH_4-N.

Soluble NH_4-N is more sensitive to changes in dry N deposition (Fig. 17.6b) than to changes in wet N deposition (Fig. 17.6a). Their appears to be a threshold (soluble NH_4-N maintained at ~5 µg FW up to wet deposition of 38 kg N ha^{-1} year^{-1}) wet deposition required to increase soluble NH_4-N, while there is no threshold for effects of gaseous NH_3 on soluble NH_4-N.

Figure 17.6b shows again that soluble NH_4-N is more sensitive than total tissue N (% N) to changes in NH_3 air concentrations.

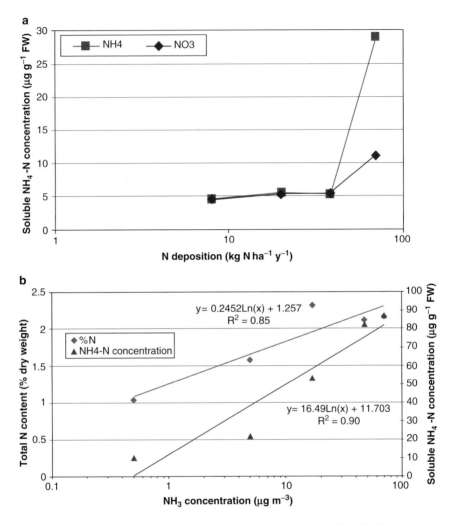

Fig. 17.6 Whim Moss experimental field: the relation ship between soluble NH_4-N concentration and (**a**) wet N deposition and (**b**) NH_3 concentration (dry N deposition) in *H. jutlandicum*

17.3.1 Applying the Method at a UK Study

After testing the simplified practical method in some field studies with N point sources, the next phase was to apply this method across the whole of the UK. Full details about the UK study can be found in Leith et al. (2005, Sections 9 and 10). Moss samples were collected from 32 sites across the whole of the UK (Fig. 17.7) with known NH_3 air concentrations.

Only a weak relationship was found between tissue content (%N) and soluble NH_4-N concentrations in the UK mosses with NH_3 air concentrations (Fig. 17.8).

This can be explained by the large amount of different moss species that were collected. Different moss species are different in sensitivity in their response to NH_3. Another reason was that the UK sites had diffuse NH_3 sources and no point sources.

Fig. 17.7 Map of the UK showing the 32 locations where the mosses were collected

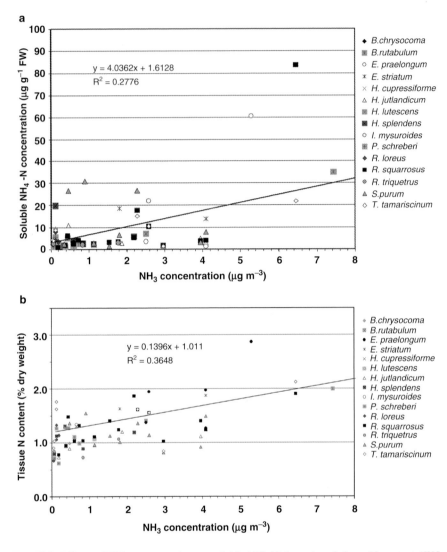

Fig. 17.8 Effects of NH₃ concentration on soluble NH₄-N (upper) and tissue N content (%N, lower graph) for pleurocarpous mosses sampled in the UK

17.4 Conclusions

- Extracting of pleurocarpous moss or grass samples for 4 h in water was found as the quickest, simplest and cheapest method tested. So this method is seen as the most practical method.
- At sites with an ammonia point source there was a strong relationship between soluble NH₄-N concentrations and NH₃ air concentrations.

- Soluble NH_4-N is very sensitive to changes in NH_3 air concentrations and less sensitive to wet N deposition.
- At sites across the UK with a diffuse NH_3 source there is a much weaker relationship between soluble NH_4-N concentrations and NH_3 air concentrations than at sites with a point NH_3 source.

Acknowledgements We would like to thank the Joint Nature Conservation Committee (JNCC), the Scotland and Northern Ireland Forum for Environmental Research (SNIFFER), Scottish Natural Heritage (SNH), Natural England and the Centre for Ecology Hydrology (CEH) for supporting this work.

References

Leith I.D., van Dijk N., Pitcairn C.E.R., Wolseley P.A., Whitfield C.P., Sutton M.A. (2005) Biomonitoring methods for assessing the impacts of nitrogen pollution: refinement and testing. JNCC Report No. 386, Peterborough, UK.

Loubet B., Milford C., Hill P.W., Tang Y.S., Cellier P., Sutton M.A. (2002) Seasonal variability of apoplastic NH_4^+ and pH in an intensively managed grassland. Plant and Soil, 238, 97–110.

Pitcairn C.E.R., Leith I.D., Sheppard L.J., Sutton M.A., Fowler D., Munro R.C., Tang Y.S., Wilson D. (1998) The relationship between nitrogen deposition, species composition and foliar nitrogen concentrations in woodland flora in the vicinity of livestock farms. Environmental Pollution, 102, 41–48.

Sutton M.A., Pitcairn C.E.R., Leith I.D., van Dijk N., Tang Y.S., Skiba U., Smart S., Mitchell R., Wolseley P., James P., Purvis W., Fowler D. (2004) Bioindicator and biomonitoring methods for assessing the effects of atmospheric nitrogen on statutory nature conservation sites. JNCC report No. 356, Peterborough, UK.

Tang Y.S., Cape J.N., Sutton M.A. (2001) Development and types of passive samplers for NH_3 and NO_x. The Scientific World 1, 513–529.

Chapter 18
Spatial Planning as a Complementary Tool to Abate the Effects of Atmospheric Ammonia Deposition at the Landscape Scale

Ulrike Dragosits, Mark R. Theobald, Chris J. Place, Helen M. ApSimon, and Mark A. Sutton

18.1 Introduction

18.1.1 Spatial Variability of NH_3 Emissions, Concentration, Deposition and Effects

Ammonia (NH_3) is emitted mainly from agricultural practices, with NH_3 concentrations decreasing rapidly away from sources (e.g. Sutton et al. 1998; Pitcairn et al. 1998, 2003). As a consequence there is a high spatial variability in N deposition and its ecological effects in agricultural landscapes (e.g. Dragosits et al. 2002, 2005, 2006), in addition to differences in sensitivity to additional nitrogen between habitat types (e.g. Fangmeier et al. 1994; Pitcairn et al. 1998, 2003; Mitchell et al. 2004).

This variability (Fig. 18.1) points to the potential to include locally tailored abatement measures as part of strategies to protect sensitive vegetation from NH_3 deposition (Dragosits et al. 2006).

18.1.2 Abatement Policies

National abatement policies typically include uniform recommendations for technical abatement measures, such as ploughing in manures after land spreading. In general, ammonia abatement strategies may be implemented in the following ways:

- All sources and source sectors reduce emissions by the same percentage (blanket approach, "common misery").

U. Dragosits
Centre for Ecology & Hydrology, Bush Estate, Penicuik, Midlothian, EH26 0QB,
United Kingdom

M. Sutton, S. Reis and S.M.H. Baker (eds), *Atmospheric Ammonia,*
© Springer Science + Business Media B.V. 2009

Fig. 18.1 Emissions, concentration and dry deposition of NH₃ at a landscape scale for an area in central England. The farms in the area are marked as F1–F4, and the nature reserves are indicated as R1–R3 (The figures have been adapted from Dragosits et al. 2006)

- Certain source sectors reduce more, because they are either cheaper/easier to abate or fall within specific regulations (e.g. IPPC 1999; also "common misery" for the selected sectors).
- Certain sources and source sectors reduce more or less depending on their source/receptor interactions, so that abatement measures are targeted to the geographic areas where they have the most significant impact.

The advantage of this last approach is that abatement is targeted precisely and only where necessary. The disadvantage is that it is less equal for farmers, and depends on the spatial location of sources relative to sensitive sink areas, which requires individual assessment. If this is followed through at a local level, the logical consequence is to look at individual farmsteads and fields near sensitive ecosystems, and prescribe local solutions (bearing in mind unequal costs for farmers and political solutions for this).

18.1.3 Spatial Planning – A Definition

Measures designed to decrease impacts of emissions from nearby sources on sensitive sink areas by taking account of relative spatial relationships (e.g. Bleeker and Erisman 1998; Lekkerkerk 1998). Examples are:

- Introducing buffer zones of low emission agriculture around NH_3 sources or sinks
- Planting tree belts around NH_3 sources or sinks
- Consideration of alternative locations for new sources at the planning stage

Some example spatial planning measures are illustrated for a study area in central England in Fig. 18.2.

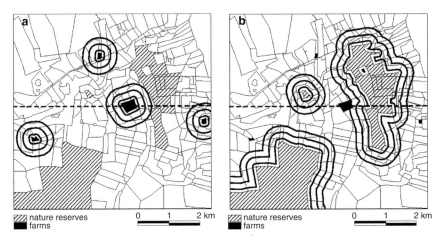

Fig. 18.2 Location of farms, nature reserves and spatial planning measures (**a**) around farms and (**b**) around nature reserves (marked as bold concentric shapes around the farms); widths of zones shown are 100, 300 and 500 m. The dashed line crossing the study area from east to west shows the location of a transect used in the analysis of the results (The figures have been adapted from Dragosits et al. 2006)

18.2   Methodology

A spatial dataset for a landscape in central England was modified in terms of landcover/land use, to allow modelling of a series of spatial planning scenarios. Several semi-natural areas in the landscape were designated as nature reserves (illustrated in Fig. 18.2), and low-emission buffer zones and tree belts of different widths were introduced around the farms and the nature reserves, respectively. In a further set of scenarios, the large poultry farm at the centre of the study area was moved to different locations (1 and 3 km west) and removed entirely.

Spatial emission inventories were then calculated at a 25×25 m grid resolution, using detailed datasets on livestock housing, manure storage, fertiliser and manure applications, grazing records etc.

The Local Area Dispersion and Deposition (LADD) model, originally developed by Hill (1998) and improved subsequently (e.g. Dragosits et al. 2002), was then used to calculate atmospheric concentrations and dry deposition of NH_3 for the landscape area. For the purpose of assessing the exceedance of critical loads (e.g. Nilsson and Grennfelt 1988; Bull 1991, 1995) for nutrient nitrogen to the nature reserves, additional data on (the much less spatially variable) NO_y and wet NH_x deposition were included from national 5×5 km grid estimates for the UK (Smith et al. 2000).

18.3   Results

Results from the modelling showed, as expected, the highest concentrations of NH_3 around the farms, with the large poultry farm at the centre of the study area being the most intensive "hot spot".

The poultry farm also dominates the patterns of NH_3 dry deposition, with critical loads exceeded for semi-natural vegetation in the vicinity of all farms (see maps for the base scenario, Fig. 18.3). The analysis of the spatial planning scenarios is summarized below.

18.3.1   Tree Belts and Low Emission Buffer Zones

The analysis shows that tree belts can reduce deposition to sensitive areas, with trees surrounding the sensitive habitats being more effective than trees around the sources (Fig. 18.3a, b, d).

Low emission buffer zones around sink areas also result in useful reductions in NH_3 deposition (Fig. 18.3c).

Smaller nature reserve sites benefit to a greater degree from such spatial planning measures, as large reserves can provide their own buffer zone to some degree.

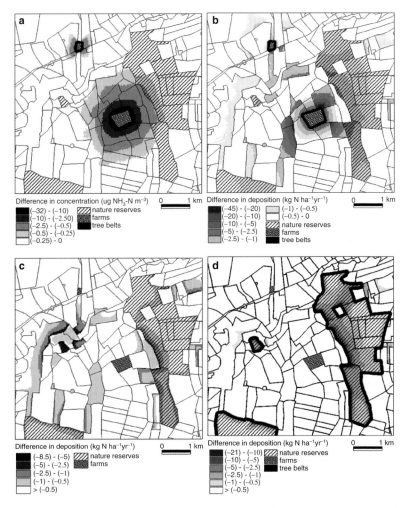

Fig. 18.3 (**a**) Absolute difference in NH₃ concentrations ($\propto$g m⁻³) between base scenario and 50 m tree belts around farms; Difference in dry deposition of NH₃ (kg NH₃-N ha⁻¹ year⁻¹) between the base scenario and (**b**) tree belts (50 m width) around the farms (**c**) buffer zone of low-emission agriculture (500 m width) around reserves, and (**d**) tree belts around the reserves (50 m width). Reprinted from Environmental Science and Policy 9/Issue 7–8, Dragosits et al., The potential for spatial planning at the landscape level to mitigate the effects of atmospheric ammonia deposition, Pages 626–638, Copyright (2006), with permission from Elsevier

18.3.2 Relocation of Sources/Planning of New Farms

Relocating point sources or using planning policies to ensure the location of large NH₃ point sources is at least 2–3 km from the sensitive habitats results in substantial reductions in NH₃ dry deposition (Fig. 18.4).

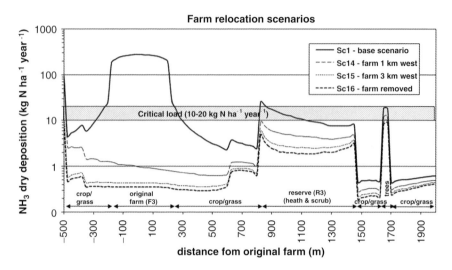

Fig. 18.4 Transect of NH$_3$ dry deposition (kg NH$_3$-N ha^{-1} year^{-1}) with (**a**) farm removed, (**b**) base scenario, (**c**) farm moved 1 km west, (**d**) farm moved 3 km west. Critical load for temperate forests and dry heathland: 10–20 kg N ha^{-1} year^{-1} (From Acherman and Bobbink 2003). Reprinted from Environmental Science and Policy 9/Issue 7–8, Dragosits et al., The potential for spatial planning at the landscape level to mitigate the effects of atmospheric ammonia deposition, Pages 626–638, Copyright (2006), with permission from Elsevier.

18.4   Discussion and Conclusions

Overall, tree belts appear more successful at decreasing NH$_3$ dry deposition than low-emission buffer zones, and spatial planning measures around sinks seem to be more effective than around sources in the study area. The latter is partially due to measures directly at the reserves decreasing incoming deposition from all surrounding sources, not just nearby farmsteads.

An additional benefit of tree belts is the increased dispersion of atmospheric NH$_3$ due to higher surface roughness of trees, and thus turbulence created by tree belts (Theobald et al. 2001, 2004a, b), compared with low-emission buffer zones.

Smaller nature reserves have been shown to benefit most from spatial planning measure, whereas the outer perimeter of larger reserves can act as their own buffer zone.

Spatial planning measures have been shown to provide a viable complementary approach to technical abatement measures. However, wider aspects and practical advantages and disadvantages for both farms and nature reserves need to be considered in detail on a case-by-case basis.

Acknowledgements   The authors are grateful to the Natural Environment Research Council (NERC) and the Department for Environment, Food and Regional Affairs (Defra) for funding of the LANAS project under the GANE Thematic Programme, and of research at Imperial College on pollution abatement strategies.

References

Acherman B., Bobbink R. (eds.) (2003) Empirical Critical Loads for Nitrogen. Proceedings of the Expert Workshop, Berne. 11–13 November 2002. Environmental Documentation No. 164. Swiss Agency for the Environment, Forests and Landscape (SAEFL), Berne, Switzerland.

Bleeker A., Erisman J.W. (1998) Spatial planning as a tool for decreasing nitrogen loads in nature areas. Environ. Pollut. 102(Suppl. 1), 649–655.

Bull K.R. (1991) The critical loads/levels approach to gaseous pollutant emission control. Environ. Pollut. 69, 105–123.

Bull K.R. (1995) Critical loads—possibilities and constraints. Water Air Soil Pollut. 85, 201–212.

Dragosits U., Theobald M.R., Place C.J., Lord E., Webb J., Hill J., ApSimon H.M., Sutton M.A. (2002) Ammonia emission, deposition and impact assessment at the field scale: a case study of sub-grid spatial variability. Environ. Pollut. 117, 147–158.

Dragosits U., Theobald M.R., Place C.J., Smith J.U., Sozanska M., Brown L., Scholefield D., Del Prado A., Angus A., Hodge I.D., Webb J., Whitehead P.G., Fowler D., Sutton M.A. (2005) Interaction of nitrogen pollutants at the landscape level and abatement strategies. In: Zhu Z., Minami K., Xing G. (eds.) 3rd International Nitrogen Conference: contributed papers, Nanjing, China, 12–16 October 2004. Science Press USA Inc., pp. 30–34 [ISBN 1-933100-10-9].

Dragosits U., Theobald M.R., Place C.J., ApSimon H.M., Sutton M.A. (2006) The potential for spatial planning at the landscape level to mitigate the effects of atmospheric ammonia deposition. Environ. Sci. Policy 9, 626–638.

Fangmeier A., Hadwiger-Fangmeier A., van der Eerden L., Jaeger H.J. (1994) Effects of atmospheric ammonia on vegetation—a review. Environ. Pollut. 86, 43–82.

Hill J. (1998) Applications of Computational Modelling to Ammonia Dispersion from Agricultural Sources. Ph.D. thesis. Imperial College, Centre for Environmental Technology, University of London, London.

Lekkerkerk L. (1998) Implications of Dutch ammonia policy on the livestock sector. Atmos. Environ. 32 (Ammonia Special Issue (3)), 581–587.

Mitchell R.J., Sutton M.A., Truscott A.M., Leith I.D., Cape J.N., Pitcairn C.E.R., van Dijk N. (2004) Growth and tissue nitrogen of epiphytic Atlantic bryophytes: effects of increased and decreased atmospheric N deposition. Funct. Ecol. 18, 322–329.

Nilsson J., Grennfelt P. (eds.) (1988) Critical Loads for Sulphur and Nitrogen. Report of a Workshop Held in Skokloster, Sweden, 19–24 March 1988. Nordic Council of Ministers, Copenhagen.

Pitcairn C.E.R., Leith I.D., Sheppard L.J., Sutton M.A., Fowler D., Tang Y.S., Munro R.C., Wilson D. (1998) The effects of ammonia on the relationship between woodland flora and nitrogen deposition in the vicinity of livestock buildings. Environ. Pollut. 102(S1), 41–48.

Pitcairn C.E.R., Fowler D., Leith I.D., Sheppard L.J., Sutton M.A., Kennedy V., Okello E. (2003) Bioindicators of enhanced nitrogen deposition. Environ. Pollut. 126, 353–361.

Smith R.I., Fowler D., Sutton M.A., Flechard C., Coyle M. (2000) Regional estimation of pollutant gas dry deposition in the UK: model description, sensitivity analysis and outputs. Atmos. Environ. 34, 3757–3777.

Sutton M.A., Milford C., Dragosits U., Place C., Singles R., Smith R.I., Pitcairn C.E.R., Fowler D., Hill J., ApSimon H., Ross C., Hill R., Jarvis S.C., Pain B.F., Phillips V.R., Harrison R., Moss D., Webb J., Espenhahn S.E., Lee D.S., Hornung M., Ullyett J., Bull K.R., Emmet B.A., Lowe J., Wyers G.P. (1998) Dispersion, deposition and impacts of atmospheric ammonia: quantifying budgets and spatial variability at local scales. Environ. Pollut. 102, 349–361.

Theobald M.R., Milford C., Hargreaves K.J., Sheppard L.J., Nemitz E., Tang Y.S., Phillips V.R., Sneath R., McCartney L., Harvey F.J., Leith I.D., Cape J.N., Fowler D., Sutton M.A. (2001) Potential for ammonia recapture by farm woodlands: design and application of a new experimental facility. In: Optimizing Nitrogen Management in Food and Energy Production and Environmental Protection: Proceedings of the Second International Nitrogen Conference on Science and Policy. The Scientific World 1(S2), 791–801.

Theobald M.R., Dragosits U., Place C.J., Smith J.U., Brown L., Scholefield D., Webb J., Whitehead P.G., Angus A., Hodge I.D., Fowler D., Sutton M.A. (2004a) Modelling nitrogen fluxes at the landscape scale. Water Air Soil Pollut.: Focus 4, 135–142.

Theobald M.R., Milford C., Hargreaves K.J., Sheppard L.J., Nemitz E., Tang Y.S., Dragosits U., McDonald A.G., Harvey F.J., Leith I.D., Sneath I.D., Williams A.G., Hoxey R.P., Quinn A.D., McCartney L., Sandars D.L., Phillips V.R., Blyth J., Cape J.N., Fowler D., Sutton M.A. (2004b) AMBER: Ammonia Mitigation by Enhanced Recapture. Impact of Vegetation and/or Other On-farm Features on Net Ammonia Emissions from Livestock Farms. Final Report on Project WA0719 to Defra (Land Management Improvement Division). CEH Edinburgh, 22 (+110, appendices).

Part IV
Regional Modelling of Atmospheric Ammonia

Chapter 19
Modelling of the Atmospheric Transport and Deposition of Ammonia at a National and Regional Scale

Addo van Pul, Ole Hertel, Camilla Geels, Anthony J. Dore,
Massimo Vieno, Hans A. van Jaarsveld, Robert Bergström, Martijn Schaap,
and Hilde Fagerli

19.1 Introduction

An overview of the current status of the modeling of the atmospheric transport and deposition of ammonia at a national and regional scale is presented. Firstly, the paper deals with the parameterizations of the transport and removal processes of ammonia used in modeling. Subsequently, an overview of the models currently in use describing the ammonia concentration and deposition at a national or (sub-) European scale is given. The emphasis lies on (a) the process parameterizations that are particularly important for ammonia such as the emission, deposition and chemical conversion and (b) the validation with measurements. Finally, discussion items as input to the Working Group 4 session are defined that follow from the model overview and validation.

Measurements of ammonia and ammonium concentrations and deposition from monitoring programmes are crucial for providing information about trends and actual loads of ammonia in the environment. Such data may together with data from field campaigns form the basis of our understanding of the physical and chemical processes governing the fate of ammonia. However, modelling of the concentrations and deposition of ammonia extends our possibilities substantially. Measurements are usually carried out at a limited number of locations. Model calculations are therefore used to obtain information with higher geographical resolutions and for estimates of loads at locations not covered by the monitoring network. Well-tested and validated models are furthermore highly useful in the interpretation of the measurements.

With modelling the relation between source and receptor is established. This means that the contribution of specific sources, economic sectors or countries to the concentration and deposition can be estimated. Modelling is the only way to

Addo van Pul
RIVM, LVM, PO Box 1, 3720 BA Bilthoven, The Netherlands

M. Sutton, S. Reis and S.M.H. Baker (eds), *Atmospheric Ammonia,*   301
© Springer Science + Business Media B.V. 2009

carry out scenario studies in environmental management. The scenario studies may answer questions like: what will happen when a source is added or a specific measure will be implemented?; how will the future situation look like given current legislation and are further emission reductions needed to meet the environmental goals?

Modelling the transport, transformation and deposition of atmospheric ammonia is a challenge due to the complexity of the governing processes (Asman et al. 1998; Hertel et al. 2006). Ammonia is generally speaking emitted from low source heights: the release takes place from the ventilation system of a stable or from volatilisation during the application of manure and from a vast amount of sources. This means that (a) ammonia, along with the fact that it deposits quickly, is deposited in larger amounts close to the source than is the case for other air pollution components and (b) the concentration and deposition patterns show large local gradients. In turn the deposited ammonia may re-emit and re-deposit. At the same time ammonia reacts (with a time scale of minutes to few hours) relatively fast with acid gases and particles in the atmosphere and forms aerosol-bound ammonium salts. The chemical transformation and the dispersion and deposition take place on similar time scales. Modelling the ammonia deposition upon ecosystems is less straightforward because of the possible re-emission of ammonia by ecosystems that have large nitrogen burden. Therefore specific parameterisations are needed for modelling the ammonia concentrations and deposition.

A number of different model types are currently in use in research as well as in environmental management in the field of ammonia. The models differ in a number of modeling concepts such as the spatial scale, the description of the emission and deposition processes. Analyses of the model performances as well as analyses of the differences in calculation results may provide useful insight into what is currently the state-of-the-art in modelling ammonia in the environment and how differences in parameterizations in models can explain observed differences between models. We focus here on the models that calculate the ammonia concentration and deposition at a national or regional scale.

So in summary the goals of this document and this Working Group at the Workshop are to:

1. Review the parameterizations used in the atmospheric dispersion and transport models of ammonia with the emphasis on the emission process, the vertical dispersion, the deposition process and the chemical conversion
2. Discuss the performance and validation of the models and discuss observed differences between modelled and measured ammonia concentrations and explain these differences in terms of (shortcomings in) the used process parameterizations
3. Define recommendations for improving the atmospheric dispersion and transport models of ammonia on a national and regional scale

First, a brief overview of the processes that describe the dispersion, transport and deposition of ammonia is given (Section 19.2. The focus in this part is on the emission parameterisations, description of the deposition processes and the chemical transformation. Subsequently, an overview of current models on a national and regional scale and their most important concepts, their performance and validation is presented (Section 19.3. In Section 19.4 we tentatively indicate the reasons for the differences in performance in terms of the modeling concepts.

19.2   Overview of Modeling of the Concentration and Deposition of Ammonia

19.2.1   Brief Overview of the Most Important Atmospheric Processes

Ammonia is emitted to the atmosphere mainly from agricultural sources. In essence the emission of ammonia is the evaporation from animal manure and is highly dependent on the specific agricultural activity and environmental circumstances. In the atmosphere ammonia is subject to dispersion and transport, removed by dry and wet deposition and transformed to aerosol-bound ammonium in reactions with acid gases and aerosols (Fig. 19.1, Hertel et al. 2006; Asman et al. 1998). Due to the relatively fast deposition and conversion process and the low emission height the atmospheric lifetime of ammonia is typically a few hours. Aerosol-bound ammonium has generally a much longer lifetime in the atmosphere and may, therefore, be transported over long distances (>1,000 km). The main removal path of the ammonium-containing aerosols is wet deposition.

19.2.2   Overview of Modelling Concepts and Parameterisations

Different approaches have been used in the modelling of the fate of ammonia in the atmosphere. The choice of the complexity of the parameterizations is often a function of among others (a) the state of knowledge of the process, (b) the availability of

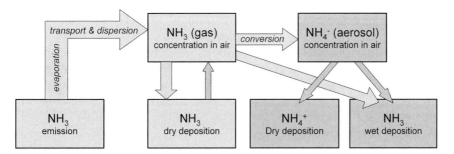

Fig. 19.1 The most important atmospheric processes that determine the fate of ammonia in the atmosphere. The thickness of the arrows indicates the relative importance of the process

input parameters, (c) the purpose of the model, (d) the available computer power. In the following subsections we elaborate the most important processes and the determining variables therein.

In the next subsections we briefly elaborate on these processes and the commonly used parameterizations for modelling these processes. Several parts of the description were taken from Hertel et al. (2006) overview of the current status of ammonia modelling where further details can be found.

19.2.2.1 Emissions

In modelling the emission amount from the sources as well as the variation in the emission factors and the spatial resolution of the data are very important. The emissions of ammonia mostly originate from animal housings and from the application of manure. The emission from animal housings depends mainly on the ventilation (Seedorf et al. 1998a) and the temperature inside the stables (Seedorf et al. 1998b; Wathes et al. 1998). The emissions of ammonia from manure application are a function of the application method, the meteorological conditions and the soil type (Huijsmans et al. 2003). Other sources of ammonia are grazing animals, storage facilities and fertilized crops (see e.g. Skjøth et al. 2006). For all sources it is common that the emission of ammonia is dependent to a large degree on environmental conditions. Therefore, the total ammonia emission varies to a large degree during the day and during the season (Battye et al. 2003; Gilliland et al. 2003).

Most transport-chemistry models deal with the seasonal and diurnal variations in a strongly simplified way. An important reason for this is the lack of input data (Hutchings et al. 2001).

The emission amount is often calculated from a combination of the activity data and the emission factor corresponding to the activity.

Most national or local scale inventories for ammonia are compiled as annual mean values for grids of varying resolution (e.g. Sutton et al. 2000; Bouwman et al. 1997; Dragosits et al. 1998). An exception is the European scale inventory (Gyldenkærne et al. 2005a, b) with a 10 × 10 km resolution using separate parameterizations for 16 source groups.

On a national level various inventories with high spatial resolution may be found. Examples are the UK 5 × 5 km inventory (Dragosits et al. 1998), and the similar Dutch inventory (Duyzer et al. 2001), which has also been compiled for resolutions of 500 × 500 m (Pul et al. 2004) and even 250 × 250 m (Duyzer et al. 2001) for 11 source groups and a Danish inventory with a 1 × 1 km (Skjøth et al. 2006) and a 100 × 100 m resolution (Geels et al. 2006; Ellermann et al. 2006) with information on 16 source groups.

At the European level, EMEP and CORINAIR collect inventories for the annual emissions on a grid with a spatial resolution of 50 × 50 km. Within the EUROTRAC GENEMIS project more detailed inventories (16.67 × 16.67 km) for the emissions from EU member countries were compiled (Schwarz et al. 2000; Wickert et al. 2001).

This inventory has been used to subdivide the annual EMEP emission inventories to 16.67 × 16.67 km (Hertel et al. 2002)

19.2.2.2   Transport and Dispersion

In describing the dispersion of the emitted ammonia in the atmosphere this process may be split into a local part at a meter to kilometre scale and a longer range part at a scale typically tens to thousand of kilometres. The latter is often referred to as the mean transport of air pollution with the mean wind flow. For ammonia, emitted at low sources, particularly the local scale dispersion is important when one wants to describe concentration and depositions at a (sub) kilometer scale (Asman 2001). The dispersion at local scale can be calculated using K-theory models in a grid framework or analytical solutions to the advection diffusion equation which deliver continuous profiles with height (Gryning et al. 1987). At what vertical grid spacing the calculation should take place depends on the spatial and temporal scale of the ammonia problem under consideration.

Describing the transport of ammonia at larger spatial scales is not different from other air pollution components. Only a brief overview of the commonly used concepts will be presented here. In modelling the transport or advection of in air pollution there are basically two types of models: the Lagrangian and the Eulerian type. In the Lagrangian models, an air parcel is traced along a trajectory computed from wind speed and wind direction. The trajectories in the models are used in a backward mode describing pollutants arriving at a number of selected receptor points and in a forward mode describing transport of pollutants from a number of selected sources in the model domain. In Eulerian models, calculations are performed simultaneously for a grid of receptor points. For each of these receptor points transport in and out of the grid cell is computed. Besides the transport, all other physical and chemical processes included in the models may in principle be identical for the two types of models; Lagrangian and Eulerian. It has been shown that a 20% uncertainty in the calculation of 96 h back-trajectories is quite common (Stohl 1998; Stohl and Koffi 1998), and it is generally considered that the transport description is more accurate in Eulerian models.

The Eulerian models are generally more computer resource demanding especially when a high geographical resolution is wanted which particularly is the case for ammonia related problems. However, with the increasing computer power becoming available, and with application of nested grid techniques solving the mathematical challenge of having a high resolution domain defined within a coarser resolution domain, this problem is becoming less significant.

It has been a general tendency over the last decade that Eulerian models have been replacing Lagrangian models where these previously were in use. However, the faster nature of Lagrangian models means that they are still in use for environmental problems which desire a relatively high spatial resolution such as for ammonia.

19.2.2.3 Dry Deposition

The dry deposition is the most important removal process of ammonia from the atmosphere. The dry deposition process is a strong function of the transport rate from the ammonia in the air to the surface and the physical, chemical and biological characteristics of the surface.

NH_3 is able to stick to almost any surface, and the dry deposition is therefore often limited by the transportation rate to the surface (Asman 1998). One of the important path ways for dry deposition of NH_3 is uptake through the stomata of plants. However, in addition there are two other major path ways for transport of NH_3 to plants: absorption of NH_3 to dew on the plants or to the thin water film on the leaves epidermis (Nemitz et al. 2004). Experimental data have shown that a co-deposition of SO_2 and NH_3 to the surface takes place. It has thus been shown that the SO_2/NH_3 concentration ratio together with relative humidity and temperature are important factors for the deposition of both SO_2 (Fowler et al. 2001) and NH_3 (Neirynck et al. 2005) to natural surfaces.

In transport-chemistry models, the dry deposition of gases and particles is often described with a deposition velocity and the concentration of the substance at a reference height. In turn the deposition velocity is often parameterized with a so-called resistance model in which the transport to the surface and the surface uptake is described with resistances (Wesely 1989). The dry deposition velocity for a gaseous compound is expressed as the reciprocal value of the total resistance to transport down to and removal on to the surface:

$$V_d = \frac{1}{R_t} = \frac{1}{R_a + R_b + R_c} \qquad (19.1)$$

where R_t is the total resistance, R_a is the aerodynamic resistance, R_b is the quasi-laminar sub-layer resistance, and R_c is the surface resistance.

Generally speaking the resistances that describe the physical or meteorological part of the transport are well known under the assumption that the roughness characteristics are known. Using this resistance scheme it is assumed that the surface concentration of the air pollutant is zero. However, for ammonia this is not the case and formulations for the surface concentration are needed. So for ammonia the deposition process is in principle the net result of an exchange process which is bi-directional. The surface concentration of ammonia is often referred to as the compensation point, being the concentration where the exchange of ammonia changes from deposition to (re-) emission or visa versa. A number of experimental studies have demonstrated the bi-directional nature of NH_3 exchange for various types of vegetation: including conifer forest (Andersen et al. 1999; Duyzer et al. 1992; Wyers and Erisman 1998), moorland (Sutton et al. 1993), grass (Phillips et al. 2004), heathland (Nemitz et al. 2004), cereal crops (Schjoerring et al. 1993) and agricultural grassland (Milford et al. 2001; Wichink et al. 2007).

Specific dry deposition sub-models for the surface resistance that include the description of a compensation point for NH_3 have been derived and implemented in connection with the analysis of different plant surfaces e.g. for beans (Farquhar et al. 1980), Oilseed rape plants (Husted et al. 2000), and *Calluna vulgaris* (Schjoerring et al. 1998). It is common to apply a two or three pathway process description (Erisman et al. 1994; Loubet et al. 2001): (a) a stomatal pathway, which is bi-directional and modelled using a stomatal compensation point, and (b) a plant surface pathway, which denotes exchange with water surfaces or waxes on the plant surface and (c) a soil surface pathway particularly important for sparse canopies or wet soil. The parameterization of the stomatal resistance is rather established and can be found in Baldocchi et al. (1987) or Wesely (1989). The stomatal compensation point may be calculated from knowledge about the aqueous phase chemistry. The equilibrium NH_3 ambient air concentration for the stomatal compensation point has been expressed as (Sorteberg and Hov 1996):

$$[NH_3(g)] = \chi_{cp} = 10^{(1.6035-4207.62/T)} \frac{[NH_4^+]}{[H^+]} \qquad (19.2)$$

Where χ_{cp} is the compensation point concentration of NH_3, and $[NH_4^+]$ and $[H^+]$ are the concentrations of ammonium and hydrogen ion in stomatal cavity, respectively. The ratio $[NH_4^+]/[H^+]$ is often referred to as Γ. This Γ is vegetation dependent (Sutton et al 1993, 1998; Nemitz et al. 2001).

It has been shown that the leaf surface may work as a capacitance for NH_3 and SO_2 uptake, and that this capacitance increase with humidity (Van Hove et al. 1989). This transport is independent of solar radiation and contrary to the uptake through stomata, this uptake will also take place during the night. However, the uptake of ammonia by the leaf surface is often parameterized by fixed low values or descriptions as a function of relative humidity (Sutton et al. 1993, 1998; Erisman et al. 1994).

A special issue is the dry deposition to marine waters. Experimental studies have shown that over sea the atmospheric fluxes of NH_3 may also be upward or downward (Sorensen et al. 2003; Quinn et al. 1988; Lee et al. 1998) depending on the meteorological conditions and the relationship between the pH and contents of NH_4^+ in the upper surface waters on the one side, and the NH_3 concentrations in ambient air just above the water surface on the other side. This bi-directional nature is similar to the above described fluxes for vegetation over land. An expression for the ammonia concentration close to the marine surface is given by Asman (1998).

19.2.2.4   Wet Deposition

Wet deposition takes place by uptake of pollutants in precipitation (rain, snow, hail) as well as in cloud droplets – termed below-cloud and in cloud scavenging, respectively. Uptake in cloud droplets may not necessarily lead to deposition, since clouds often evaporate without producing precipitation; in average every 10th cloud encountered by an air parcel precipitates (Raes et al. 1993). A cloud droplet has a considerably longer atmospheric residence time compared with a rain droplet. In-cloud scavenging is therefore generally a more efficient removal process for pollutants than the process of below-cloud scavenging. The wet deposition is a very important removal process for ammonia since ammonia is highly soluble in water. For ammonia both in-cloud and below-cloud scavenging are of importance. The uptake in rain and cloud droplets is limited by the diffusion into the droplet rather than the equilibrium concentration in the droplet.

In recent years many models include complex wet phase chemistry. However, in many transport-chemistry models, the overall principle behind the description of wet deposition is still considerably more simplified and based on a fixed relationship between the concentration in the droplet and the concentration in ambient air. Given this relationship is known, the rate of removal may be determined by a so-called scavenging coefficient:

$$\Lambda = \frac{S \times I}{H} \qquad (19.3)$$

where Λ is the scavenging coefficient, S is the scavenging ratio (the ratio between the concentration in air and droplet), I is the precipitation intensity and H is the height at which the wet scavenging takes place.

In many cases the dependency of precipitation intensity is neglected and a constant scavenging rate is applied. The scavenging ratio for gases depends also on the solubility in water of the gas in question, the water content in the cloud and the ambient temperature. The scavenging ratio has been expressed in various ways, but one example is the one applied in the ACDEP model (Hertel et al. 1995):

$$S = \frac{1}{\dfrac{(1-cl)}{(H_{eff} \times R \times T)} + cl} \qquad (19.4)$$

where cl is the cloud water content, H_{eff} is the effective Henry's law coefficient, R is the gas constant and T is the ambient temperature.

The accuracy of precipitation data is of course crucial for producing reliable wet deposition estimates. Precipitation amounts and intensity is highly heterogeneous and may vary strongly within short distances (Badas et al. 2006).

The below-cloud scavenging of ammonia may be of importance in source regions. It has been shown from experimental results in America and South Korea that the wet deposition of NH_4^+ is correlated to the local NH_3 emission density (Aneja et al. 2003; Park and Lee 2002). Also van Jaarsveld et al. (2000) clearly show a correlation between spatial distribution of the wet deposition and the ammonia emissions over the Netherlands. When considering the contribution from a single farm, the wet deposition of NH_3 will, however, be very limited. This is due to the short periods with precipitation compared with the dry periods, and at the same time a result of the short residence time of the pollutants in the nearby region of the farm. In the dutch OPS model a description of the below-cloud scavenging is incorporated (van Jaarsveld 2004).

19.2.2.5 Chemical Conversion

In the atmosphere NH_3 is quickly transformed into particulate NH_4^+ in the reaction with acid gases and aerosol particles (Seinfeld and Pandis 1998). If sulphuric acid (H_2SO_4) is present in the atmosphere, gaseous NH_3 will practically always react with H_2SO_4 in gas or aerosol phase. The H_2SO_4 is formed from oxidation of SO_2 by OH radical or by ozone (O_3). The latter process is pH dependent, and may be catalyzed by NH_3 (Junge and Ryan 1958; Apsimon et al. 1994); since NH_3 increases the pH when it is taken up by the aerosols. Presence of nitric acid (HNO_3) and/or hydrochloric acid (HCl) together with NH_3 will lead to an equilibrium between these gases and their aerosol phase reaction products – the ammonium salts: ammonium nitrate (NH_4NO_3) and ammonium chloride (NH_4Cl). In the reactions between gas phase NH_3 and gas phase acids, new aerosol particles are formed. However, NH_3 may also condense onto existing atmospheric particles.

In transport-chemistry models, the reaction between NH_3 and H_2SO_4 forming ammonium bisulphate (NH_4HSO_4) and ammonium sulphate ($(NH_4)_2SO_4$) is considered as an irreversible process. For most atmospherically relevant conditions, the humidity is high enough so that most inorganic aerosol particles exist as highly concentrated salt solutions rather than solid crystals. Several field and model process studies of cloud processing have thus indicated that once ammonium sulphate is incorporated into a cloud, re-mixing with other ions in the aqueous phase can effectively achieve NH_3 degassing on subsequent evaporation of the cloud (Bower et al. 1995; Wells et al. 1997; Milford et al. 2000). The rate of the reaction between NH_3 and H_2SO_4 has been analysed in details in a number of laboratory studies (Baldwin and Golden 1979; Huntzicker et al. 1980; McMurry et al. 1983). At high relative humidity the limiting factor for the transformation is the molecular diffusion of NH_3 to the acid particles, whereas at low humidity only 10–40% of the collisions between NH_3 gas molecules and H_2SO_4 containing particles lead to reaction (Huntzicker et al. 1980; McMurry et al. 1983). For small particles the large surface area makes the diffusion process more efficient. Organic material on the surface of the particles may, however, limit the uptake of NH_3 (Daumer et al. 1992). In the traditional model formulation the reaction takes place over two steps:

$$NH_3 + H_2SO_4 \rightarrow NH_4HSO_4$$
$$NH_3 + NH_4HSO_4 \rightarrow (NH_4)_2SO_4 \quad (19.5)$$

The reaction with HNO_3 is on the other hand a reversible process (Harrison and Pio 1983; Seinfeld and Pandis 1998). Experimental studies have shown that to a good approximation an equilibrium product of the gas phase concentrations of NH_3 and HNO_3 at saturation of the air, may be expressed by a function depending solely on temperature and humidity (Stelson et al. 1979; Stelson and Seinfeld 1982):

$$NH_3 + HNO_3 \leftrightarrow NH_4NO_3 \quad (19.6)$$

Besides the reactions with H_2SO_4 and HNO_3, NH_3 may also take part in a reaction with HCl and form NH_4Cl (Pio and Harrison 1987). Usually HCl appear in very low ambient concentrations, but it may be released from sea spray particles when these take up HNO_3:

$$HNO_3 + NaCl \rightarrow NaNO_3 + HCl$$ (19.7)
$$NH_3 + HCl \leftrightarrow NH_4Cl$$

Similarly as for the reaction between NH_3 and HNO_3, experimental studies have determined an equilibrium product at saturation of the air with these two gases (Pio and Harrison 1987).

In a number of transport models the conversion of ammonia to ammonium is parameterized in a simplified way using pseudo first order reaction rates of the conversion.

19.3   Overview Current Models

The models included in this overview are listed in Table 19.1.

19.3.1   Overview Main Features of Each Model

In this section the main features, validation and uncertainty sources of the models are presented.

19.3.1.1   FRAME

A summary of the model description is given below. More detailed descriptions and analysis of results are given in Singles et al. (1998), Fournier et al. (2004, 2005a, b), Vieno (2005) and Dore et al. (2007).

Model Description

- $5 \times 5 \, km^2$ resolution over the British Isles (incorporating the Republic of Ireland) grid dimensions: 244×172 with a $1°$ angular resolution in the trajectories.
- Input gas and aerosol concentrations at the edge of the model domain are calculated with FRAME-Europe, using European emissions and running on the EMEP 150 km scale grid.
- Air column divided into 33 layers moving along straight-line trajectories in a Lagrangian framework with a $1°$ angular resolution. The air column advection speed and frequency for a give wind direction is statistically derived from radiosondes measurements. Variable layer thickness from 1 m at the surface to 100 m at the top of the mixing layer.

Table 19.1 Overview of ammonia models at a national or regional scale

Name	Owner/reference	Type	Vertical structure; lowest troposhere	Emissions spatial scale = horiz. Resolution Eulerian models	Emissions sub sector split	Emissions temporal and meteorological variations
FRAME	CEH, Univ. Edin. (Singles et al. 1998; Vieno 2005; Dore et al. 2006)	Lagrangian	Variable; starting first layer from 1 m	5 × 5 km	Detailed agricultural activities	Annual
OPS	MNP/RIVM Van Jaarsveld 2004	Lagrangian	Variable; calculation of profile	1 × 1 km (500 × 500 m)	Detailed agricultural activities	function of ambient temperature
MATCH	SMHI	Eulerian	Variable, first layer 60 m, 6–11 layers below 1,500 m (1999–2003)	44 × 44 km (Europe) 11×11 km (Sweden)	(Agriculture) as one sector; minor emissions in other SNAP sectors	Simple schematic daily and seasonal variations
DAMOS/ DEHM	NERI (Christensen 1997; Frohn 2001, 2002, 2004)	Nested grid 3-D Eulerian	20 vertical layers with highest resolution in the PBL.	Europe on 50km × 50km, DK and nearby areas 16.67km × 16.67km.	The applied emission inventories are based on the snap coding of EMEP.	For the inner domain ammonia emission inventories on hourly basis using local agricultural practice and meteorological conditions
LOTOS/ EUROS	TNO/MNP/RIVM	Eulerian	First layer 25 m		As one sector	Simple schematic daily and seasonal variations
EMEP-unified	UNECE/EMEP	Eulerian	First layer 90 m	50 × 50km	(Agriculture) as one sector; minor emissions in other SNAP sectors	Simple schematic daily and seasonal variations

- Annual emissions of NH_3 are gridded separately for cattle, pigs, poultry, sheep, fertiliser and non-agricultural sources and mixed into the lowest surface layers with a source-dependent emissions height.
- Vertical diffusion in the air column is calculated using K-theory eddy diffusivity and solved with the Finite Volume Method.
- Wet deposition is calculated using a diurnally varying scavenging coefficient depending on mixing layer depth and a 'constant drizzle' approximation. A precipitation model is used to calculate wind-direction-dependent orographic enhancement of wet deposition. A single scavenging coefficient is used to represent in-could and below-cloud processes.
- Dry deposition for NH_3 is ecosystem specific and includes five land classes: forest, moorland, grassland, arable, urban and water. A canopy resistance parameterisation is employed including an optional canopy compensation point module for representation of bi-directional exchange of NH_3 (Vieno 2005). A fixed deposition velocity is used for ammonium aerosol.
- The model chemistry includes gas phase and aqueous phase reactions of oxidised sulphur and oxidised nitrogen and conversion of NH_3 to ammonium sulphate and ammonium nitrate aerosol.
- The modelled chemical species treated include: NH_3, NH_4^+ aerosol, NO, NO_2, HNO_3, PAN, NO_3^- aerosol, SO_2, H_2SO_4 and SO_4^{2-} aerosol.
- Current model run time: 25 min on CEH Edinburgh Beowulf cluster using 100 processors.

Performance and Validation of FRAME

The output from the model includes maps of annual average surface concentration of NH_3 (. 19.2) which may be used to assess exceedance of the critical level. Maps of annual wet deposition and vegetation-specific dry deposition of reduced nitrogen (Figs. 19.2a–c) are used for calculation of exceedance of critical loads for acid deposition and nitrogen deposition

Assessment of the accuracy of FRAME in estimating atmospheric concentrations and deposition rates of reduced nitrogen was made by comparison with measurements. For this purpose, data from the UK national ammonia monitoring network was employed comprising over 100 DELTA samplers and ALPHA samplers (http://www.cara.ceh.ac.uk/nh3network). The network uses monthly sampling from the CEH DELTA system, (DEnuder for Long Term Atmospheric sampling, Sutton et al. 2001). ALPHA samplers are passive diffusion samplers, developed for long term monitoring and suitable for use in remote areas with low ammonia concentrations (Tang et al. 2001). Wet deposition data were obtained from the secondary acid precipitation monitoring network, comprising fortnightly collections of precipitation from 38 sites with ion concentrations analysed by ion chromatography (NEGTAP 2001).

Figures 19.3a–c illustrate the correlation of the model with measurements. The correlation of modelled concentrations of NH_3 with measurements (Fig. 19.3a) shows considerable scatter. The principal reason for this is the highly localised

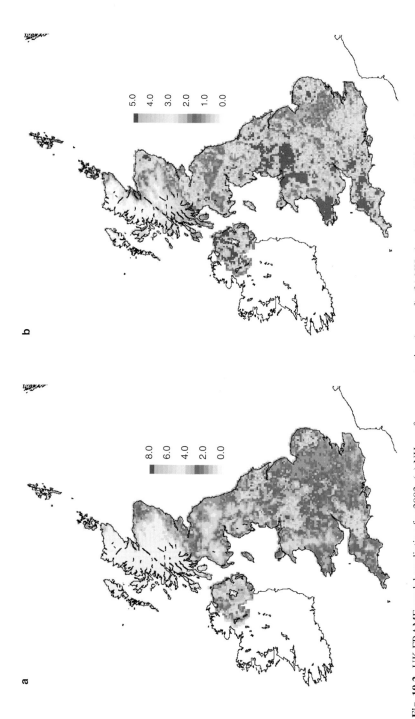

Fig. 19.2 UK FRAME model prediction for 2002: (**a**) NH$_3$ surface concentration ($\propto$g m^{-3}), (**b**) NH$_x$ dry deposition (kg N ha^{-1}) and (**c**) NH$_x$ wet deposition (kg N ha^{-1}) (Fournier et al. 2005; Dore et al. 2006)

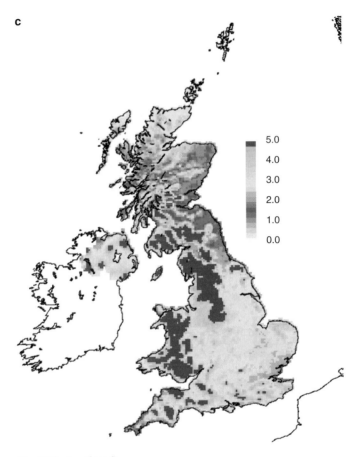

Fig. 19.2 (continued)

nature of NH$_3$ emissions, such that the modelled average concentration from a 5 × 5 km^2 model grid cell may differ significantly from that measured at a specific location within the grid cell (Dragosits et al. 2002). The graph shows evidence that, particularly at low concentrations, the model overestimates NH$_3$ surface concentrations. There is a need for finer scale national modelling of ammonia concentrations, preferably at a 1 km resolution, in order to perform a more accurate model-measurement comparison. A better correlation is observed between modelled and measured NH$_4^+$ concentrations (Fig. 19.3b) and wet deposition (Fig. 19.3c). This is due to the more slowly changing pattern in NH$_4$+ aerosol concentrations, which are not expected to vary on a scale smaller than the 5 km model grid resolution. Figure 19.3b shows that the model generally underestimates NH$_4^+$ aerosol concentrations which may indicate that the rate of production of NH$_4^+$ aerosol from NH$_3$ gas is underestimated in the model.

A sensitivity study was conducted to assess which model parameters were responsible for the greatest uncertainty in total dry and wet deposition of reduced nitrogen to the

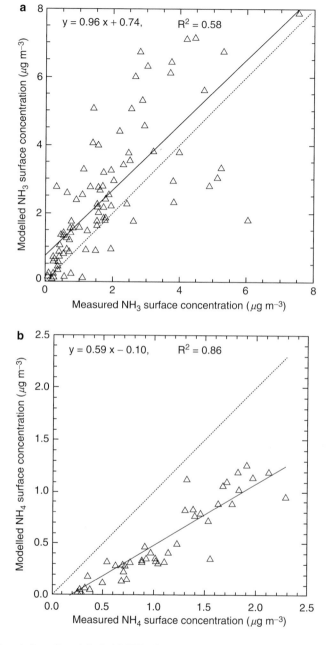

Fig. 19.3 Correlation of modelled: (**a**) NH$_3$, (**b**) NH$_4^+$ aerosol concentrations and (**c**) NH$_4^+$ wet deposition with measurements from the national monitoring network for the year 2002 over the UK (Fournier et al. 2005; Dore et al. 2006)

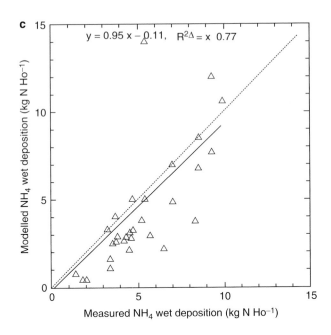

$$y = 0.95 x - 0.11, \quad R^{2\triangle} = x \; 0.77$$

Fig. 19.3 (continued)

UK. For the dry deposition of NH_3, the two most significant model parameters in introducing uncertainty were the canopy resistance, R_c (for which a 100% increase resulted in a 23% reduction in dry deposition) and the emissions of NH_3 (for which a 30% increase resulted in a 32% increase in dry deposition). For wet deposition of NH_x, the two most significant parameters were the washout coefficient (for which a 100% increase resulted in a 32% increase in wet deposition) and the vertical diffusivity (for which a 100% increase resulted in a 14% increase in wet deposition).

Observation sites of the UK national ammonia monitoring network can be grouped into three categories representative of: mixed agricultural, nature reserve and woodland. Figure 19.4 shows the correlation between the FRAME model predictions and the site observations for land-use specific sites. A strong difference in the gradient of the line of best fit is evident for the different groups of land categorisation. The model appears to be significantly over-estimating ammonia concentrations at low semi-natural sites. This occurs because nature sites tend to be 'havens' of low ammonia concentration within a model grid square which may have average emissions that are associated with intensive agricultural activity. On the other hand, it is noticeable that this division between different site types much improves the correlation between measurement and modelling for woodland and semi-natural areas (with R^2 of 0.84–0.91) compared with all the sites combined ($R^2 = 0.48$). Sites in such woodland and semi natural areas will be less influenced by local sources than the sites in mixed agricultural landscapes ($R^2 = 0.58$), demonstrating that natural spatial variability within each $5 \times 5 \, km^2$ grid square is a key reason for the modest R^2 values obtained between measured and modelled NH_3 concentrations.

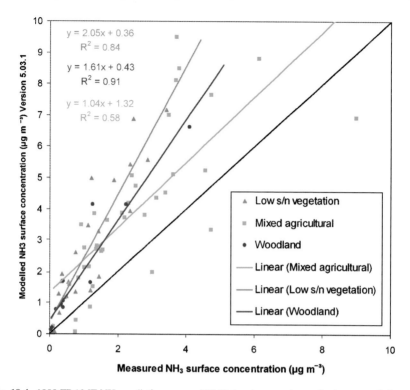

Fig. 19.4   1999 FRAME NH_3 predictions versus UK National ammonia monitoring network ($\mu g\ m^{-3}$) (Vieno 2005)

19.3.1.2   OPS

Model Description

General

The OPS model represents a combination of a Gaussian plume model for local-scale application and a trajectory model for long-range transport (Van Jaarsveld 1995, 2004). Especially in the case of ammonia the local scale plume model allows for a detailed approach of the low level release height in combination with near-source deposition. Dry and wet deposition for both NH_3 and the secondary product, NH_4^+ are calculated with a spatial resolution mainly dependent on the resolution of the emission data. The model is used for issues on acidification/ eutrophication as well as in heavy metals and persistent organic pollutants. Asman and van Jaarsveld (1992) applied the model to the calculating the NH_x distribution in Europe. Furthermore, the model was successfully applied to deduce SO_2 and NO_x trends from measurements of ambient concentrations. An early version of the OPS model (called TREND model) has taken part in a number of model intercomparison studies (Derwent et al. 1989; Sofiev et al. 1999). More recently, results of the OPS model were compared with those of the EMEP unified model on the issue of, among others, ammonia and ammonium concentration and deposition in the Netherlands (Velders et al. 2003).

In the next three sections the processes are described which are important particularly for ammonia.

Specific Processes

Emission process. The emissions from land spreading of manure (EC_{spread}) are adjusted to the meteorological conditions. The factor relative to the average emission strength reads:

$$EC_{spread} = 1 + 1.55 \ 10^{-5} \ [(100/R_a)^{0.8}(T+23)^{2.3}]^{1.2} \qquad (19.8)$$

in which T is the ambient temperature (in °C) and R_a the aerodynamic resistance over the lower 4 m of the boundary layer (see Eq. 19.1). In the latter the effect of wind speed and atmospheric stability is taken into account. The relation between emission and meteorological conditions is a fit on the results of the numerical air-soil exchange model DEPASS (Van Jaarsveld 1996). The factor determined in this way amounts to 1.8 on sunny days to 0.07 on very stable conditions. On average the factor varies from about 0.4 in January to 1.5 in July.

The emissions from animal housings (EC_{house}) are parameterized taking into account the outdoor temperature T:

$$EC_{house} = 1 + 0.04^*(T - T_{gem}) \qquad (19.9)$$

Where T_{avg} is the long-term average outdoor temperature, for the Netherlands of 10°C. The average factor for emissions from animal housings is about 0.7 in January and 1.3 in July. The factor 0.04 is based on relations with indoor temperatures and mechanically ventilated cattle-housings. It is assumed that the temperature variations for indoor and outdoor are similar. There is no distinction between the housing systems for cattle, pigs and poultry nor for natural and mechanically ventilated systems. Also there is no wind speed dependency included.

Vertical Dispersion

The vertical dispersion is described for a number of regimes in the atmospheric boundary layer each characterised by distinct scaling parameters (Holtslag and Nieuwstadt 1986). The parameterization of the vertical dispersion is modelled dependent on the height of the emission or centre of the plume and the atmospheric stability. In this way a dedicated modelling of the local dispersion close to (low level) sources is obtained of which results are verified using data of the so called Prairie grass experiment (Van Ulden 1978).

Chemical Conversion

The formation of ammonium is simulated using a one-dimensional model, including the relevant chemical reactions as applied in the MPA model (De Leeuw et al. 1990) and also deposition processes. This model is used on the basis of actual meteorological data and supplied with background concentrations of SO_2, NO_x, NH_3, O_3 and OH radicals. The conversion rate follows from the production of ammonium sulphate and ammonium nitrate over a (long) period, divided by the mean ammonia concentration. The conversion rates are then translated into a parameterisation for this rate using regression analyses. This resulted in the following relation between the $NH_3 > NH_4$ conversion rate K_{NH3}:

$$K_{NH3} = 0.67 + 1.36\,C_1 + 10.7\,C_2 + 3.06\,(C_2)^4 - 0.29\,(C_2)^6 \qquad (19.10)$$

Where $C_1 = NO_2/NH_3$ and $C_2 = SO_2/NH_3$ both in ppb/ppb.

Background concentrations on a $10 \times 10\,km$ scale for the period 1980–2002 are available for calculating the conversion. Averaged over the Netherlands the conversion in 1980 was about 16%/h and in 1997 about 5%/h.

Deposition

Dry deposition of ammonia is parameterized following the well known resistance modelling. This is built in the module DEPAC and is described in detail by Erisman et al. (1994). The deposition to the surface is described for three pathways: stomata, external leaf surface and soil. The resistances are a function of the wetness of the surface indicated with a switch wet/dry and of the ratio between the NH_3 and SO_2 concentrations indicated with a switch high/low. The latter switch is always set to high for the Dutch situation.

Aerodynamic resistances are calculated as a function of atmospheric conditions and roughness length. The dry deposition of particles is also modelled with a resistance scheme. It consists of the aerodynamic resistance and a R_{part} in which all surface related processes are incorporated. The R_{part} is split in a resistance parameterization for surfaces with a roughness length below 0.5 m in which the description is a function of turbulence alone and above 0.5 m in which the description is a function of turbulence, wind speed at canopy height and a particle collection efficiency. The latter has a value for dry and wet conditions.

Wet deposition is described as a combination of in-cloud and below cloud scavenging. Near sources a clear relation between the ammonia concentrations in rain and air are found (Van Jaarsveld et al. 2000). This means that below cloud scavenging is an important process for ammonia. The process of wet scavenging is considered as a irreversible process for a well soluble component as ammonia. The scavenging coefficient is parameterized as a function of the molecular diffusion, rain intensity and the drop-size distribution.

Performance and Validation

Ammonia concentrations were measured using passive samplers during 1 year at 159 locations equally distributed over the Netherlands. This dataset is used here as a reference for the capability of the model to describe spatial differences. Model calculations were carried out using 500 × 500-m resolution emission data as well as 5,000 × 5,000 m data. Results given in Fig. 19.5 show that the model simulates the spatial distribution nicely but that there is still an influence of the emission resolution. Of the 159 locations, 4 were situated within 50 m from animal housings. Yet, these locations do not pop up as outliers in the comparison indicating that the OPS model is capable of describing concentration gradients at a very local scale as long as the emissions are represented at the appropriate spatial scale.

Nevertheless, the concentrations have been found to be underestimated (about 30%) by the model on the basis of these emissions. The underestimation found for the concentrations at the passive sampler sites is similar to the underestimation found for the LML network sites. This again indicates that the LML sites are representative for the range of ammonia concentrations in the Netherlands.

A similar comparison was made on a much more local scale in the so called VELD project. In this project local emissions were inventoried with the help of

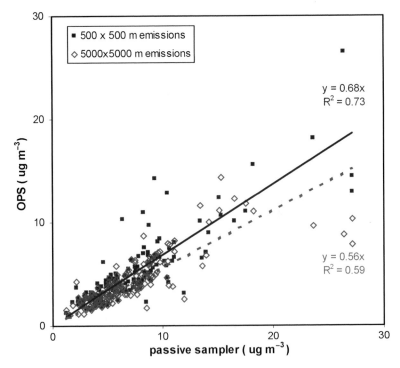

Fig. 19.5 Ammonia concentrations for 159 locations in the Netherlands calculated with the OPS model based on emissions on a 500 m resolution (black squares) and 5,000 m resolution (open diamonds) against ammonia concentrations measured with passive samplers (Van Pul et al. 2004)

local farmers. On the basis of these detailed emissions more than 75% of the (annual average) concentration differences (50 locations with passive samplers) within an area of $3 \times 3\,km^2$ could be explained by the model. Also this result indicates that simulations of atmospheric ammonia concentrations can be significantly improved when actual and detailed emission data is available.

In Fig. 19.6 similar results are given for ammonium concentrations and wet deposition. Here the model results are compared to observations of the LML network for the year 2000. This comparison shows that the model not only underestimates the ammonia concentration but also ammonium and wet deposition. Moreover, the underestimation appears to be systematic since the relative underestimation is nearly constant over the years (Fig. 19.7). Trends in atmospheric levels are well simulated by the model. This means that the change over the years may be attributed to a general change (decrease) in emissions.

Uncertainties

Most sensitive (and uncertain) process in the model is the surface exchange process. Sensitivity analysis shows that the parameterization of the surface resistance is critical. Furthermore, the simulation of dispersion, transport and dry deposition in the lower part of the boundary layer during stable atmospheric conditions has a large impact on local concentrations. Also the emission term for land spreaded

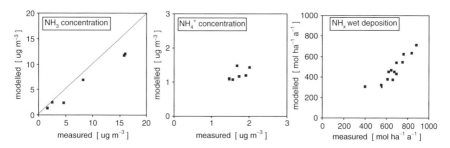

Fig. 19.6 Comparison of the (2000) modelled spatial distributions of NH_3, NH_4^+ concentrations and NH_x wet deposition with observations of the Dutch Monitoring Network (LML)

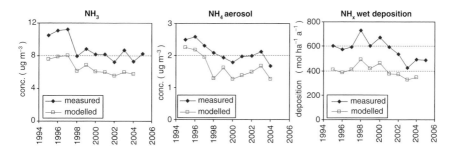

Fig. 19.7 Trends in measured and modelled NH_3 and NH_4^+ concentrations and wet deposition

manure plays an important role. In terms of source categories it is shown that the contribution of land spreaded manure is significantly underestimated. This points most probably to too low emissions for this category.

Main Conclusions

- A systematic difference of approximately 30% on a yearly basis is still found between measured and modelled NH_x species in the Netherlands. The underestimation is largest in the early spring situation.
- Trends in NH_x species in the Netherlands can be explained to a large extent by changes in anthropogenic NH_3 emissions.
- Spatial variations in ammonia concentrations can be well simulated when actual emission data is available.

19.3.1.3   The MATCH Model

Model Description

The Multi-scale Atmospheric Transport and Chemistry (MATCH) model is a three-dimensional, Eulerian model developed at the Swedish Meteorological and Hydrological Institute (SMHI). It is used in a range of applications from urban scale studies (e.g., Gidhagen et al. 2005) on ca. 5 km, or higher, resolution to regional/ continental scale studies on acidifying/eutrophying deposition and photochemistry (e.g. Andersson et al. 2007; Langner et al. 2005; Siniarovina and Engardt 2005). MATCH is also used for air pollution assessment in Sweden and the Baltic Sea region. The air pollution budgets of nitrogen and sulphur compounds for Sweden are calculated annually, using a system combining the MATCH model calculations and monitoring data from Sweden and the neighbouring countries. The model is also used operationally to provide forecasts of radioactivity in case of nuclear emergencies in Europe (Langner et al. 1998).

Emissions

Anthropogenic emissions of NH_3, NO_x, SO_x, NMVOC and CO (and natural emissions of SO_x) are taken from the UNECE/EMEP database (50 × 50 km resolution, Vestreng 2003) and regridded to the MATCH model grid (usually using a 0.4° × 0.4° (ca. 44 km) resolution). For Sweden, higher resolution emission data (ca. 1–5 km resolution) are available and these are used in national applications, e.g., the annual air pollution assessments, for the Swedish EPA, which are done at 11 km horizontal resolution. Emissions are handled in a simple way with emission heights (or profiles) and temporal variations specified for the different emission sectors used. Standard plume-rise calculations can be performed based on stack parameters (stack diameter, effluent temperature, volume flux). Individual point sources can be included but this is rarely done in the large scale studies with the photochemistry model.

For ammonia the agricultural emission sector dominates. For this sector all emissions are released in the lowest model level. Usually the same vertical resolution is used as for the meteorological data; this means that the surface emissions are spread in a ca. 60 m thick layer, when using the operational HIRLAM data from SMHI for Europe (for the years 1999–2003). Emissions of ammonia are not dependent on meteorological parameters. Only variations depending on month, weekday and hour are used. The temporal variations are based on results from the GENEMIS project and on data from TNO.

Transport

The basic transport model includes modules describing emissions, advection, turbulent diffusion and dry and wet deposition. Depending on the application specific modules describing, e.g., chemistry can be added to the basic transport model. MATCH is an "off-line" model. This means that atmospheric weather data are taken from some external source, usually a numerical weather prediction (NWP) model, and fed into the model at regular time intervals, currently every 3 or 6 h. Such data are then interpolated in time to yield hourly data. Special attention is given to interpolation of the horizontal wind where vector increments are applied. The vertical wind is calculated internally to assure mass consistency of the atmospheric motion after the time interpolation of the horizontal winds. The model design is flexible with regard to the horizontal and vertical resolution, principally defined by the input weather data, and allows for an arbitrary number of chemical compounds. The advection scheme is Bott-type (Bott 1989 a, b), using fourth-order scheme in the horizontal and a second-order scheme in the vertical. A complete description of the transport model can be found in Robertson et al. (1999).

The vertical resolution of the model is usually based on the resolution of the meteorological data. Typically the lowest model level has a thickness of ca. 60 m when the operational HIRLAM NWP model of SMHI is used. For the years 1999–2001 the total number of model levels was 14 and for 2003 this was increased to 22 levels. In both cases the model vertical extent is ca. 5,500 m.

Deposition

The dry deposition of gaseous and particulate species is calculated using a resistance approach depending on land-use. For MATCH-Europe a simple scheme is used with only four different land-use classes (Water, Forest, Low vegetation and No vegetation). The dry deposition flux is proportional to the concentration of each component and the inverse of the sum of the aerodynamic resistance and a species specific surface resistance. For simplicity the same aerodynamic resistance is used for all surfaces within a grid square. For species with stomatal uptake as a major deposition route, surface resistance is calculated taking into account soil moisture, soil type, vegetation type, leaf area index, photosynthetic active radiation and temperature. For other species a simpler approach is used with only monthly varying surface resistances. For NH_3 and some other species lower deposition velocities are used for snow covered surfaces.

In the air pollution assessments for Sweden a more detailed dry deposition scheme is used with 10 land use classes and a dry deposition parameterization based on Erisman et al. (1994) and Bartnicki et al. (2001).

Wet scavenging of NH_3 is assumed to be proportional to the precipitation intensity and a species-specific scavenging coefficient:

$$\frac{dc_i}{dt} = - c_i \Lambda_i P \tag{19.11}$$

where c_i is the concentration of species i, Λ_i is the scavenging coefficient (s^{-1} mm^{-1} h) and P is the precipitation rate (in mm h^{-1}). A vertical variation of the scavenging coefficient is often used. For example, the NH_3 the scavenging coefficient may be set to $0.000195\,s^{-1}$ mm^{-1} h in the lowest model level, increasing to $0.000389\,s^{-1}$ mm^{-1} h above the boundary layer z_i. For particulate ammonium slightly lower values are used: 0.000028–$0.000195\,s^{-1}$ mm^{-1} h.

Chemical Conversion

Only a few chemical reactions are considered for the ammonia-ammonium conversion in MATCH:

$$2\ NH_3 + SO_4 2- \rightarrow 2(NH_4)_2 SO_4 \text{ (irreversible reaction)} \tag{19.12a}$$

$$NH_3(g) + HNO_3(g) \leftrightarrow NH_4 NO_3 \text{ (s, aq)} \tag{19.12b}$$
$$\text{(equilibrium; humidity and temperature dependent)}$$

The sulphate/sulphuric acid, participating in the first reaction, can be directly emitted or formed in the model by gas phase oxidation of SO_2, by OH or CH_3O_2, or oxidation in cloud droplets, by H_2O_2 or O_3, (a constant cloud water pH is used, usually set to 5).

As an alternative to the first reaction ammonium sulphate production can be considered as leading to an equal mixture of NH_4HSO_4 and $(NH_4)_2SO_4$:

$$NH_3 + SO_4^{2-} \rightarrow \tfrac{1}{2} NH_4 HSO_4 + \tfrac{1}{2}(NH_4)_2 SO_4 \tag{19.13}$$

Performance/Validation

The MATCH-Europe model has, until recently, not been very thoroughly validated for ammonia or ammonium. The model has participated in some European model intercomparisons, e.g., within EUROTRAC-2 (Hass et al. 2003) and, more recently, the EuroDelta project (http://aqm.jrc.it/eurodelta/). Much of the focus has been on aerosols. Thus, ammonium has received much more attention than gaseous ammonia.

In the EuroDelta project five different regional scale models participate (Chimere, RCG, EMEP, MATCH and LOTOS/EUROS; in the validation/intercomparison part DEHM and the global TM5 models also took part). The models were run for two different meteorological years (1999 and 2001) using a number of different emission scenarios.

A validation of the models performance for 2001 for aerosol components, including ammonium, will be published (Schaap et al. 2008). The validation is mainly against EMEP measurement stations in the western/central/southern part of Europe (the domain common to all participating models). Some results from this validation are given below.

For measurement stations that measure total NH_x ($NH_4^+ + NH_3$) MATCH, on the average, underestimates the concentrations by ca. 15% and the average station correlation coefficient (for daily concentrations) is ca. 0.57 (the ensemble of all the participating models has the correlation coefficient 0.54 and essentially zero average bias). The RRMSE is 50% both for the MATCH model and the ensemble of all models. The spatial correlation coefficient is 0.77 for the MATCH model, which is about equal to the EMEP model and slightly lower than the ensemble of models (0.80).

The MATCH results are of similar quality for stations that measure aerosol ammonium (NH_4^+); average correlation coefficient 0.61, RRMSE 44%, spatial correlation coefficient 0.85. The average bias is small but MATCH shows a tendency to overestimate spring and late summer concentrations and underestimate winter concentrations.

Results for gas phase ammonia are in general poorer. The concentration is usually heavily underestimated and correlation coefficients are low, when comparing to daily measurements at EMEP stations. There are probably several reasons for this. It seems likely that some of the most important are the relatively coarse resolution of the model (both vertically and horizontally) and overestimated deposition of NH_3 in the MATCH-Europe version used in the EuroDelta project.

Comparisons of MATCH-Europe results with measurements at EMEP stations are shown in Figs. 19.8–19.11 (monthly average results for the EMEP stations for the years 2000, 2001 and 2003) and Figs. 19.12–19.14 (station average results for the periods with measurements in the same years) (Table 19.2).

Ammonia is underestimated, on the average, by a factor of 3–4 and the temporal correlation, for daily average concentrations, at individual stations is very low. At the northernmost stations the model concentration of ammonia is very close to zero indicating that losses during transport from the source regions are too fast. In Fig. 19.15 calculated and observed timeseries of ammonia (and NH_4^+) are shown for Birkenes (southern Norway) and Vredepeel (the Netherlands).

Results for aerosol NH_4^+ and total NH_x are in much better agreement with observations; at some stations the model results are actually fairly good with correlation coefficients around 0.8 for daily average concentrations. The model bias (−10% to −20%) is also much smaller than for NH_3. For the high ammonia station Vredepeel MATCH actually tends to overestimate the NH_4^+ concentration, especially in

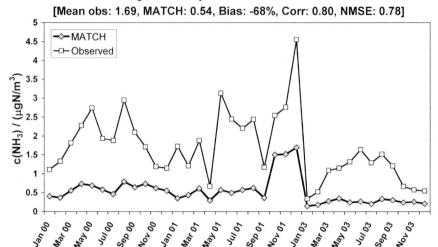

Fig. 19.8 Monthly concentrations of ammonia in air (averaged over EMEP sites with measurements); MATCH model results and EMEP measurements for the years 2000–2001 and 2003. Note that the measurement sites are not the same for all years. (μg N m^{-3})

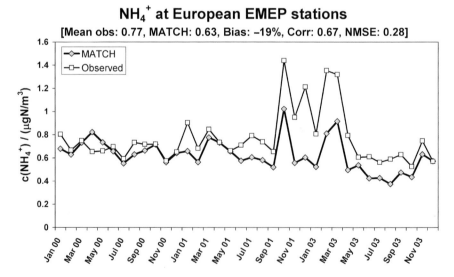

Fig. 19.9 Monthly concentrations of particulate ammonium in air (averaged over EMEP sites with measurements); MATCH model results and EMEP measurements for the years 2000–2001 and 2003. Note that the measurement sites are not the same for all years. (μg N m^{-3})

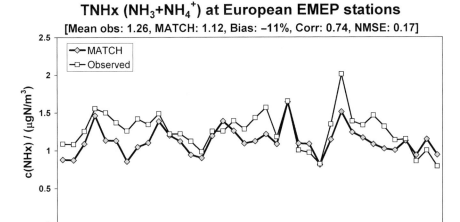

Fig. 19.10 Monthly concentrations of total NH$_x$ (NH$_3$ + NH$_4^+$) in air (averaged over EMEP sites with measurements); MATCH model results and EMEP measurements for the years 2000–2001 and 2003. Note that the measurement sites are not the same for all years. (μg N m^{-3})

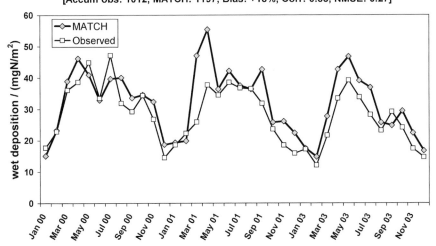

Fig. 19.11 Monthly wet deposition of NH$_x$ (averaged over EMEP sites with measurements); MATCH model results and EMEP measurements for the years 2000–2001 and 2003. Note that the measurement sites are not the same for all years (mg N m^{-2})

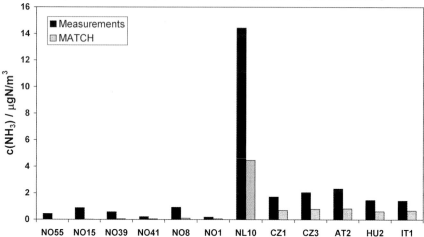

Fig. 19.12 Average concentrations of ammonia in air (averaged over periods with measurements for the years 2000–2001 and 2003); MATCH model results and EMEP measurements. Note that the all sites do not have measurements for all years (μg N m^{-3})

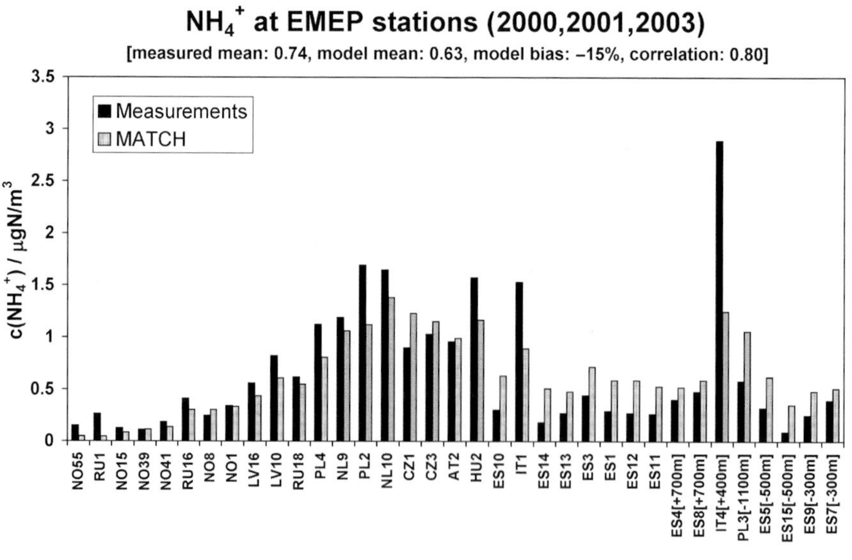

NH$_4^+$ at EMEP stations (2000,2001,2003)

[measured mean: 0.74, model mean: 0.63, model bias: −15%, correlation: 0.80]

Fig. 19.13 Average concentrations of particulate ammonium in air (averaged over periods with measurements for the years 2000–2001 and 2003); MATCH model results and EMEP measurements. Note that the all sites do not have measurements for all years. For the last eight stations the model topography deviates more than 250 m from the measurement site altitude. The model altitude deviation is given within brackets (μg N m^{-3})

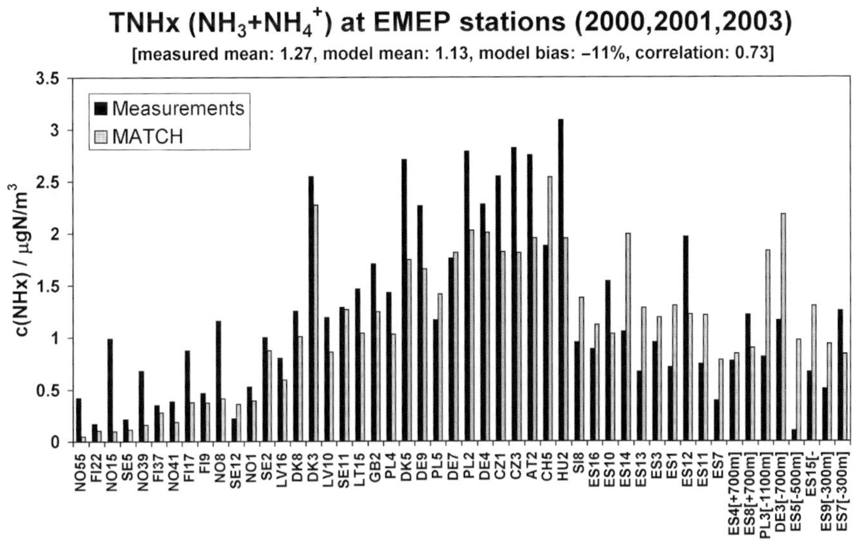

TNHx (NH₃+NH₄⁺) at EMEP stations (2000,2001,2003)

[measured mean: 1.27, model mean: 1.13, model bias: –11%, correlation: 0.73]

Fig. 19.14 Average concentrations of total NH_x ($NH_3 + NH_4^+$) in air; MATCH model results and EMEP measurements (averaged over periods with measurements for the years 2000–2001 and 2003). Note that the all sites do not have measurements for all years. For the last eight stations the model topography deviates more than 250 m from the measurement site altitude. The model altitude deviation is given within brackets (μg N m^{-3})

Table 19.2 Comparison between annual average concentrations of NH_3 at a few sites in Europe

Site	Code	Measured	Modelled
Braunschweig	DE97	5.4	1.4
K-puszta	HU02	1.6	1.0
Montelibretti	IT01	1.7	1.5
Vredepeel	NL10	19.0	8.6

April–May (see Fig. 19.15). This may indicate that conversion of NH_3 to NH_4+ is a bit too efficient in MATCH.

It is also likely that the wet scavenging of NH_x is too efficient in the present version of the model. In spite of the underestimated NH_x concentrations in air, the wet deposition of NH_x is actually somewhat overestimated compared to measurements at EMEP stations (see Fig. 19.11).

19.3.1.4   DEHM/DAMOS

Model Description

The Danish Ammonia Modelling System (DAMOS) is a combination of the Eulerian long-range transport model DEHM (Christensen 1997; Frohn et al. 2002; Geels et al. 2005) and the Gaussian local scale transport-deposition model OML-DEP

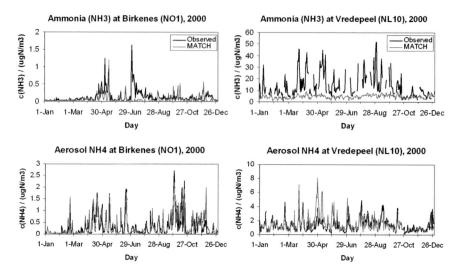

Fig. 19.15 Examples of MATCH results for NH_3 and NH_4^+. Calculated and observed concentrations at Birkenes (Norway) and Vredepeel (the Netherlands) for the year 2000 (μg N m^{-3})

(Olesen 1995). The system has been developed at the National Environmental Research Institute in order to assess detailed ammonia distributions resulting from the combined field of local to national and international emissions. DAMOS is used to support the nation wide air pollution monitoring programme in Denmark and for detailed studies of the nitrogen load at specific sensitive nature areas.

The Danish Eulerian Hemispheric Model (DEHM) is a full 3D transport-chemistry model covering the majority of the Northern Hemisphere and including multiple two-way nesting capabilities. The model is set up with an outer domain (150 × 150 km resolution) a nest over Europe (50 × 50 km resolution) and a second nest over Denmark and nearby areas (16.67 × 16.67 km resolution). In the vertical the model is divided into 20 sigma levels extending up to ca. 15 km in the troposphere and with a lowest level of about 60 m. Meteorological input is generated by the MM5 model run in a corresponding setup (Grell et al. 1995).

DEHM has through the years been used for many different applications within the field of air pollution and several versions of the model including different chemical species and tracers exists today (SO_2, SO_4 and lead, Christensen 1997; POPs, Hansen et al. 2008; CO_2, Geels et al. 2004). The chemistry version of DEHM includes an explicit chemical mechanism with 63 chemical compounds and 120 chemical reactions. The mechanism is based on the work by Strand and Hov (1994), but has been extended with a detailed description of ammonia chemistry through the inclusion of gas phase ammonia and related species, ammonium nitrate, ammonium bisulphate, ammonium sulphate and particulate nitrate formed from nitric acid. The dry deposition velocities are in DEHM performed with a deposition module based on the methodology in the EMEP model (Simpson et al. 2003). The deposition is calculated for nine land use categories. For ammonia the surface resistance Rc is divided into a stomatal and a nonstomatal part, where the latter depends on the SO_2 level in the atmosphere. The compensation point for ammonia is not taken into account. Wet deposition is based on a scavenging ratio formulation.

The local-scale model OML-DEP is a modern Gaussian plume model, which makes use of the surface depletion method from Horst (1977) for the dry deposition and uses a pseudo first order reaction velocity for the conversion of NH_3 to NH_4^+ (Asman et al. 1989). The dry deposition velocity is calculated with the same module as in DEHM. The meteorological input is also obtained from MM5. Regulation of ammonia from agriculture in Denmark is today based on standard calculations with the OML-DEP model.

Within DAMOS the OML-DEP model is applied for a 16×16 km domain that covers the nature area for which detailed deposition mapping is needed. The calculations are performed for 40×40 receptor points evenly distributed over the domain each representing a 400×400 m area. Background concentrations of ammonia and sulphur dioxide are obtained from the DEHM model for each hour by interpolation between up to three grid cells upwind from the OML-DEP domain. In this way DAMOS is used for calculation of ammonia deposition to local ecosystems from single farm units on a 400×400 m receptor net (Hertel et al. 2006).

Emissions

The emission of primary pollutants (NO_x, SO_x, CO, CH_4, NMVOC and NH_3) is in the DEHM model obtained from a combination of global and regional emission inventories (EMEP, EDGAR, GEIA, GENEMIS) and regridded into the three domains. Emissions are categorized in different sectors (SNAP categories), where each category has its own temporal (monthly, weekly, daily) variations as well as vertical distribution. For the Danish area more detailed (100 m to 1 km) emissions of NO_x and NH_3 are included. The NH_3 emissions in both DEHM and OML-DEP are computed using newly developed parameterisations with high spatial and temporal resolution based on local agricultural practice and meteorological conditions (Gyldenkaerne et al. 2005; Skjoth et al. 2004). These emissions are aggregated from an inventory on single farm and field level (100×100 m) to the 16.67×16.67 km and 400×400 m grids of the two models. The high resolution in the inventories has shown to be very important for the model performance (see the discussion in Hertel et al. 2006) at the Danish sites measuring NH_3. Next step will be to extend the inventory to cover all of Europe. Parameterisations of the temporal variability then have to be based on available information on for example the agricultural practice of the specific country or region.

Performance and Validation

The comparison of results of the DEHM model with EMEP measurements of reduced nitrogen components (concentration and wet deposition) is presented as a mean of all EMEP stations for the period 1993–2002 in Fig. 19.16. In order to cover most stations, the

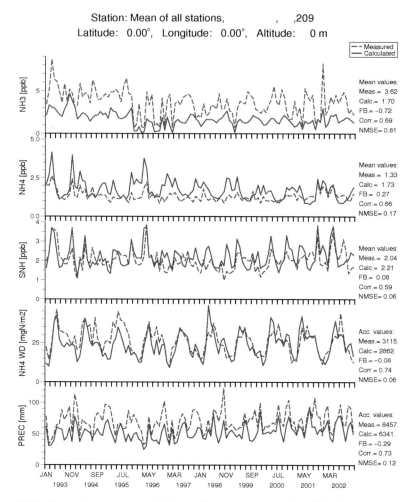

Fig. 19.16 Comparison of measured (dashed line) and calculated (solid line) concentrations of NH$_3$ (first panel), NH$_4^+$ (second panel) and the sum of NH$_3$ and NH$_4^+$ (SNH, third panel), wet deposition of NH$_4^+$ (fourth panel) and precipitation (fifth panel). Results are obtained as a mean value of all available measurements for the EMEP stations in the period 1993–2002 (note that the number of stations can vary throughout the period). The calculated results are from the European domain (50 × 50 km resolution) in the DEHM model

results from the European domain (resolution of 50 × 50 km) are shown here. Specific results for the Danish stations Anholt, Keldsnor and Tange are presented in Figs. 19.17–19.19, also from the 50 × 50 km model run. NH$_3$ and NH$_4^+$ are also measured specifically for these Danish stations, however, the results are not reported to EMEP and therefore only the sum of NH$_3$ and NH$_4^+$ (denoted SNH) is shown in these plots.

From Fig. 19.16 it is clear that DEHM underestimates the NH$_3$ concentrations throughout this 10 year period. The measured mean for the period and across all stations is 3.62 ppb, while the simulated mean is 1.70 ppb. This underestimation is

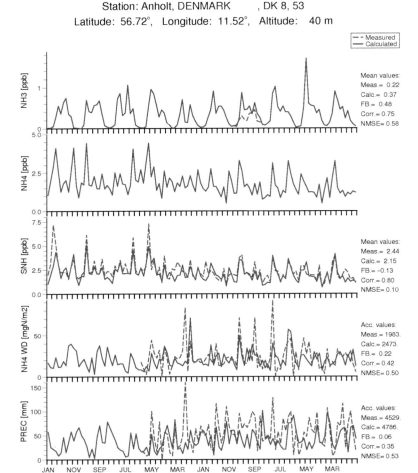

Fig. 19.17 Comparison of measured (dashed line) and calculated (solid line) concentrations of NH$_3$ (first panel), NH$_4^+$ (second panel) and the sum of NH$_3$ and NH$_4^+$ (SNH, third panel), wet deposition of NH$_4^+$ (fourth panel) and precipitation (fifth panel). All results are obtained for the Danish EMEP station Anholt for the period 1993–2002. The calculated results are from the European domain (50 × 50 km resolution) in the DEHM model

mainly due to an underestimation of the concentration at high emission locations like the Vredepeel (NL10) station in the Netherlands. Based on this time series of monthly means it is seen that the seasonal cycle across Europe is reasonable well simulated with a correlation coefficient of 0.69. Also the overall seasonal cycle of the NH$_4^+$ concentration is captured by the DEHM model, while the level is somewhat overestimated (ca. 30%). This shows that a long-rang transported pollutant like ammonium is easier to model than for example ammonia. The latter is much more dependent on local characteristics of the emission regions driven by differences in

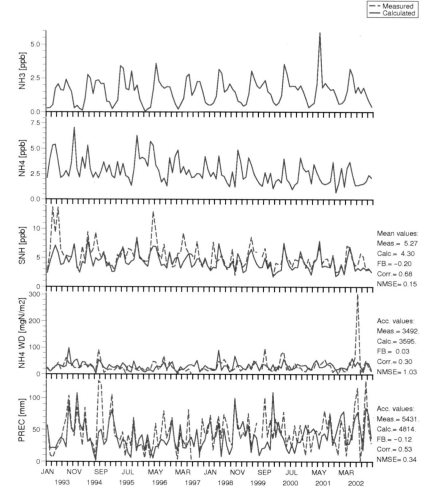

Fig. 19.18 Comparison of measured (dashed line) and calculated (solid line) concentrations of NH₃ (first panel), NH₄⁺ (second panel) and the sum of NH₃ and NH₄⁺ (SNH, third panel), wet deposition of NH₄⁺ (fourth panel) and precipitation (fifth panel). All results are obtained for the Danish EMEP station Keldsnor for the period 1993–2002. The calculated results are from the European domain (50 × 50 km resolution) in the DEHM model

agricultural activity and local-scale meteorological processes that is not resolved in this mesoscale model setup and with the current emission inventory for Europe. The under- and overestimation of NH₃ and NH₄⁺, respectively, could also indicate that the conversion between the two components happens too fast in the DEHM model. As seen in Fig. 19.16 the level of the sum (SNH) is captured within approximately 10% in this period. A good correspondence between measurements of SNH and model results is also seen at the Danish sites Anholt, Keldsnor and Tange.

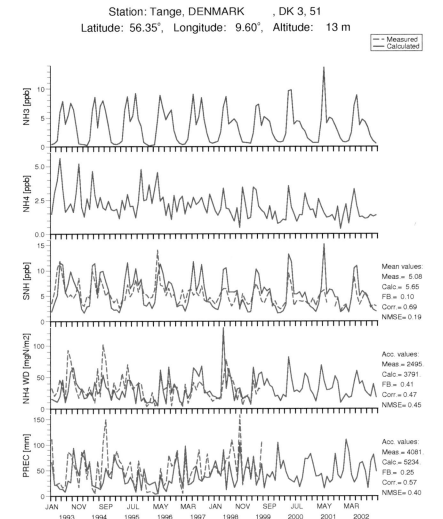

Fig. 19.19 Comparison of measured (dashed line) and calculated (solid line) concentrations of NH_3 (first panel), NH_4^+ (second panel) and the sum of NH_3 and NH_4^+ (SNH, third panel), wet deposition of NH_4^+ (fourth panel) and precipitation (fifth panel). All results are obtained for the Danish EMEP station Tange for the period 1993–2002. The calculated results are from the European domain (50 × 50 km resolution) in the DEHM model

Monthly concentrations of wet depositions of ammonium are also seen in Figs. 19.16–19.19. As a mean over all the EMEP sites both level and seasonal variability is seen to be well captured by the model. However, the model tends to underestimate the deposition in some periods. The overall overestimation of the air concentration by the model then seems to be counteracted by a general under-estimation of the monthly precipitation (also shown in the figures) when the wet deposition is calculated.

An analysis of yearly time series for the period 1989–2001, have shown that DEHM captures the observed year-to-year variability and the overall decreasing trend of the ammonium concentration throughout Europe (Geels et al. 2005).

As part of NOVANA i.e. the National Monitoring and Assessment Programme for the Aquatic and Terrestrial Environment both DEHM and DAMOS are applied every year with updated NH_3 emissions for Denmark. In Fig. 19.20 the model results are compared to the measured yearly mean NH_3 concentration at five Danish stations in 2005. The DEHM calculations represent the mean value for the 16.67 × 16.67 km cell where the station is placed. The results of the DAMOS system are for the 400 × 400 m grid cell in OML-DEP, where the station is placed. Measurements of NH_3 on Anholt and Tange are performed with filter pack samplers, whereas measurements at Keldsnor, Lindet and Ulborg are performed using the denuder method.

The local ammonia emissions are very low in the area where the Anholt and Keldsnor stations are located. At these stations both DEHM and DAMOS are close to the measured level and it is long-rang transported ammonia that dominates the signal. The stations at Lindet, Tange and Ulborg are all placed in agricultural active regions and the DEHM results hence represents the concentration resulting from the sum of emissions in the 16.67 × 16.67 km (~280 km²) grid cell. The stations are, however, placed some distance from local sources and the actual ammonia level measured is much lower than calculated with the regional model. The DAMOS system including the high resolution emission inventory and locale scale transport captures the measured levels significantly better, but still overestimates the level at especially the Ulborg site. A detailed evaluation and discussion of DEHM and DAMOS will be given by Geels et al. A coupled model system (DAMOS) improves the accuracy of simulated atmospheric ammonia over Denmark (in preparation) (Figs. 19.21, 19.22).

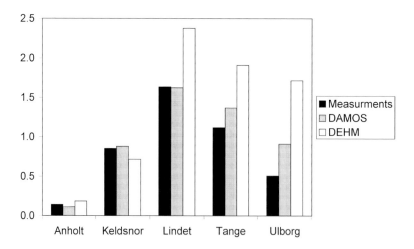

Fig. 19.20 Measured and calculated NH_3 concentrations ($\propto$g NH_3-N m^{-3}) at five Danish rural stations in 2005. The model results are from the DAMOS system and from the regional model, DEHM, alone (Data from Ellermann et al. 2006)

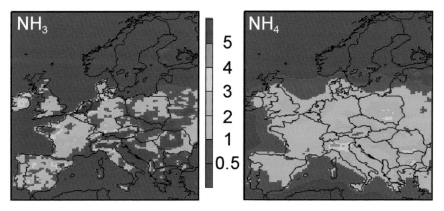

Fig. 19.21 Annual modelled concentrations of ammonia and ammonium for 2001

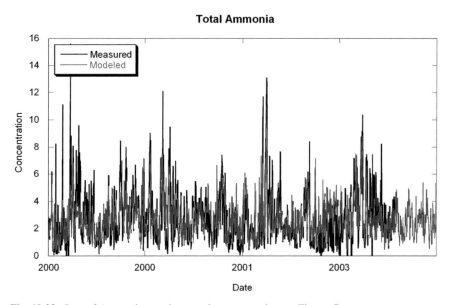

Fig. 19.22 Sum of Ammonium and ammonia concentrations at Zingst, Germany

19.3.1.5 LOTOS/EUROS

Model Description

LOTOS-EUROS is a three-dimensional (3D) chemistry transport model for
Europe. The domain of LOTOS-EUROS is the area between 35° and 70° North and
10° West and 40° East. The projection is normal longitude-latitude and the standard
grid resolution is 0.50° longitude × 0.25° latitude, approximately 25 × 25 km. In

the vertical there are three dynamic layers and an optional surface layer. The model extends in vertical direction 3.5 km above sea level. The lowest dynamic layer is the mixing layer, followed by two reservoir layers. The height of the mixing layer is part of the diagnostic meteorological input data. The heights of the reservoir layers are determined by the difference between the mixing layer height and 3.5 km. Both reservoir layers are equally thick with a minimum of 50 m. Simulations usually performed including the optional surface layer of a fixed depth of 25 m. Hence, this layer is always part of the dynamic mixing layer.

The transport consists of advection in three dimensions, horizontal and vertical diffusion, and entrainment/detrainment. The recently improved and highly-accurate, monotonic advection scheme developed by Walcek (2000) is used to solve the advection. Each hour the vertical structure of the model is adjusted to the new mixing layer depth. After the new structure is set the pollutant concentrations are redistributed using linear interpolation. Vertical diffusion is described using standard eddy diffusivity (K_z) theory. Vertical exchange is calculated employing the new integral scheme by Yamartino et al. (2004).

LOTOS-EUROS uses the TNO CBM-IV (Schaap et al. 2005) scheme which is a modified version of the original CBM-IV (Whitten et al. 1980). The scheme includes 28 species and 66 reactions, including 12 photolytic reactions. N_2O_5 hydrolysis is computed following Dentener and Crutzen (1993) and Jacob (2000). The formation of ammonium nitrate and ammonium sulphate is represented using ISORROPIA (Nenes et al. 1998). The dry deposition in LOTOS-EUROS is parameterised following a resistance approach (Erisman et al. 1994). No compensation point is taken into account for the deposition/emission of ammonia. Below cloud scavenging is described using simple scavenging coefficients for gases (Schaap et al. 2005) and following Simpson et al. (2003) for particles. In-cloud scavenging is neglected due to the limited information on clouds. Neglecting in-cloud scavenging results in too low wet deposition fluxes but has a very limited influence on ground level concentrations of ammonia and other components (see Schaap et al. 2004).

The standard meteorological data for Europe are produced at the Free University of Berlin employing a diagnostic meteorological analysis system based on an optimum interpolation procedure on isentropic surfaces (Kerschbaumer and Reimer 2003). The anthropogenic emissions used here are a combination of the TNO emission database (Visschedijk and Denier van der Gon 2005) and CAFE baseline emissions for 2000. For each source category and each country, we have scaled the country totals of the TNO emission database to those of the CAFE baseline emissions. Hence, we use the official emission totals as used within the LRTAP protocol but we benefit from the higher resolution of the TNO emission database (0.25×0.125 lon-lat). For a detailed description of the model and the input data we refer to Schaap et al. (2005).

Discussion on Treatment Ammonia Emissions

The seasonal variation in ammonia emissions is highly variable and differs regionally as function of farming procedures and climatic conditions. The seasonal variation in the ammonia emissions is modelled based on experimental data representative for the Netherlands (Bogaard and Duyzer 1997). The seasonal variation shows a distinct maximum in March and a slight maximum in August due to the application of manure on top of a function that roughly scales with duration of daylight. Following Asman (2001) we assumed a diurnal cycle in the emission with half the average value at midnight and twice the average at noon. We recognize that these functions may only represent practices in nortwestern Europe.

Exchange, emission or deposition, of ammonia depends on the compensation point, which refers to the situation in which the ammonia concentration in air is in equilibrium with the vegetation. Assessing the compensation point of ammonia is not possible for many surfaces (Asman 2001). Furthermore, the sub-grid variability in this parameter is expected to be very high. Hence, we do not take it into account.

Due to the emissions there is a large vertical gradient of ammonia concentrations in the source areas with highest concentrations near the ground. However, in our model the emissions are completely vertically mixed over the first mixing layer. We may therefore underestimate the effective dry deposition of ammonia close to the sources. To account for this effect Asman and Janssen (1987) and Dentener and Crutzen (1994) lowered the 'effective' emissions in their model by 25%, assuming that this part of the emission was removed on sub-grid scales. Janssen and Asman (1988) argued that by uniformly lowering the ammonia emission, ammonium formation could be underestimated and more sophisticated correction factors were proposed. These correction factors would be highly variable depending on region, the surface roughness downwind of the sources, availability of acidic precursors, meteorological conditions and the history of the air parcel (e.g. Asman 1998). Much of this information is not available in our model and therefore no correction factors are used in our model.

The deposition is treated following Erisman et al. (1994). The deposition velocity is calculated for nine land use classes within a grid cell and then a weighted average (as function of area) is used to assess the average dry deposition velocity.

Performance and Validation

The LOTOS-EUROS model calculates the concentration in air and dry deposition fluxes on an hour by hour basis. As we operate a regional model we compare our results mainly to the EMEP network, which has only a few stations for NH_3. However, a larger set of measurements is available for ammonium plus ammonia.

In Fig. 19.23, we compare modelled and measured annual average concentrations for ammonium and ammonia for 2001. The nice spatial correlation shows that the model is able to capture the large scale variability in ammonia levels in Europe. The temporal correlation is on average 0.44 (−0.17 to 0.77), which is not very good.

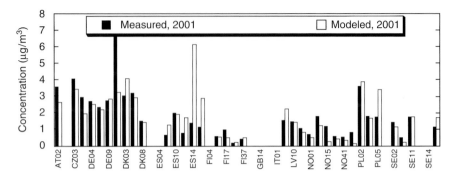

Fig. 19.23 Comparison of measured and modelled annual average of sum of ammonium and ammonia concentrations for 2001 in Europe

In source areas the ammonia concentrations are underestimated significantly by the model. For example, At Vredepeel or Braunschweig we underestimate by a factor 3. Also, the temporal correlation is low. This is not a surprise given the resolution of the model and the siting of the stations. Simulations on a higher resolution are needed and planned for next year.

The model is able to simulate the large scale variability and temporal variation at background sites. In source areas the levels are largely underestimated and the temporal correlation is not good. Reasons are the inherent fast mixing due to the Eulerian approach in combination with the averaging in the timing of the emissions which may in fact be highly variable.

19.3.1.6 EMEP Unified Model

Model Description

The Eulerian EMEP model is a multi-layer atmospheric dispersion model designed for simulating the long-range transport of air pollution over several years. The model domain is centered over Europe and also includes most of the North Atlantic and the polar region. The model has 20 vertical layers in σ-coordinates below 100 hPa, with a surface layer of approximately 90 m. It is primarily intended for use with a horizontal resolution of ca. $50 \times 50\,km^2$ (at $60°$ N) in the EMEP polar stereographic grid. The EMEP Unified model use meteorological data from PARLAM (PARalell version of HIRLAM) (Benedictow 1999), a dedicated version of the operational HIRLAM model (High Resolution Limited Area Model) maintained and verified at met.no. The numerical solution of the advection terms is based upon the scheme of Bott (1989a, b). The fourth order scheme is utilized in the horizontal directions. In the vertical direction a second order version applicable to variable grid distances is employed.

Chemical Conversion

The scheme is based upon the ozone chemistry from the Lagrangian photo-oxidant model (Simpson et al. 1993; Simpson 1995; Andersson-Sköld and Simpson 1999; Kuhn et al. 1998), but with additional reactions introduced to extend the model's coverage to acidification and eutrophication issues. These additions include ammonium chemistry, gas and aqueous oxidation of SO_2 to sulphate, and night-time production of nitrate. Additionally, a coarse particle nitrate species has been introduced. In total, the chemical scheme uses about 140 reactions between 70 species. The module EQSAM (Metzger et al. 2002a, b) is used to calculate the partitioning between gas and aerosol phase of HNO_3 and NO_3-aerosol and NH_3 and NH_4^+ aerosol, respectively.

Dry Deposition

The dry deposition module is based on the resistance analogy and the dry deposition velocity depends on the aerodynamic resistance, the quasi-laminar resistance and the surface (canopy) resistance. The deposition velocity is calculated independently over each land-cover and applied as a net deposition rate (weighted by the fraction of each land-cover type within the grid) to the modeled concentration at the reference height.

R_c is divided into a stomatal (R_{sto}) and a nonstomatal component (R_{ns}), where the R_{sto} parameterization is described in Simpson et al. (2003).

Surface Resistance, R_c

Surface (or canopy) resistance is the most complex variable in the deposition model, as it depends heavily on surface characteristics and the chemical characteristics of the depositing gas. Our approach makes use of bulk canopy resistances and conductances (R and G terms, where $G_x = 1/R_x$ for any x), and of unit-leaf-area (one-sided) resistances and conductances, which we denote with lower-case letters r, g. The general formula for bulk canopy conductances, G_c, is:

$$G_c = LAI \cdot g_{sto} + G_{ns} \qquad (19.14)$$

where LAI is the leaf-area index ($m^2\ m^{-2}$, one sided), g_{sto} is the stomatal conductance, and G_{ns} is the bulk non-stomatal conductance. For non-vegetative surfaces only the last term is relevant.

At sub-zero temperatures many of the following formulas use a low-temperature resistance. We use the formulation of Wesely (1989), where T_s is here in °C:

$$R_{low} = 1000\ e^{-(T_s+4)} \qquad (19.15)$$

Non-Stomatal Resistances

Gns is calculated specifically for O_3, SO_2, and NH_3. Values for other gases are obtained by interpolation of the O_3 and SO_2 values.

The non-stomatal resistance R_{ns} for NH_3 is assumed to depend upon surface (2 m) temperature, T_s (°C), humidity levels, RH (%), and on the molar 'acidity ratio':

$$a_{SN} = 0.6 \times [SO_2] / [NH_3] \qquad (19.16)$$

This acidity ratio is a first attempt to account for the observed changes in resistance in areas with different pollution climates (Erisman et al. 2001; Fowler and Erisman 2003). Other possible ratios include $[NH_3 + NH_4^+]/[SO_2 + SO_4^{2-}]$, but there is insufficient data upon which to choose between these ratios for modelling purposes at this time. The factor 0.6 is used to allow for the fact that the ratio of these gases at the surface should be higher than predicted by the EMEP model, due to the large vertical gradients of NH_3 above source areas.

The parameterisation of Smith et al. (2000) has been modified in order to take into account the effects of a_{SN}, based upon discussions with the Centre for Ecology & Hydrology (Smith et al. 2003). The resulting scheme can be expressed as:

$$
\begin{array}{llll}
R_{ns} & = & \beta F_1(T_s, RH) \, F_2(a_{SN}) & (T_s > 0) \\
 & & 200 & (-5 < T_s \leq 0) \\
 & & 1000 & (T_s \leq -5) \qquad (19.17)
\end{array}
$$

Where β is a normalising factor (0.0455), and $F_1 = 10 \log(T_s + 2)e^{((100-RH)/7)}$ as well as $F2 = 10^{(-1.1099aSN + 1.6769)}$.

The F1 term is identical to that of Smith et al. (2000) and provides a relationship of R_{ns} with temperature and relative humidity. The second function, F2, is an equation derived from observations presented in Nemitz et al. (2001), and relates the value at 95% relative humidity and 10°C to the molar ratio of SO_2/NH_3. The two terms are equal for molar SO_2/NH_3 ratio 0.3. The factor β is introduced in order to normalize one equation to the other, i.e. to ensure that the combined parameterisation is equal to the two separate terms for 95% relative humidity, 10°C and molar ratio 0.3.

For above-zero temperatures R_{ns} is constrained to lie between 10 and 200 s/m.

Finally, we do not distinguish wet or dry surfaces in this formulation (they are included in the RH dependency used above), so the conductances are:

$$G_{ns,dry} = G_{ns,wet} = 1/R_{ns} \qquad (19.18)$$

Aerosol Dry Deposition

Aerosol dry deposition velocity at height z_{ref} is calculated as:

$$Vg = \frac{1}{Ra + Rb + RaRbvs} + vs \qquad (19.19)$$

where vs is the gravitational settling velocity. Other terms are as for gases. An assumption is made that all particles stick to the surface, so that the surface resistance R_c is set to zero.

Wet Deposition

Parameterisation of the wet deposition processes in the Unified EMEP model includes both in-cloud and sub-cloud scavenging of gases and particles.

In-Cloud Scavenging

The in-cloud scavenging of a soluble component C is given by the expression:

$$\Delta C_{wet} = -C \frac{W_{in} \cdot P}{\Delta z \cdot \rho_w} \tag{19.20}$$

where W_{in} is the in-cloud scavenging ratio, P (kg m^{-2} s^{-1}) is the precipitation rate, Δz is the scavenging depth (assumed to be 1,000 m) and ρw is the water density (1,000 kg m^{-3}). We do not account for the effect that dissolved material may be released if clouds or rain water evaporate.

Below-Cloud Scavenging

For below cloud scavenging a distinction is made between scavenging of particulate matter and gas phase components. The sub-cloud scavenging of the gases is calculated as:

$$\Delta C_{wet} = -C \frac{W_{sub} \cdot P}{\Delta z \cdot \rho_w} \tag{19.21}$$

where W_{sub} is the sub-cloud scavenging ratio.

Wet deposition rates for particles are calculated, based on Scott (1979), as:

$$\Delta C_{wet} = -C \frac{A \cdot P}{V_{dr}} \cdot \overline{E} \tag{19.22}$$

where V_{dr} is the the raindrop fall speed (V_{dr} = 5 m s^{-1}), A = 5.2 m^3 kg^{-1} s^{-1} is the empirical coefficient (a Marshall-Palmer size distribution is assumed for rain drops), and E is the size-dependent collection efficiency of aerosols by the raindrops.

The emissions input required by the EMEP model consists of gridded annual national emissions. These emissions are provided for 10 anthropogenic source-sectors denoted by so-called SNAP codes (*Selected Nomenclature for Air Pollution Sectors*, Vestreng et al. 2003). In addition, an 11th source-sector consisting almost entirely of emissions from natural and biogenic sources exists.

The emissions are distributed vertically according to a default distribution based upon the SNAP codes. Emissions are distributed temporally according to monthly (Jan.–Dec.) and daily (Sun.–Sat.) factors derived from data provided by the University of Stuttgart (IER). These factors are specific to each pollutant, emission sector, and country. Simple day-night factors are also applied.

The model is fully documented in Simpson et al. (2003) and Fagerli et al. (2004) and applications of the model can be found in e.g. Fagerli et al. (2003), Simpson et al. (2006a, b), Jonson et al. (2006) and Fagerli et al. (2007).

Performance, Validation and Uncertainties

The results from the EMEP Unified model have been compared to ammonia measurements available in the EMEP network between 1995 and 2004. Seventeen sites have reported data for at least 1 year during this period. Of these sites, eight sites are situated in Norway, one in Denmark, one in Austria, one in the Netherlands, one in Italy, one in Turkey, two in Czech Republic, one in Lithuania and two in Latvia. These sites, although few in number, represent very different pollution climates with respect to ammonia, with yearly average concentrations ranging from about $0.05\,\mu g$ (N) m^{-3} at the remote site Jergul in northern Norway to ~$15\,\mu g$ (N) m^{-3} at the Dutch site Vredepeel. It should be noted that only three of these sites use denuders, whilst at the other sites the gas-particle separation is achieved using filter packs, thus the measurements may be biased. The large gradient in ammonia concentrations over Europe is well captured by the model (spatial correlation coefficient of 0.99–1.0), reflecting that the spatial distribution of the ammonia emissions is reasonable.

The EMEP model systematically underestimates the ammonia concentrations by a factor of 2. This might be related to the large height of the lowest layer (90 m), which makes it difficult to simulate the large vertical gradient of ammonia above sources.

The modeled seasonal cycle in ammonia concentrations are in reasonable agreement with the measurements, with most correlation coefficients (based on monthly averages) of 0.5 or larger, reflecting that the factors used for disaggregating the ammonia emissions from yearly to monthly emissions are reasonable. Correlation coefficients between model results and measurements for day-to-day variations range between 0 (most of the Norwegian sites except NO01, NO08 and NO42) and 0.5 (NO01, DK08, HU02, CZ03, NL10). At present, ammonia emissions in the EMEP model are not coupled to meteorological conditions, and even if they were, many of the local scale processes (e.g. soil moisture, leaf wetness, pH) which control ammonia emissions are difficult to parameterize even on a field-scale and cannot be captured within large scale models. Moreover, ammonia measurements are seldom characteristic for a $50 \times 50\,km^2$ grid square. Thus, it is difficult for a large scale model to capture short term variations in ammonia concentrations.

The model results of ammonium aerosol concentrations are in better agreement with measurements both with respect to absolute levels and short term variations. This secondary component dry deposits only slowly and is more determined by long range transport and therefore easier to model on a large scale. There are approximately 20 EMEP sites that report ammonium aerosol measurements. In general, the absolute levels of ammonium aerosol is captured within 30% at the different sites, most of the sites being overestimated. The model has a tendency to overestimate ammonium aerosol concentrations in winter time (30–40%) and slightly underestimate summer concentrations (~20%), with temporal (daily) correlation coefficients of $R^2 \sim 0.6$–0.8. For the sum of ammonia and ammonium (NH_x), almost 50 sites report measurements to EMEP. A comparison of model results and measurements of the seasonal cycle of reveals a similar pattern as for ammonium aerosol, confirming the systematic underestimation of ammonia at all seasons (Figs. 19.24 and 19.25).

For aerosol nitrate in air (measured at ~20 EMEP sites), there is a similar seasonal performance as for aerosol ammonium, indicating that the somewhat different seasonal pattern in the model results compared to the measurements are caused by a too efficient formation of ammonium nitrate in the cold periods (the seasonal cycle of sulfate agree well with the measurements). However, from a comparison with the few HNO_3 measurements available, no systematic underestimation of HNO_3 in the cold season is found. Thus, it is not clear whether or not the overestimation of NH_4NO_3 is related to the equilibrium chemistry between the NH_3, HNO_3 and NH_4NO_3.

For wet deposition of reduced nitrogen, more than 40 EMEP sites report measurements regularly. On average, modeled reduced nitrogen depositions and concentrations in precipitation are around 10% lower than the measurements. The gradient in reduced nitrogen wet deposition over Europe is rather well represented ($R^2 \sim 0.7$). The underestimation is largest during the early summer months June and July, whilst depositions in March are often overestimated. This systematic pattern might indicate that (1) the seasonal variation imposed on the ammonia emissions needs improvement or (2) the low modeled ammonium concentrations in summer leads to a too short residence time of reduced nitrogen in air and thereby too little wet deposition of

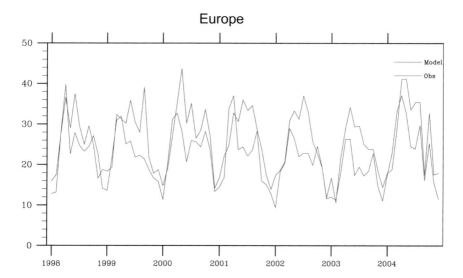

Fig. 19.24 Monthly concentrations of ammonia (μg (N) m^{-3}) in air (averaged over all EMEP sites with measurements), EMEP model results (scaled with 1.67) and EMEP measurements 1998–2004. Note that the number of measurement sites are not the same for every year

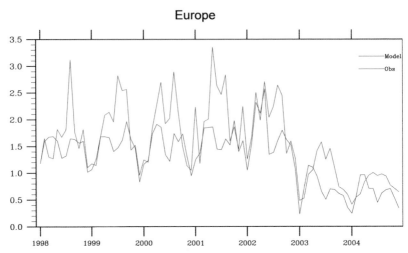

Fig. 19.25 Monthly concentrations of ammonium (μg (N) m^{-3}) in air (averaged over all EMEP sites with measurements), EMEP model results and EMEP measurements 1998–2004. Note that the number of measurement sites are not the same for every year

reduced nitrogen. Above we discussed that ammonia concentrations were underestimated with approximately the same amount throughout the whole year, which oppose (1). Ammonium concentrations in the cold months January and February are overestimated, but wet deposition of ammonium in the same months are frequently underestimated. Thus, it is not clear why the model underestimate wet deposition of reduced nitrogen.

No attempt has been made towards simulating trends of ammonia as not enough measurements have been available suitable for validating ammonia trends on a European scale.

In Fagerli and Aas (2008, in press), modeled trends of the sum of ammonia and ammonium in air (1990–2003) and trends of reduced nitrogen depositions (1980–2003) were compared to measurements at EMEP sites in Europe. In areas with high ammonia emissions, the decrease in the concentrations was similar to the decrease in ammonia emissions, whilst the decrease in the background areas (e.g. Norway, Sweden, Finland) was larger than expected based on the changes in emissions. In general, the model results were found to give similar trends as the measurements (Figs. 19.26–19.30).

19.3.1.7 EMEP4UK

- The Unified EMEP model (Tarrasón et al. 2003) is currently applied with increased spatial resolution over the UK. The new application is intended to allow mesoscale atmospheric transport model calculations over the British Isles.
- The new EMEP Unified model application is called EMEP4UK (Vieno et al. 2009, this volume) and is the Unified EMEP model implemented at the much finer horizontal resolution of $5 \times 5\,km^2$. Pollutants such as reactive nitrogen

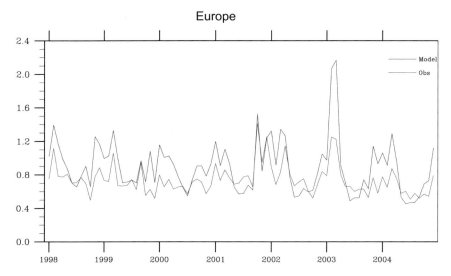

Fig. 19.26 Monthly concentrations of ammonium wet deposition (mg (N) /m⁻²) (averaged over all EMEP sites with measurements), EMEP model results and EMEP measurements 1998–2004. Note that the number of measurement sites are not the same for every year

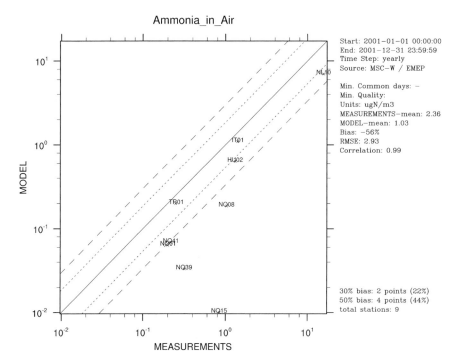

Fig. 19.27 Scatter plot of EMEP model results versus EMEP measurements of ammonia in air for 2001

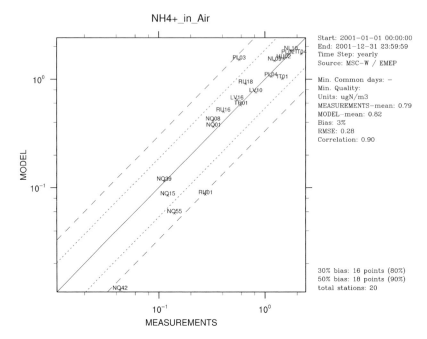

Fig. 19.28 Scatter plot of EMEP model results versus EMEP measurements of ammonium in air for 2001

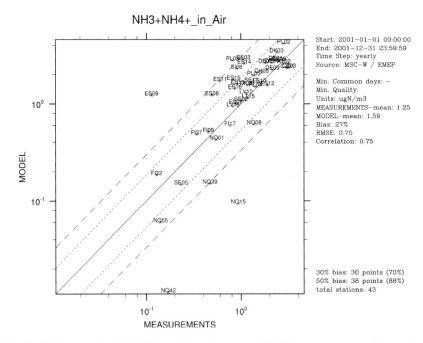

Fig. 19.29 Scatter plot of EMEP model results versus EMEP measurements of ammonia + ammonium in air for 2001

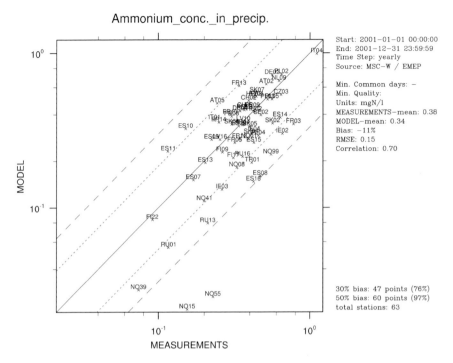

Fig. 19.30 Scatter plot of EMEP model results versus EMEP measurements of ammonium in precipitation for 2001

and sulphur have a high spatial variability in the emissions and a short life time; therefore the associated dry deposition also has a high spatial variability (Vieno 2005). This is very important when critical loads of nitrogen are calculated for specific ecosystems. To address these issues in the UK, the EMEP Unified model is being developed, using a nested approach, to run at high resolution over the UK.

- The $1 \times 1\,km^2$ National Atmospheric Emissions Inventory (NAEI) is used to create the emission input for EMEP4UK. The NAEI emissions use the British National Grid (Transverse Mercator) coordinate system therefore an Arc info script is used to convert the emissions from $1 \times 1\,km^2$ Transverse Mercator into the $5 \times 5\,km^2$ polar stereographic projection used by both EMEP Unified and EMEP4UK models. The horizontal resolution can be freely chosen, but, in the current version of EMEP4UK is set to be $5 \times 5\,km^2$.

- The emission input required by the EMEP4UK model consists of gridded annual national emissions of sulphur dioxide (SO_2), nitrogen oxides ($NO_x = NO + NO_2$), ammonia (NH_3), non-methane volatile organic compounds (NMVOC), carbon monoxide (CO), and particulates (PM2.5, PM10).

- EMEP4UK can in principle use meteorological data from various sources using a pre-processor which converts and re-grids the required meteorological data. For initial simulations two datasets were used to drive the EMEP4UK model:

interpolated ERA40 and the Weather Research Forecast (WRF) model outputs (http://www.wrf-model.org/). The WRF model presents the state-of-the-art in weather forecast modelling and is widely used by the academic community for research and weather forecast purposes. WRF uses a nesting domain approach to provide metrological data at the required horizontal and vertical resolution. Initial and boundary conditions (chemistry) are obtained from the EMEP Unified model.

19.4 Discussion and Conclusion

The models that are presented in Section 19.3 differ in a number of aspects. The differences in modeling concepts lead to a difference in model performance.

Issues that emerge from the model presentations and the validation against measurements:

1. Spatial scale of the model; particularly the models that calculate the concentrations at a relatively large scale (several tens of kilometers) underestimate the ammonia concentrations in air. The reason is twofold: (a) these models do often have a large depth of the first (surface) layer which leads to a too rapid mixing of the emitted ammonia over this first layer and (b) the local sub grid gradients in the emissions are not caught by the models. So sub-grid concentration gradients cannot be calculated and often leads to an underestimation of the concentration for a site. However, this is also strongly dependent on the vicinity of the measurement site.
2. Dry deposition parameterization; in general the models use dry deposition parameterizations that do not include the compensation point of ammonia. The ammonia concentration in air by the models may be underestimated caused by a too large dry deposition.
3. Chemical conversion of ammonia to ammonium; it is found by some of the models that a too high conversion may be the reason for the underestimation of ammonia concentrations.
4. Emissions; the lack or errors in the spatial and temporal representation of the ammonia emissions may lead to over- and underestimations of the ammonia concentrations. It is shown by models that using more spatial detailed emissions improved the correlation between measured and modelled concentrations considerably.

Acknowledgements The author gratefully acknowledges financial support for this review from COST Action 729, the ESF NinE programme and the NitroEurope Integrated Project, together with underpinning support for the model development from respective national funding sources.

References

Andersen H. V., Hovmand M. F., Hummelshoj P. and Jensen N. O. (1999) Measurements of ammonia concentrations, fluxes and dry deposition velocities to a spruce forest 1991–1995. Atmospheric Environment, 33, 1367–1383.

Andersson C., Langner J. and Bergström R. (2007) Inter-annual variation and trends in air pollution over Europe due to climate variability during 1958–2001 simulated with a regional CTM coupled to the ERA40 reanalysis. Tellus B, 59, 77–98, doi: 10.1111/j.1600–0889.2006.00196.x.

Andersson-Sköld Y. and Simpson D. (1999) Comparison of the chemical schemes of the EMEP MSC-W and the IVL photochemical trajectory models. Atmospheric Environment, 33, 1111–1129.

Aneja V. P., Nelson D. R., Roelle P. A., Walker J. T. and Battye W. (2003) Agricultural ammonia emissions and ammonium concentrations associated with aerosols and precipitation in the southeast United States. Journal of Geophysical Research-Atmospheres, 108.

Apsimon H. M., Barker B. M. and Kayin S. (1994) Modeling studies of the atmospheric release and transport of ammonia in anticyclonic episodes. Atmospheric Environment, 28, 665–678.

Asman W. A. H. (1998) Factors influencing local dry deposition of gases with special reference to ammonia. Atmospheric Environment, 32, 415–421.

Asman W. A. H. (2001) Modelling the atmospheric transport and deposition of ammonia and ammonium: an overview with special reference to Denmark. Atmospheric Environment, 35, 1969–1983.

Asman W. A. H. and Janssen A. J. (1987) A long-range transport model for ammonia and ammonium for Europe. Atmospheric Environment, 21, 2099–2119.

Asman W. A. H. and van Jaarsveld J. A. (1992) A variable-resolution transport model applied for NH_x in Europe. Atmospheric Environment, 26A, 445–464.

Asman W. A. H., Pinksterboer E. F., Maas H. F. M., Erisman J. W., Waijers Yperlaan A., Slanina J. and Horst T. W. (1989) Gradients of the ammonia concentration in a nature reserve - model results and measurements. Atmospheric Environment, 23, 2259–2265.

Badas M. G., Deidda R. and Piga E. (2006) Modulation of homogeneous space-time rainfall cascades to account for orographic influences. Natural Hazards and Earth System Sciences, 6, 427.

Baldwin A. C. and Golden D. M. (1979) Heterogeneous atmospheric reactions - sulfuric-acid aerosols as tropospheric sinks. Science, 206, 562–563.

Bartnicki J., Olendrzynski K., Jonson J. E., Berge E. and Unger S. (2001) Description of the Eulerian acid deposition model. http://projects.dnmi.no/ emep/acid/eudm.pdf

Battye W., Aneja V. P. and Roelle P. A. (2003) Evaluation and improvement of ammonia emissions inventories. Atmospheric Environment, 37, 3873–3883.

Benedictow A. (1999) Meteorological fields produced by PARLAM-PS and used as input for Eulerian EMEP model. Documentation and characterization, Technical Report. The Norwegian Meteorological Institute, Oslo, Norway, 2002.

Bogaard A. and Duyzer J. (1997) Een vergelijking tussen resultaten van metingen en berekeningen van de concentratie van ammoniak in de buienlucht op een schaal kleiner dan5 kilometer, TNO-report, TNO-MEP-R97/423, Apeldoorn, the Netherlands.

Bott A. (1989a) A positive definite advection scheme obtained by non-linear re-normalization of the advection fluxes. Monthly Weather Review, 117, 1006–1015.

Bott A. (1989b) Reply. Monthly Weather Review, 117, 2633–2636.

Bouwman A. F., Lee D. S., Asman W. A. H., Dentener F. J., VanderHoek K. W. and Olivier J. G. J. (1997) A global high-resolution emission inventory for ammonia. Global Biogeochemical Cycles, 11, 561–587.

Bower K. N., Wells M., Choularton T. W. and Sutton M. A. (1995) A model of ammonia/ammonium conversion and deposition in a hill cap cloud. Quarterly Journal of the Royal Meteorological Society, 121, 569–591.

Christensen J. H. (1997) The Danish Eulerian hemispheric model - a three-dimensional air pollution model used for the Arctic. Atmospheric Environment, 31, 4169–4191.

Daumer B., Niessner R. and Klockow D. (1992) Laboratory studies of the influence of thin organic films on the neutralization reaction of H_2SO_4 aerosol with ammonia. Journal of Aerosol Science, 23, 315–325.

De Leeuw F. A. A. M., van Rheineck L. H. J. and Builtjes P. J. H. (1990) Calculation of long term averaged ground level ozone concentrations. Atmospheric Environment, 24A, 185–193.

Dentener F. J. and Crutzen P. J. (1993) Reaction of N_2O_5 on tropospheric aerosols: impact on the global distributions of NO_x, O_3, and OH. Journal of Geophysical Research, 7149–7163.

Dentener F. J. and Crutzen P. J. (1994) A three-dimensional model of the global ammonia cycle. Journal of Atmospheric Chemistry, 19, 331–369.

Derwent R. G., Hov Ø., Asman W. A. H., van Jaarsveld J. A. and de Leeuw F. A. A. M. (1989) An intercomparison of long-term atmospheric transport models; the budgets of acidifying species for the Netherlands. Atmospheric Environment, 23, 1893–1909.

Dore A. J., Vieno M., Tang Y. S., Dragosits U., Dosio A., Weston K. J. and Sutton M. A. (2007) Modelling the atmospheric transport and deposition of sulphur and nitrogen over the United Kingdom and assessment of the influence of SO2 emissions from international shipping. Atmospheric Environment, 41(11), 2355–2367, doi:10.1016/j.atmosenv.2006.11.013.

Dragosits U., Sutton M. A., Place C. J. and Bayley A. A. (1998) Modelling the spatial distribution of agricultural ammonia emissions in the UK. Environmental Pollution, 102.

Dragosits U., Theobald M. R., Place C. J., Lord E., Webb J., Hill J., ApSimon H. M. and Sutton M. A. (2002) Ammonia emission, deposition and impact assessment at the field scale: a case study of sub-grid spatial variability. Environmental Pollution, 117, 147–158.

Duyzer J. H., Verhagen H. L. M., Weststrate J. H. and Bosveld F. C. (1992) Measurement of the dry deposition flux of NH_3 on to Coniferous forest. Environmental Pollution, 75, 3–13.

Duyzer J., Nijenhuis B. and Weststrate H. (2001) Monitoring and modelling of ammonia concentrations and deposition in agricultural areas of the Netherlands. Water Air and Soil Pollution: Focus, 1, 131–144.

Ellermann T., Andersen H. V., Bossi R., Brandt J., Christensen J., Frohn L. M., Geels C., Kemp K., Løfstrøm P., Mogensen B. and Monies C. (2006) Atmospheric Deposition. NOVANA (In Danish: Atmosfærisk deposition. NOVANA), 595, pp. 1–66. National Environmental Research Institute. Technical Report.

Erisman J. W., van Pul A. and Wyers P. (1994) Parametrization of surface resistance for the quantification of atmospheric deposition of acidifying pollutants and ozone. Atmospheric Environment, 28, 2595–2607.

Erisman J. W., Hensen A., Fowler D., Flechard C. R., Grüner A., Spindler G., Duyzer J. H., Weststrate H., Römer F., Vonk A. W. and Jaarsveld H. V. (2001) Dry deposition monitoring in europe. Water Air and Soil Pollution: Focus, 1, 17–27.

Fagerli H. and Aas W. (2008) Trends of nitrogen in air and precipitation. Model results and observations at EMEP sites in Europe, 1980–2003, doi : 10.1016/j.envpol.2008.01.024 (in press).

Fagerli H., Simpson D. and Aas W. (2003) Model performance for sulphur and nitrogen compounds for the period 1980 to 2000. In: Transboundary Acidification, Eutrophication and Ground Level Ozone in Europe. EMEP Status Report 1/2003, Part II Unified EMEP Model Performance, pp. 1–66. The Norwegian Meteorological Institute, Oslo, Norway, 2003. Available from http://www.emep.int

Fagerli H., Simpson D. and Tsyro S. (2004) Unified EMEP model: updates. In: Transboundary Acidification, Eutrophication and Ground Level Ozone in Europe. Status Report 1/2004, pp. 11–18. The Norwegian Meteorological Institute, Oslo, Norway, 2004. Available from http://www.emep.int

Fagerli H., Legrand M., Preunkert S., Vestreng V., Simpson D. and Cerqueira M. (2007) Modeling historical long-term trends of sulfate, ammonium, and elemental carbon over Europe: a comparison with ice core records in the Alps. Journal of Geophysical Research, 112, D23S13, doi:10.1029/2006JD008044.

Farquhar G. D., Firth P. M., Wetselaar R. and Weir B. (1980) On the gaseous exchange of ammonia between leaves and the environment - determination of the ammonia compensation point. Plant Physiology, 66, 710–714.

Fournier N., Dore A. J., Vieno M., Weston K. J., Dragosits U. and Sutton M. A. (2004) Modelling the deposition of atmospheric oxidised nitrogen and sulphur to the United Kingdom using a multi-layer long-range transport model. Atmospheric Environment, 38(5), 683–694.

Fournier, N., Weston K. J., Dore A. J. and Sutton M. A. (2005a) Modelling the wet deposition of reduced nitrogen over the British Isles using a Lagrangian multi-layer atmospheric transport model. Quarterly Journal of Royal Met. Society, 131, 703–722.

Fournier N., Tang Y. S., Dragosits U., de Kluizenaar Y. and Sutton M. A. (2005b) Regional atmospheric budgets of reduced nitrogen over the British Isles assessed using an atmospheric transport model. Water Air and Soil Pollution, 162, 331–351.

Fowler D. and Erisman J. W. (2003) Biosphere/atmosphere exchange of pollutants. Overview of subproject BIATEX-2. In: Midgley P. M. and Reuther M. (eds.) Towards Cleaner Air for Europe – Science, Tools and Applications, Part 2. Overviews from the Final Reprots of the EUROTRAC-2 Subprojects. Margraf Verlag, Weikersheim, Germany.

Fowler D., Sutton M., Flechard C., Cape J. N., Storeton-West R. L., Coyle M. and Smith R. I. (2001) The control of SO_2 dry deposition on to natural surfaces by NH3 and its effects on regional deposition. Water Air and Soil Pollution: Focus, 1, 39–48.

Frohn L. M. (2004) A study of long-term high-resolution air pollution modelling. Ph.D. thesis reports. Ministry of the Environment, National Environmental Research Institute, Roskilde, Denmark.

Frohn L. M., Christensen J. H., Brandt J. and Hertel O. (2001) Development of a high resolution integrated nested model for studying air pollution in Denmark. Physics and Chemistry of the Earth Part B-Hydrology Oceans and Atmosphere, 26, 769–774.

Frohn L. M., Christensen J. H. and Brandt J. (2002) Development and testing of numerical methods for two-way nested air pollution modelling. Physics and Chemistry of the Earth, 27, 1487–1494.

Geels C., Doney S. C., Dargaville R., Brandt J. and Christensen J. (2004) Investigating the sources of synoptic variability in atmospheric CO_2 measurements over the Northern Hemisphere continents: a regional model study. Tellus, 56B(1), 35–50.

Geels C., Brandt J., Christensen J. H., Frohn L. M. and Hansen K. (2005) Long-term calculations with a comprehensive nested hemispheric air pollution transport model. In: Farago I. et al. (eds.) Advances in Air Pollution Modeling for Environmental Security, 185–196.

Geels C., Bak J., Callesen T., Frohn L., Frydendall J., Gyldenkaerne S., Hansen A. G., Hutchings N., Jacobsen A. S., Pedersen P., Schneekloth M., Winther S., Hertel O. and Moseholm L. (2006) Guideline for approval of livestock farms (In Danish: Vejledning om godkendelse af husdyrbrug), 568, 83 p. National Environmental Research Institute, Roskilde, Denmark.

Gidhagen L., Johansson C., Langner J. and Foltescu V. L. (2005) Urban scale modeling of particle number concentration in Stockholm. Atmospheric Environment, 39, 1711–1725.

Gilliland A. B., Dennis R. L., Roselle S. J. and Pierce T. E. (2003) Seasonal NH_3 emission estimates for the eastern United States based on ammonium wet concentrations and an inverse modeling method. Journal of Geophysical Research-Atmospheres, 108.

Grell G. A., Dudhia J. and Stauffer D. R. (1995) A description of the fifth-generation Penn State/NCAR mesoscale model (MM5) [NCAR/TN-398 + STR], 117 pp. NCAR Technical Note.

Gryning S. E., Holtslag A. A. M., Irwin J. S. and Sivertsen B. (1987) Applied dispersion modelling based on meteorological scaling parameters. Atmospheric Environment, 21, 79–89.

Gyldenkærne S., Skjøth C. A., Christensen J., Ellermann T., Frohn L. M., Brandt J. and Hertel O. (2005a) A high resolution ammonia emission inventory for regional scale air pollution models. Journal of Geophysical Research, 110, D07108, doi:10.1029/2004JD005459.

Gyldenkærne S., Skjøth C. A., Hertel O. and Ellermann T. (2005b) A dynamical ammonia emission parameterization for use in air pollution models. Journal of Geophysical Research-Atmospheres, 110.

Hansen K. M., Halsall C. J., Christensen J. H., Brandt J., Geels C., Frohn L. M. and Skjøth C. A. (2008) The effect of snow on the fate of α-HCH in a dynamic multimedia model. Environmental Science & Technology, 42(8), 2943–2948.

Harrison R. M. and Pio C. A. (1983) An investigation of the atmospheric HNO_3-NH_3-NH_4NO_3 equilibrium relationship in a cool, humid climate. Tellus Series B-Chemical and Physical Meteorology, 35, 155–159.

Hass H., van Loon M., Kessler C., Stern R., Matthijsen J., Sauter F., Zlatev Z., Langner J., Foltescu V. and Schaap M. (2003) Aerosol modelling: results and intercomparison from European regional-scale modelling systems EUROTRAC ISS, Munich, EUROTRAC-2 Special Report, 77.

Hertel O., Christensen J., Runge E. H., Asman W. A. H., Berkowicz R., Hovmand M. F. and Hov O. (1995) Development and testing of a new variable scale air-pollution model - Acdep. Atmospheric Environment, 29, 1267–1290.

Hertel O., Skjoth C. A., Frohn L. M., Vignati E., Frydendall J., de Leeuw G., Schwarz U. and Reis S. (2002) Assessment of the atmospheric nitrogen and sulphur inputs into the North Sea using a Lagrangian model. Physics and Chemistry of the Earth, 27, 1507–1515.

Hertel O., Skjoth C. A., Lofstrom P., Geels C., Frohn L. M., Ellermann T. and Madsen P. V. (2006) Modelling nitrogen deposition on a local scale - a review of the current state of the art. Environmental Chemistry, 3, 317–337.

Holtslag A. A. M. and Nieuwstadt F. T. M. (1986) Scaling the atmospheric boundary layer. Boundary-Layer Meteorology, 36, 201–209.

Horst T. W. (1977) Surface depletion model for deposition from a gaussian plume: atmospheric environment, 11, 41–46.

Huijsmans J. F. M., Hol J. M. G. and Vermeulen G. D. (2003) Effect of application method, manure characteristics, weather and field conditions on ammonia volatilization from manure applied to arable land. Atmospheric Environment, 37, 3669–3680.

Huntzicker J. J., Cary R. A. and Ling C. S. (1980) Neutralization of sulfuric-acid aerosol by ammonia. Environmental Science & Technology, 14, 819–824.

Husted S., Schjoerring J. K., Nielsen K. H., Nemitz E. and Sutton M. A. (2000) Stomatal compensation points for ammonia in oilseed rape plants under field conditions. Agricultural and Forest Meteorology, 105, 371–383.

Hutchings N. J., Sommer S. G., Andersen J. M. and Asman W. A. H. (2001) A detailed ammonia emission inventory for Denmark. Atmospheric Environment, 35, 1959–1968.

Jacob D. J. (2000) Heterogeneous chemistry and tropospheric ozone. Atmospheric Environment, 34, 2131–2159.

Janssen A. J. and Asman W. A. H. (1988) Effective removal parameters in long-range air pollution transport models. Atmospheric Environment, 22, 359–367.

Jonson J. E., Simpson D., Fagerli H. and Solberg S. (2006) Can we explain the trends in European ozone levels? Atmospheric Chemistry and Physics, 6(1), 51–66.

Junge E. and Ryan T. G. (1958) Study of the SO_2 oxidation in solution and its role in atmospheric chemistry. Quarterly Journal of the Royal Meteorological Society, 84, 46–55.

Kerschbaumer A. and Reimer E. (2003) Preparation of Meteorological Input Data for the RCG-Model. UBA-Rep. 299 43246, Free University. Berlin Institute for Meteorology (in German).

Kuhn M., Builtjes P. J. H., Poppe D., Simpson D., Stockwell W. R., Andersson-Skoeld Y., Baart A., Das M., Fiedler F., Hov Ø., Kirchner F., Makar P. A., Milford J. B., Roemer M. G. M., Ruhnke R., Strand A., Vogel B. and Vogel H. (1998) Intercomparison of the gas-phase chemistry in several chemistry and transport models. Atmospheric Environment, 32(4), 693–709.

Langner J., Robertson L., Persson C. and Ullerstig A. (1998) Validation of the operational emergency response model at the Swedish meteorological and hydrological institute using data from ETEX and the Chernobyl accident. Atmospheric Environment, 32, 4325–4333.

Langner J., Bergström R. and Foltescu V. L. (2005) Impact of climate change on surface ozone and deposition of sulphur and nitrogen in Europe. Atmospheric Environment, 39, 1129–1141.

Lee D. S., Halliwell C., Garland J. A., Dollard G. J. and Kingdon R. D. (1998) Exchange of ammonia at the sea surface - a preliminary study. Atmospheric Environment, 32, 431–439.

Loubet B., Milford C., Sutton M. A. and Cellier P. (2001) Investigation of the interaction between sources and sinks of atmospheric ammonia in an upland landscape using a simplified dispersion-exchange model. Journal of Geophysical Research-Atmospheres, 106, 24183–24195.

McMurry P. H., Takano H. and Anderson G. R. (1983) Study of the ammonia (gas) sulfuric-acid (aerosol) reaction-rate. Environmental Science & Technology, 17, 347–352.

Metzger S. M., Dentener F. J., Jeuken A., Krol M. and Lelieveld J. (2002a) Gas/aerosol partitioning 2. global modeling results. Journal of Geophysical Research, 107(D16), ACH 17.

Metzger S. M., Dentener F. J., Lelieveld J. and Pandis S. N. (2002b) Gas/aerosol partitioning 1. a computionally efficient model. Journal of Geophysical Research, 107(D16), ACH 17.

Milford C., Sutton M. A., Allen A. G., Karlsson A., Davison B. M., James J. D., Rosman K., Harrison R. M. and Cape J. N. (2000) Marine and land-based influences on atmospheric ammonia and ammonium over Tenerife. Tellus Series B-Chemical and Physical Meteorology, 52, 273–289.

Milford C., Theobald M. R., Nemitz E. and Sutton M. A. (2001) Dynamics of ammonia exchange in response to cutting and fertilising in an intensively-managed grassland. Water Air and Soil Pollution: Focus, 1, 167–176.

NEGTAP (2001) Transboundary Air Pollution: Acidification, Eutrophication and Ground Level ozone in the UK. Report of the National Expert Group on Transboundary Air Pollution, DEFRA, London.

Nemitz E., Milford C. and Sutton M. A. (2001) A two-level canopy compensation point model for describing bi-directional biosphere-atmosphere exchange of ammonia. Quarterly Journal of the Royal Meteorological Society, 127, 815–833.

Nemitz E., Sutton M. A., Wyers G. P. and Jongejan P. A. C. (2004) Gas-particle interactions above a Dutch heathland: I. Surface exchange fluxes of NH_3, SO_2, HNO_3 and HCl. Atmospheric Chemistry and Physics, 4, 989–1005.

Nenes A., Pilinis C. and Pandis S. N. (1998) Isorropia: a new thermodynamic model for multiphase multicomponent inorganic aerosols. Aquatic Geochemistry, 4, 123–152.

Neirynck J., Kowalski A. S., Carrara A. and Ceulemans R. (2005) Driving forces for ammonia fluxes over mixed forest subjected to high deposition loads. Atmospheric Environment, 39, 5013–5024.

Olesen H. R. (1995) Regulatory dispersion modeling in Denmark. International Journal of Environment and Pollution, 5, 412–417.

Park S. U. and Lee Y. H. (2002) Spatial distribution of wet deposition of nitrogen in South Korea. Atmospheric Environment, 36, 619–628.

Phillips S. B., Arya S. P. and Aneja V. P. (2004) Ammonia flux and dry deposition velocity from near-surface concentration gradient measurements over a grass surface in North Carolina. Atmospheric Environment, 38, 3469–3480.

Pio C. A. and Harrison R. M. (1987) The equilibrium of ammonium-chloride aerosol with gaseous hydrochloric-acid and ammonia under tropospheric conditions. Atmospheric Environment, 21, 1243–1246.

Quinn P. K., Charlson R. J. and Bates T. S. (1988) Simultaneous observations of ammonia in the atmosphere and ocean. Nature, 335, 336–338.

Raes F., Van Dingenen R., Wilson J. and Saltelli A. (1993) Dimethyl Sulphide, Oceans, Atmosphere and Climate (Restelli G. and Angeletti G. (eds.). Kluwer, Dordrecht, The Netherlands, pp. 311–322.

Robertson L., Langner J. and Engardt M. (1999) An Eulerian limited-area atmospheric transport model. Journal of Applied Meteorology, 38, 190–210.

Schaap M., van Loon M., ten Brink H. M., Dentener F. D. and Builtjes P. J. H. (2004) Secondary inorganic aerosol simulations for Europe with special attention to nitrate. Atmospheric Physics and Chemistry, 4, 857–874.

Schaap M., Roemer M., Sauter F., Boersen G., Timmermans R. and Builtjes P. J. H. (2005) LOTOS-EUROS Documentation, TNO report B&O 2005/297. TNO, Apeldoorn, The Netherlands.

Schaap M., Vautard R., Bergström R., van Loon M., Bessagnet B., Brandt J., Christensen J. H., Cuvelier K., Foltescu V., Graff A., Jonson J. E., Kerschbaumer A., Krol M., Langner J., Roberts P., Rouïl L., Stern R., Tarrasón L., Thunis P., Vignati E., White L., Wind P. and Builtjes P. J. H. (2008) Atmospheric Environment (submitted).

Schjoerring J. K., Kyllingsbaek A., Mortensen J. V. and Byskovnielsen S. (1993) Field investigations of ammonia exchange between barley plants and the atmosphere.1. Concentration profiles and flux densities of ammonia. Plant Cell and Environment, 16, 161–167.

Schjoerring J. K., Husted S. and Poulsen M. M. (1998) Soil-plant-atmosphere ammonia exchange associated with Calluna vulgaris and Deschampsia flexuosa. Atmospheric Environment, 32, 507–512.

Schwarz U., Wickert B., Obermeier A. and Friedrich R. (2000) Generation of Atmospheric Emission Inventories in Europe with High Spatial and Temporal Resolution. Southampton SO40 7AA, WIT, Southampton, UK.

Scott B. C. (1979) Parameterization of sulphate removal by precipitation. Journal of Applied Meteorology, 17, 11375–11389.

Seedorf J., Hartung J., Schroder M., Linkert K. H., Pedersen S., Takai H., Johnsen J. O., Metz J. H. M., Groot Koerkamp P. W. G. and Uenk G. H. (1998a) A survey of ventilation rates in livestock buildings in Northern Europe. Journal of Agricultural Engineering Research, 70, 39–47.

Seedorf J., Hartung J., Schroder M., Linkert K. H., Pedersen S., Takai H., Johnsen J. O., Metz J. H. M., Groot Koerkamp P. W. G. and Uenk G. H. (1998b) Temperature and moisture conditions in livestock buildings in Northern Europe. Journal of Agricultural Engineering Research, 70, 49–57.

Seinfeld J. H. and Pandis S. N. (1998) Atmospheric Chemistry and Physics: From Air Pollution to Climate Change. Wiley, New York.

Simpson D. (1995) Biogenic emissions in Europe 2: Implications for ozone control strategies. Journal of Geophysical Research, 100(D11), 22891–22906.

Simpson D., Andersson-Skoeld Y. and Jenkin M. E. (1993) Updating the Chemical Scheme for the EMEP MSC-W Oxidant Model: Current Status. Norwegian Meteorological Institute, EMEP MSC-W Note 2/93.

Simpson D., Fagerli H., Jonson J. E., Tsyro S., Wind P. and Tuovinen J-P. (2003) Transboundary Acidification, Eutrophication and Ground Level Ozone in Europe, PART I, Unified EMEP Model Description. 1, pp. 1–104. 1-8-2003.

Singles R. J., Sutton M. A. and Weston K. J. (1998) A multi-layer model to describe the atmospheric transport and deposition of ammonia in Great Britain. Atmospheric Environment, 32, 393–399.

Siniarovina U. and Engardt M. (2005) High-resolution model simulations of anthropogenic sulphate and sulphur dioxide in Southeast Asia. Atmospheric Environment, 39, 2021–2034.

Skjøth C. A, Ellermann T., Gyldenkærne S., Hertel O., Geels C., Frohn L., Frydendall J. and Løfstrøm P. (2006) Footprints on Ammonia Concentrations from Emission Regulations. Presented at Bolger Conference Center, Potomac, MD, 5–8 June 2006.

Smith R., Fowler D., Sutton M. A., Flechard C. and Coyle M. (2000) Regional estimation of pollutant gas dry deposition in the UK: model description, sensitivity analyses and outputs. Atmospheric Environment, 34, 3757–3777.

Smith R., Fowler D. and Sutton M. A. (2003) The External Surface Resistance in the EMEP Eulerian Model. Unpublished Note, Centre for Ecology and Hydrology, Penicuik, Scotland.

Sorensen L. L., Hertel O., Skjoth C. A., Lund M. and Pedersen B. (2003) Fluxes of ammonia in the coastal marine boundary layer. Atmospheric Environment, 37, S167–S177.

Sorteberg A. and Hov O. (1996). Two parametrizations of the dry deposition exchange for SO_2 and NH_3 in a numerical model. Atmospheric Environment, 30, 1823–1840.

Stelson A. W. and Seinfeld J. H. (1982) Relative-humidity and temperature-dependence of the ammonium-nitrate dissociation-constant. Atmospheric Environment, 16, 983–992.

Stelson A. W., Friedlander S. K. and Seinfeld J. H. (1979) Note on the equilibrium relationship between ammonia and nitric-acid and particulate ammonium-nitrate. Atmospheric Environment, 13, 369–371.

Stohl A. (1998) Computation, accuracy and applications of trajectories - a review and bibliography. Atmospheric Environment, 32, 947–966.

Stohl A. and Koffi N. E. (1998) Evaluation of trajectories calculated from ECMWF data against constant volume balloon flights during ETEX. Atmospheric Environment, 32, 4151–4156.

Sutton M. A., Fowler D. and Moncrieff J. B. (1993) The exchange of atmospheric ammonia with vegetated surfaces.1. Unfertilized vegetation. Quarterly Journal of the Royal Meteorological Society, 119, 1023–1045.

Sutton M. A., Burkhardt J. K., Guerin D., Nemitz E. and Fowler D. (1998) Development of resistance models to describe measurements of bi-directional ammonia surface-atmosphere exchange. Atmospheric Environment, 32, 473–480.

Sutton M. A., Dragosits U., Tang Y. S. and Fowler D. (2000) Ammonia emissions from non-agricultural sources in the UK. Atmospheric Environment, 34, 855–869.

Sutton M. A., Tang Y. S., Miners B. and Fowler D. (2001) A new diffusion denuder system for long-term, regional monitoring of atmospheric ammonia and ammonium. Water Air and Soil Pollution: Focus, 1, 145–156.

Tang Y. S., Cape J. N. and Sutton M. A. (2001) Development and types of passive samplers for monitoring atmospheric NO_2 and NH_3 concentrations. The Scientific World, 1, 513–529.

Van Hove L. W. A., Adema E. H., Vredenberg W. J. and Pieters G. A. (1989) A study of the adsorption of NH3 and SO_2 on leaf surfaces. Atmospheric Environment, 23, 1479–1486.

Van Jaarsveld J. A. (1995) Modelling the long-term atmospheric behaviour of pollutants on various spatial scales. Ph.D. thesis Utrecht University.

Van Jaarsveld J. A. (1996) The dynamic exchange of pollutants at the air-soil interface and its impact on long range transport. In: Sven-Erik G. and Schiermayer F. (eds.) Air Pollution Modeling and Its Application XI. Plenum, New York.

Van Jaarsveld J. A. (2004) The Operational Priority Substances model. RIVM 500045001.

Van Jaarsveld J. A., Bleeker A. and en Hoogervorst N. J. P. (2000) Evaluatie ammoniakredukties met behulp van metingen en modelberekeningen (Evaluation of ammonia reductions on the basis of measurements and model calculations). RIVM report (in Dutch) 722108025, RIVM, Bilthoven.

Van Pul A., Jaarsveld H. V., Meulen T. V. D. and Velders G. (2004) Ammonia concentrations in the Netherlands: spatially detailed measurements and model calculations. Atmospheric Environment, 38, 4045–4055.

Van Ulden A. P. (1978) Simple estimates for vertical diffusion from sources near the ground. Atmospheric Environment 12, 2125–2129.

Velders G. J. M., de Waal E. S., van Jaarsveld J. A. and de Ruiter J. F. (2003) The RIVM-MNP contribution to the evaluation of the EMEP Unified (Eulerian) model. RIVM-rapport 500037002/2003.

Vestreng V. (2003) EMEP/MSC-W Technical Report. Review and Revision. Emission data reported to CLRTAP, MSC-W Status Report 2003. EMEP/MSC-W Note 1/2003. ISSN 0804–2446.

Vieno M. (2005) The Use of an Atmospheric Chemistry-Transport Model (FRAME) over the UK and the Development of Its Numerical and Physical Schemes. Ph.D. thesis, University of Edinburgh, Edinburgh, UK.

Visschedijk A. J. H. and Denier van der Gon H. A. C. (2005) Gridded European anthropogenic emission data for NO_x, SO_x, NMVOC, NH_3, CO, PPM_{10}, $PPM_{2.5}$ and CH_4 for the year 2000. TNO-Report B&O-A R 2005/106.

Walcek C. J. (2000) Minor flux adjustment near mixing ratio extremes for simplified yet highly accurate monotonic calculation of tracer advection. Journal of Geophysical Research, 105(D7), 9335–9348.

Wells M., Bower K. N., Choularton T. W., Cape J. N., Sutton M. A., Storeton-West R. L., Fowler D., Wiedensohler A., Hansson H. C., Svenningsson B., Swietlicki E., Wendisch M., Jones B., Dollard G., Acker K., Wieprecht W., Preiss M., Arends B. G., Pahl S., Berner A., Kruisz C., Laj P., Facchini M. C. and Fuzzi S. (1997) The reduced nitrogen budget of an orographic cloud. Atmospheric Environment, 31, 2599–2614.

Wesely M. L. (1989) Parameterization of surface resistances to gaseous dry deposition in regional-scale numerical-models. Atmospheric Environment, 23, 1293–1304.

Whitten G., Hogo H. and Killus J. (1980) The carbon bond mechanism for photochemical smog. Environmental Science & Technology, 14, 14690–14700.

Wichink K. R. J, van Pul W. A. J., Otjes R. P., Hofschreuder P., Jacobs A. F. G. and Holtslag A. A. M. (2007) Ammonia fluxes and derived canopy compensation points over non-fertilized agricultural grassland in the Netherlands using the new gradient ammonia—high accuracy—monitor (GRAHAM). Atmospheric Environment, 41, 1275–1287.

Wickert B., Heidegger A. and Friedrich R. (2001) Calculations of Emissions in Europe with CAREAIR. Springer, Berlin/Heidelberg/New York.

Wyers G. P. and Erisman J. W. (1998) Ammonia exchange over coniferous forest. Atmospheric Environment, 32, 441–451.

Yamartino R. J., Flemming J. and Stern R. M. (2004) Adaption of analytic diffusivity formulations to eulerian grid model layers finite thickness. 27th ITM on Air Pollution Modelling and its Application. Banff, Canada, October 24–29, 2004.

Chapter 20
Application of a Lagrangian Model FRAME to Estimate Reduced Nitrogen Deposition and Ammonia Concentrations in Poland

Maciej Kryza, Anthony J. Dore, Marek Błaś, and Mieczysław Sobik

20.1 Introduction

This paper presents the preliminary results of modelling the annual average NH_3 concentrations and reduced nitrogen depositions in Poland. The Lagrangian Fine Resolution Atmospheric Multi-pollutant Exchange (FRAME) model is applied here, working with $5 \times 5\,km$ grid size. The input data are briefly described, including emission inventory and meteorological information used. The results are compared with the EMEP Unified model estimates of NH_3 concentrations and NH_x dry and wet depositions and checked against the available point measurements of NH_3 air concentrations. FRAME dry and wet deposition budget for Poland is calculated, showing good agreement with the EMEP model estimates and measurement-based wet deposition budget calculated by Chief Inspectorate of Environmental Protection.

Emissions of reduced nitrogen in Poland have fallen by 41% since 1985, compared to a 72% reduction of SO_2 and 46% reduction of NO_x emissions (Olendrzyński et al. 2004, 2007). While SO_2 and, to the less extent, NO_x emissions still show a downward trend, the NH_3 emission level has stabilized at about 320 Gg since the year 2000. The substantial reduction of SO_2 and NO_x emissions is a result of successful application of abatement strategies in Poland and economic transformations in 1990s (Mill 2006). The reduction of emissions resulted not only in decreased nitrogen and acid deposition but also the relative contribution of chemical species in acid and eutrophic deposition has changed, with the NH_x deposition gaining in importance. The EMEP estimates for the year 2004 show that the nitrogen input from NH_x deposition in Poland is about 20% larger than from oxidised nitrogen.

Until recently, the main source of spatial information on atmospheric pollutants depositions and concentrations was the Unified EMEP model, which is briefly discussed in Chapter 19 and the detailed description can be found in Tarrasón

Maciej Kryza
Department of Meteorology and Climatology, Institute of Geography and Regional Development, Wrocław University, Poland

M. Sutton, S. Reis and S.M.H. Baker (eds), *Atmospheric Ammonia,*
© Springer Science + Business Media B.V. 2009

et al. (2003). The resolution of the EMEP model (50 × 50 km) is often insufficient for regional scale analysis and might affect e.g. the critical levels and loads calculations (Mill 2006). The coarse spatial resolution of the atmospheric transport model is of special importance when reduced nitrogen is considered, because of its high spatial variation of emissions, concentrations and dry deposition (Dore et al. 2007a).

This study presents the preliminary results of applying the regional scale model, FRAME (Fine Resolution Atmospheric Multi-pollutant Exchange), for modelling spatial patterns of NH_3 concentrations and deposition of reduced nitrogen for Poland with 5 × 5 km grid. The FRAME model was described in Chapter 19. However, for the readability of this chapter we will repeat a brief description of the model. Furthermore the modelled results of annual average concentrations and a comparison with the available measurements are presented. The FRAME dry and wet deposition budgets for Poland are calculated and compared with EMEP reports and CIEP (Chief Inspectorate of Environmental Protection 2006) measurement-based estimates.

20.2   FRAME Model Outlook

FRAME is a Lagrangian model used for modelling the long-range transport and annual deposition of atmospheric pollutants with high horizontal (5 × 5 km) and vertical (33 layers) resolution, which makes it especially useful for modelling ammonia and ammonium concentrations and depositions on a regional scale. In fact, FRAME was originally developed as the Fine Resolution AMmonia Exchange model and was used to simulate transport and deposition of reduced nitrogen over the UK (Singles et al. 1998). Brief description of the model can be found in Chapter 19 while details are provided by Singles et al. (1998), Fournier et al. (2005a, b) and Dore et al. (2007b).

FRAME simulates an air column moving along straight-line trajectories. Trajectories are run at a 1° resolution for all grid squares at the edge of the model domain. The air column advection speed and frequency for a given wind direction is statistically derived from radio-sonde measurements (Dore et al. 2006). Here the frequency and wind speed roses use 6-hourly operational radiosonde data from the stations of Wrocław, Legionowo, Łeba, Greifswald, Lindenberg, Prague, Poprad and Kiev, spanning the whole 2002 year period (Fig. 20.1).

The dry deposition is calculated individually at each grid square using a canopy resistance model. The resistances included are: the atmospheric surface layer resistance (R_a), molecular sub layer resistance (R_b) and surface resistance (R_c) respectively which is dependent on surface characteristics.

Wet deposition of chemical species is calculated using scavenging coefficients based on those used in the EMEP model. The model employs a constant drizzle approach using precipitation rates calculated from a climatological map of annual precipitation for Poland (Kryza 2006). The FRAME model contains a seeder-feeder

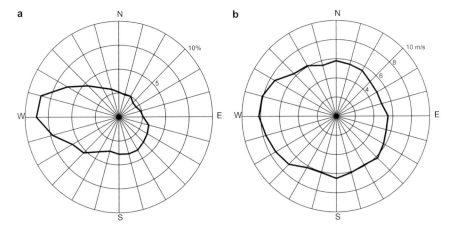

Fig. 20.1 (**a**) Wind frequency rose (percentage per 15° radial band) and (**b**) wind speed rose used for FRAME simulations for year 2002

module and enhanced washout rate is assumed over mountainous areas, due to the scavenging of cloud droplets by the seeder-feeder effect, to calculate local scale orographic enhancement of precipitation and concentration (Dore et al. 1999). The washout rate for the orographic component of rainfall is assumed to be twice that calculated for the non-orographic component.

20.3   Emission Data

Total emission of NH_3 from Poland is estimated to be 325 Gg in the year 2002, according to the national emission inventory (Olendrzyński et al. 2004). The year 2002 is chosen here because the detailed census data on animal number, necessary to calculate spatial patterns of NH_3 emissions, are available for this year from the Regional Data Bank (National Statistical Office 2006).

According to the official inventory of NH_3 emissions in Poland (Olendrzyński et al. 2004), agriculture contributes almost 97% (313.8 Gg of NH_3) and the remainder comes from waste treatment and production processes (Fig. 20.2). It should be mentioned however that recently Pietrzak (2006) suggested that the emission factors that are in use for Western Europe, and which were used by Olendrzyński et al. (2004), might not be valid for the NH_3 emission assessment in Poland due to the differences in agricultural practices. The total ammonia emission estimated by Pietrzak (2006) is about 10% lower than calculated by Olendrzyński et al. (2004) with the emission factors specific to Western Europe. The differences in the estimated emissions from fertilizer application are even greater and reach 30%. In this paper, the emission factors proposed by Olendrzyński et al. (2004) are applied, as they are also used in official national reports, including the

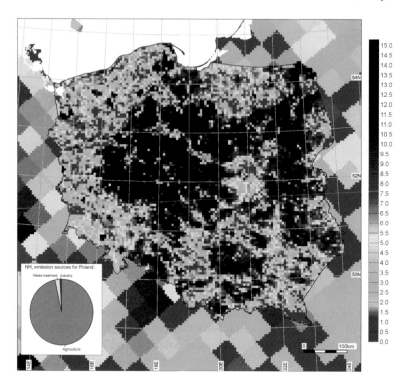

Fig. 20.2 Ammonia emission sources (%) and spatial emission (kg N ha⁻¹year⁻¹) in Poland in 2002

EMEP reports. As the FRAME model results are compared here with the spatial patterns of NH_3 concentrations and NH_x depositions calculated with the EMEP model, there is a need to provide the similar spatial emission inventory to the model. Spatial information on NH_3 emissions for Poland was prepared with 5×5 km spatial resolution (Fig. 20.2) using the method proposed by Dragosits et al. (1998) and data provided by the National Statistical Office (2006). The emissions from the remaining area of the model domain were taken from the EMEP expert emission inventory (Vestreng et al. 2005).

20.4 NH_3 Air Concentration

Both the FRAME and EMEP models show close agreement with measurements of NH_3 concentrations, although the data only from three EMEP network sites are available for comparison (Table 20.1). Moreover, two stations are located in specific places. Śnieżka is located on the top of the highest summit of the Karkonosze Mountains in SW Poland (Mt. Śnieżka 1603 m a.s.l.) in the centre of the

Karkonosze National Park, while Łeba station is on the sea shore. For the Śnieżka station, the fivefold overestimation in modelled NH_3 concentration, produced by the EMEP model, might be related to the coarse grid size. The RMSE (Root Mean Square Error) for the FRAME modelled concentrations is 0.34 and for EMEP is greater than 0.73.

The highest modelled NH_3 concentrations are in central Poland (Fig. 20.3). This is the region with intensive agriculture and large ammonia emission from low sources, particularly from the fertilizer application (Fig. 20.2). The FRAME modelled spatial pattern of NH_3 air concentrations is very complex, often showing a large local gradient. The concentrations modelled by FRAME are locally higher than those calculated with the EMEP model, though similar general spatial features are apparent (Fig. 20.3). The EMEP model shows the highest concentrations close to the fertilizer production sites (e.g. 21.5° E, 51.5° N).

20.5   Dry and Wet NH_x Deposition

Dry deposition, modelled with FRAME, is the highest close to the source areas (Fig. 20.3). As for the FRAME modelled air concentrations, NH_x dry depositions show large local gradients. In general, FRAME modelled dry deposition is locally higher than that estimated by the EMEP model. This can be attributed to the finer grid resolution of the FRAME model, resulting in a larger spread in the range of modelled concentrations, particularly where high emissions are concentrated in small areas.

In general, spatial patterns of wet deposition, calculated with the FRAME and EMEP model are similar (Fig. 20.3). Both models estimate the highest wet depositions over the mountainous areas in the south, which is caused by high precipitation supported by the enhanced washout rate in FRAME where the seeder-feeder effect is concerned. Central Poland is the second region of increased wet deposition, as this is the source region of emission and with high air concentrations of both NH_3 and NH_4^+.

20.6   Dry and Wet NH_x Deposition Budget

Total mass of reduced nitrogen deposited in the year 2002 in Poland, as estimated by the FRAME model, is close to 227 Gg of N (Table 20.2) and this is about 40 Gg of N less than was emitted. The FRAME estimated dry, wet and total deposition budgets of reduced nitrogen are in close agreement with the EMEP data. The FRAME dry deposition budget is smaller than that estimated by the EMEP model, while wet deposition is larger. Simultaneously, FRAME wet

Table 20.1 Measured and modelled NH_3 air concentration (μg N/m^3)

Station	Measured	FRAME	EMEP
Jarczew	1.38	1.97	1.54
Śnieżka	0.29	0.37	1.54
Łeba	0.64	0.62	0.54

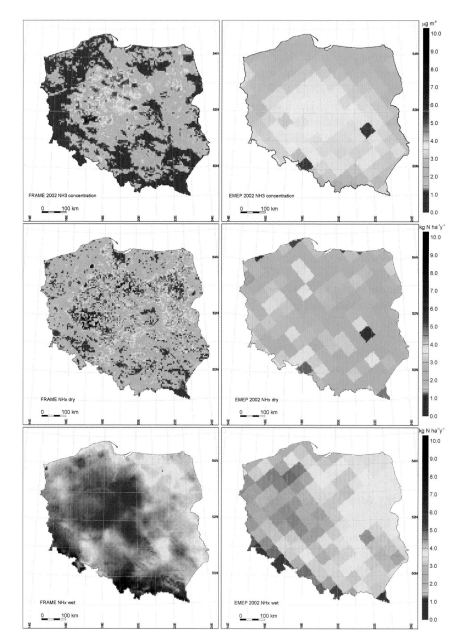

Fig. 20.3 FRAME (left column) and EMEP (right) modelled NH_3 air concentration (μg N m^{-3} year^{-1} and dry and wet deposition of NH_x (kg N ha^{-1} y^{-1})

Table 20.2 Dry, wet and total deposition budget of reduced nitrogen in Poland 2002 (Gg of N)

	FRAME	EMEP	CIEP
Dry	80.4	85.9	Not available
Wet	146.5	125.1	151.3
Total	226.9	211.1	Not available

deposition budget is close to the measurement-based estimates presented by CIEP (GIOŚ 2007).

20.7 Conclusions

Preliminary results of modelling the concentrations and depositions of ammonia for Poland with the FRAME model are presented. The model, which was originally developed for the UK, needs further development before it can become an operational tool for supporting decision making processes in Poland, but the preliminary results presented here are encouraging.

The FRAME model results for ammonia concentrations and dry and wet deposition of reduced nitrogen show close agreement with the EMEP model estimates. Due to the fine resolution of the FRAME model the spatial distribution of NH_x shows large local scale gradients, which are not present in the coarse resolution EMEP model. There is also a good spatial agreement between FRAME, EMEP and CIEP estimates, but the FRAME model calculates higher deposition over the mountainous areas due to the seeder-feeder effect.

The differences in deposition budgets for FRAME, EMEP and CIEP (wet only) estimates are below 6% for dry and 17% for wet deposition. The differences can be considered as small, taking into account different input data and model formulations. High resolution FRAME model locally shows higher concentrations and depositions, if compared with the EMEP estimates. This might be of special importance if critical levels and loads are taken into consideration.

The next steps should be focused on the model validation. The measurements for the model verification are however very limited in Poland, especially in case of air concentration and dry deposition.

Acknowledgement We gratefully acknowledge financial support from COST Action 729, the ESF NinE programme and CEH.

References

Chief Inspectorate of Environmental Protection (2006) National Monitoring of the Environment, www.gios.gov.pl (Accessed: 07/02/2008).

Dore A.J., Sobik M., Migała K. (1999) Patterns of precipitation and pollutant deposition in the Western Sudety Mountains, Poland. Atmospheric Environment 33, 3301–3312.

Dore A.J., Vieno M., Fournier N., Weston K., Sutton M.A. (2006) Development of a new wind-rose for the British Isles using radiosonde data, and application to an atmospheric transport model. Quarterly Journal of the Royal Meteorological Society 132, 2769–2784.

Dore A.J., Theobald M.R., Kryza M., Vieno M., Tang S.Y., Sutton M.A. (2007a) Modelling the deposition of Reduced Nitrogen at different scales in the United Kingdom. Proceedings of the 29th NATO/SPS International Technical Meeting on Air Pollution Modelling and its Application, Aveiro, Portugal, 24–28 September 2007 (in press).

Dore A.J., Vieno M., Tang Y.S., Dragosits U., Dosio A., Weston K.J., Sutton M.A. (2007b) Modelling the atmospheric transport and deposition of sulphur and nitrogen over the United Kingdom and assessment of the influence of SO_2 emissions from international shipping. Atmospheric Environment 41, 2355–2367.

Dragosits U., Sutton M.A., Place C.J., Bayley A. (1998) Modelling the spatial distribution of ammonia emissions in the United Kingdom. Environmental Pollution 102(S1), 195–203.

Fournier N., Tang Y.S., Dragosits U., De Kluizenaar Y., Sutton M.A. (2005a) Regional atmospheric budgets of reduced nitrogen over the British Isles assessed using a multi-layer atmospheric transport model. Water Air and Soil Pollution 162, 331–351.

Fournier N., Weston K.J., Dore A.J., Sutton M.A. (2005b) Modelling the wet deposition of reduced nitrogen over the British Isles using a multi-layer atmospheric transport model. Quarterly Journal of the Royal Meteorological Society 131, 703–722.

GIOŚ (2007) Monitoring jakości powietrza, www.gios.gov.pl (Accessed: 07/02/2008).

Kryza M. (2006) Zastosowanie GIS do przestrzennego modelowania miesięcznych sum opadu atmosferycznego w Polsce, [in:] Migała K., Ropuszyński P. (eds.), Współczesna meteorologia i klimatologia w geografii i ochronie środowiska, PTG Wrocław, 77–87.

Mill W. (2006) Tempral and spatial development of critical loads exceedance f acidity to Plish forest ecosystems in view of econoic transformation and national environmental policy. Environmental Science & Policy 9, 563–567.

National Statistical Office (2006) Regional Data Bank, www.stat.gov.pl (Accessed: 07/02/2008).

Olendrzyński K., Dębski B., Skośkiewicz J., Kargulewicz I., Fudała J., Hławiczka S., Cenowski M. (2004) Inwentaryzacja emisji do powietrza SO2, NO2, NH3, CO, pyłów, metali ciężkich, NMLZO i TZO w Polsce za rok 2002. IOŚ.

Olendrzyński K., Dębski B., Kargulewicz I., Skośkiewicz J., Cieslinska J., Fudała J., Hławiczka S., Cenowski M. (2007) Emission Inventory of SO2, NOx, NH3, CO, PM, NMVOCs, HMs, and POPs in Poland in 2005. UNECE-EMEP/Poland Report, National Emission Centre, IOŚ, 42.

Pietrzak S. (2006) Inventory method for ammonia emissions from agricultural sources in Poland and its practical application. Water-Environment-Rural Areas 6(16), 319–334.

Singles R., Sutton M.A., Weston K.J. (1998) A multi-layer model to describe the atmospheric transport and deposition of ammonia in Great Britain. Atmospheric Environment 32, 393–399.

Tarrasón L., Fagerli H., Eiof Jonson J., Klein H., van Loon M., Simpson D., Tsyro S., Vestreng V., Wind P., Posch M., Solberg S., Spranger, T., Cuvelier K., Thunis P., White L. (2003) Transboundary acidification, eutrophication and ground level ozone in Europe. PART I Unified EMEP Model description. EMEP Status Report 2003.

Vestreng V., Breivik K., Adams M., Wagener A., Goodwin J., Rozovskkaya O., Pacyna J. M. (2005) Inventory Review 2005, Emission Data reported to LRTAP Convention and NEC Directive, Initial review of HMs and POPs, Technical report MSC-W 1/2005, ISSN 0804–2446.

Chapter 21
Application of the EMEP Unified Model to the UK with a Horizontal Resolution of $5 \times 5\,km^2$

Massimo Vieno, Anthony J. Dore, Peter Wind, Chiara Di Marco,
Eiko Nemitz, Gavin Phillips, Leonor Tarrasón, and Mark A. Sutton

21.1 Introduction

The EMEP Unified model (Simpson et al. 2003; http://www.emep.int) is an Eulerian model that is driven by real-time meteorology. The model is applied over Europe for multiple years on a $50 \times 50\,km^2$ grid, with meteorological fields updated every 3 h. While comparisons with measurements have shown generally robust performance of the EMEP model on a European scale (e.g. Simpson et al. 2006), pollutants such as reactive nitrogen and sulphur have a high spatial variability in their emissions and a short life time. Therefore, the associated dry deposition also has a high spatial variability (Vieno 2006; van Pul et al. 2009, this volume). This is very important when critical loads of nitrogen are calculated for specific ecosystems. For this reason a number of models have been developed for high resolution operation at a national scale. To address these issues for the UK, the EMEP Unified Model is being developed, using a nested approach. This model application, referred to as EMEP4UK, has been developed at a 5 × 5 km² resolution covering the whole of the British Isles. By comparison with existing statistical models of atmospheric chemistry and transport over the UK (e.g. Singles et al. 1998; Lee et al. 2000; Metcalfe et al. 2001; Fournier et al. 2005; Vieno 2006; Dore et al. 2007), the EMEP4UK model therefore has the advantage of using real time meteorology, enabling the interactions between emissions, meteorology, concentrations and deposition to be addressed at a fine spatial scale.

21.2 Model Description

The EMEP4UK model framework is a collection of model pre-processors and post-processors which work together to produce a detailed representation of the physical and chemical state of the atmosphere over Europe and in particular over the UK.

Massimo Vieno
Institute Atmospheric and Environmental Science, Crew Building, University of Edinburgh, King's Buildings, Edinburgh EH9 3JN, UK and Centre for Ecology & Hydrology, Penicuik, Midlothian EH26 9HF, United Kingdom

M. Sutton, S. Reis and S.M.H. Baker (eds), *Atmospheric Ammonia,*
© Springer Science + Business Media B.V. 2009

The chemical scheme is identical to the EMEP Unified Model and includes the prognostic variables NH_3, NH_4NO_3, $(NH_4)_2SO_4$, NO, NO_2, NO_3^-, HNO_3, PAN, SO_2, H_2SO_4, SO_4^{2-}; a full description can be found in Simpson et al. (2003). The chemical scheme includes ammonium chemistry, gas and aqueous phase oxidation of SO_2 to sulphate, providing a comprehensive chemistry for both photo-oxidant and acidification studies.

EMEP4UK can, in principle, use meteorological drivers from various sources, using a pre-processor that converts and re-grids the required meteorological data. In the present implementation, the Weather Research Forecast (WRF) model (http://www.wrf-model.org/) is run specifically to provide the $5 \times 5\,km^2$ meteorological inputs necessary to drive the EMEP4UK model. WRF is run in back-cast mode including data assimilation (Newtonian nudging) of the United States National Center for Environmental Prediction (NCEP)/National Center for Atmospheric Research (NCAR) Global Forecast System (GFS) numerical weather prediction model reanalysis of meteorological observations applied for 2003–2006.

The WRF model represents the state-of-the-art in weather forecast modelling. It is easily available and widely used by the academic community for research and weather forecast purposes. WRF is applied here with a nesting domain approach of $50 \times 50\,km^2$, $10 \times 10\,km^2$ and $5 \times 5\,km^2$, to provide meteorological data at the required horizontal and vertical resolution. A coarse domain of $50 \times 50\,km^2$ is used to drive the EMEP Unified Model across the European domain to calculate the chemical initial conditions and boundary conditions for the EMEP4UK model.

21.3 Results and Discussion

EMEP4UK preliminary results for December 2005 are shown in Fig. 21.1; (a) ammonia surface concentration, (b) ammonium surface concentration, reduced nitrogen (c) dry deposition and (d) wet deposition. The model is able to capture the spatial variability of ammonia surface concentration when qualitatively compared with other studies such as Singles et al. (1998) and Dore et al. (2006, 2007). An interesting feature of Fig. 21.1 a and b is the modelled depletion of gaseous ammonia concentrations and matching enhancement of particulate ammonium concentrations at several point locations in central and eastern England. These positions match the location of major point sources of sulphur dioxide emissions, indicating the occurrence of locally rapid formation of ammonium sulphate for the modelled period (1–20 December 2005). This feature is not so clearly revealed in the annual average maps from, for example, the FRAME model (Singles et al. 1998; Vieno 2006; Dore et al. 2006, 2007), and deserves further investigation in relation to field observations.

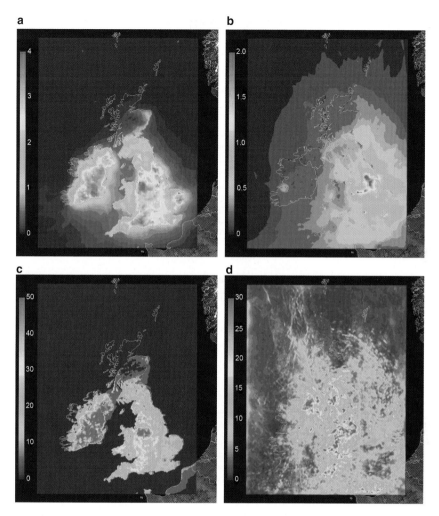

Fig. 21.1 Example outputs of the EMEP4UK model (v.244/WRF 2.1) showing simulations for 1–20 December 2005: (**a**) NH$_3$ surface concentration (µg m^{-3} NH$_3$-N); (**b**) Aerosol NH$_4^+$ surface concentration (µg m^{-3} NH$_4^+$ -N) (**c**) NH$_x$ dry deposition (mg m^{-2} NH$_x$-N for the period) and (**d**) NH$_x$ wet deposition (mg m^{-2} NH$_x$-N for the period). In this simulation only the emissions from the British Isles are included

Wet deposition may be underestimated by the EMEP4UK model for the simulation 1–20 December 2005, as the rain drop velocity used to convert the WRF rain mixing ratio into mm was too low. For the run shown for 2006 (Fig. 21.2), a larger rain drop velocity was used to be more representative of the rainfall velocity distribution. However, the spatial pattern reproduces well the higher wet deposition over mountain terrains as suggested by Dore et al. (1992) and Fournier et al. (2005).

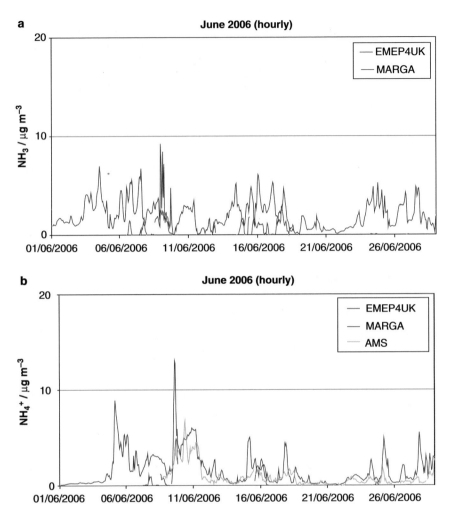

Fig. 21.2 Ammonia and ammonium surface air concentrations for June 2006 as simulated by the EMEP4UK model and compared with observations from the June 2006 EMEP Intensive Measurement Period at Auchencorth and Bush, Scotland: (**a**) NH$_3$ (MARGA) and (**b**) NH$_4^+$ (MARGA and AMS instruments). This simulation includes emissions from the parts of Ireland, France, Holland, Germany, Denmark, Belgium and Norway which are included in the EMEP4UK domain, as well as import from the EMEP model for other parts of Europe using WRF meteorology

It should be noted that the performance of the EMEP Lagrangian Model and the Unified Model, in estimating the spatial patterns of wet deposition of sulphur and nitrogen in the UK has long been a matter of discussion (e.g. NEGTAP 2001). The present analysis shows that the key challenge is to be able to capture the spatial scale of orographic precipitation. This is difficult when the EMEP model is run with a 50 × 50 km^2 grid resolution. By contrast, using the scale of

$5 \times 5 \, km^2$ in EMEP4UK, the main features of UK orographic precipitation and wet deposition are resolved (see e.g. Fig. 21.2d).

The 2006 EMEP4UK model predictions for ammonia and ammonium have been compared with hourly observations from the June 2006 Intensive Period of the EMEP air chemistry monitoring program. The EMEP4UK model estimates of gaseous ammonia and particulate ammonium surface air concentrations are compared with measured estimates shown in Fig. 21.2a and b, respectively. The observations were made with a wet-chemistry monitor for inorganic species in aerosols and gases in ambient air (MARGA; Thomas et al., under review) and by quadrupole aerosol mass spectrometry (AMS, Jayne et al. 2000). The MARGA was located at the Auchencorth Moss EMEP Super-Site (N55:47:35; W3:14:34), while the AMS was located approximately 7 km north east at Bush (N55:51:56; W3:12:25). Model simulations and other long-term observations indicate broadly similar submicron aerosol concentrations at these two sites.

The EMEP4UK model performs reasonably well in simulating the hourly ammonium surface concentrations, especially if compared with the AMS. Less data are available in the June 2006 Intensive Period for ammonia, therefore it is difficult to assess model performance. The model, however, seems to overestimate ammonia surface concentration. This may reflect the importance of sub-grid variability for ammonia (within the 5 km grid square) not resolved by EMEP4UK, as discussed elsewhere in this volume by Loubet et al. (2009, this volume).

The results shown here are preliminary, but demonstrate that the EMEP4UK model can be successfully applied to the UK at higher resolution than the standard resolution of $50 \times 50 \, km^2$ applied in the EMEP Unified model. The model has the advantage of simultaneously simulating high resolution deposition processes in both space and time.

Acknowledgements This work is supported jointly by the UK Department for Environment Food and Rural Affairs (Defra), Centre for Ecology & Hydrology (CEH) and the Norwegian Meteorological Institute (Met.No), and provides a contribution to the work of the NitroEurope Integrated Project. We gratefully acknowledge travel support from the COST Action 729 and the European Science Foundation NinE programme.

References

Dore A.J., Choularton T.W. and Fowler D. (1992) An improved wet deposition map of the United Kingdom incorporating the seeder-feeder effect over mountainous terrain. Atmos. Environ. 26A, 1375–1381.

Dore A.J., Vieno M., Fournier N., Weston K.J. and Sutton M.A. (2006) Development of a new wind rose for the British Isles using radiosonde data and application to an atmospheric transport model. Q. J. Roy. Meteor. Soc. 132, 2769–2784.

Dore A.J., Vieno M., Tang Y.S., Dragosits U., Dosio A., Weston K.J. and Sutton M.A. (2007) Modelling the atmospheric transport and deposition of sulphur and nitrogen over the United Kingdom and assessment of the influence of SO2 emissions from international shipping. Atmos. Environ. 41, 2355–2367.

Fournier N., Weston K.J., Dore A.J. and Sutton M.A. (2005) Modelling the wet deposition of reduced nitrogen over the British Isles using a Lagrangian multi-layer atmospheric transport model. Q. J. Roy. Meteor. Soc. 131, 703–722.

Jayne J.T., Leard D.C., Zhang X., Davidovits P., Smith K.A., Kolb C.E. and Worsnop D.R. (2000) Development of an aerosol mass spectrometer for size and composition analysis of submicron particles. Aerosol Sci. Tech. 33, 49–70.

Lee D.S., Kingdon R.D., Jenkin M.E. and Garland J.A. (2000) Modelling the atmospheric oxidised and reduced nitrogen budgets for the UK with a Lagrangian multi-layer long-range transport model. Environ. Model. Assess. 5, 83–104.

Metcalfe S.E., Whyatt J.D., Broughton R., Derwent R.G., Finnegan D., Hall J., Mineter M., O'Donoghue M. and Sutton M.A. (2001) Developing the hull acid rain model: its validation and implications for policy makers. J. Environ. Sci. Policy 4, 25–37.

NEGTAP (2001) Transboundary Air Pollution: Acidification, Eutrophication and Ground Level Ozone in the UK. Report of the National Expert Group on Transboundary Air Pollution. Defra, London (www.nbu.ac.uk/negtap).

Simpson D., Fargerli H., Jonson J.E., Tsyro S., Wind P. and Tuovinen J.P. (2003) Transboundary Acidification, Eutrofication and Ground Level Ozone in Europe. Part 1. Unified EMEP Model Description. (Eds.) EMEP/MSC-W Report 1/03. Norwegian Meteorological Institute, Blindern, Norway.

Simpson D., Butterbach-Bahl K., Fagerli H., Kesik M., Skiba U. and Tang S. (2006) Deposition and emissions of reactive nitrogen over European forests: a modelling study. Atmos. Environ. 40, 5712–5726.

Singles R.J., Sutton M.A. and Weston K.J. (1998) A multi-layer model to describe the atmospheric transport and deposition of ammonia in Great Britain. Atmos. Environ. 32, 393–399.

Thomas R., Trebs I., Otjes R., Jongejan J.P.C., ten Brink H., Phillips G., Kortner M., Meixner F.X. and Nemitz E (2008) A continuous analyser to measure exchange fluxes of water soluble inorganic aerosol compounds and reactive trace gases. Environ. Sci. Technol. (submitted).

Vieno M. (2006) The Use of an Atmospheric Chemistry-Transport Model (FRAME) over the UK and the Development of Its Numerical and Physical Schemes. Ph.D. thesis University of Edinburgh, Edinburgh, UK.

Part V
Conclusions and Outlook

Chapter 22
Critical Levels for Ammonia

John Neil Cape, Ludger van der Eerden, Andreas Fangmeier, John Ayres,
Simon Bareham, Roland Bobbink, Christina Branquinho, Peter Crittenden,
Christina Cruz, Teresa Dias, Ian Leith, Maria Amélia Martins-Loução,
Carole Pitcairn, Lucy Sheppard, Till Spranger, Mark Sutton, Netty van Dijk,
and Pat Wolseley

22.1 Summary

The issues involved in assessing and evaluating Critical Levels for atmospheric NH_3 and vegetation are discussed, recognizing that measurable effects of exposure to NH_3 may be observed in plant species without necessarily any adverse effects on that species. Consequently, impacts of NH_3 are best measured in terms of community response, where there is a measurable change in species composition and/or loss of individual species. Such data are currently available only from field measurements, either from measurements close to point sources, where NH_3 concentrations have also been measured, or from controlled field fumigation experiments. The relationship between Critical Levels and Critical Loads is explored, and the redundancy of the 1993 Critical Level ($8 \, \mu g \, m^{-3}$ as an annual average), when compared with the Critical Loads for N deposition, is demonstrated. Statistical techniques for estimating 'no effect' concentrations from field measurements are presented, and then used to review the available data. For sensitive vegetation types (lichens and bryophytes) field evidence shows changes in species composition at NH_3 concentrations (as an annual average) of $1 \, \mu g \, m^{-3}$, which is proposed as the new Critical Level for annual average NH_3 concentrations for sensitive ecosystems (i.e. those in which lichens and bryophytes are important components). There are fewer data for higher plants, but a Critical Level of $3 \, \mu g \, m^{-3}$ is proposed, as a long-term average NH_3 concentration. The existing Critical Level of $23 \, \mu g \, m^{-3}$ for the monthly average concentration is retained, given the recent changes in seasonal patterns of NH_3 emissions from agricultural practices such as slurry spreading.

J.Neil Cape
Centre for Ecology & Hydrology Bush Estate,Penicuik, Midlothian, EH26 0QB,
United Kingdom

L.van der Eerden
Foundation OBRAS, Centre for Art and Science, Evoramonte, Portugal

M. Sutton, S. Reis and S.M.H. Baker (eds), *Atmospheric Ammonia,*
© Springer Science+Business Media B.V. 2009

22.2   History and Background

The current NH_3 Critical Levels (CLEs) for vegetation were based on measurements and observations from the 1980s, mostly from the Netherlands, and are set as:

hour: 3,300/day: 270/month: 23/year $8 \mu g \ m^{-3}$.

Critical Loads (CLOs) for total N deposition are specified per habitat and have been updated several times in the past decade. Since then, discussions have clarified that CLEs and CLOs have different uses and roles in environmental assessment and regulations, for several reasons:

- In CLOs no differentiation is made between different reactive N species, while in the CLE of NH_3 no differentiation between habitats is made.
- CLOs are applied at a regional or national scale, and have limitations for local use, specific habitat protection, etc.
- NH_3 concentrations can be measured more easily than N deposition, so CLEs are more easy to use in air quality regulation.

For the present annual CLE, dry deposition of N from NH_3 alone would far exceed the relevant CLO for N deposition in most, if not all habitats, implying that the current CLEs are redundant. Nevertheless, the working group (WG) decided that the CLE for NH_3 needs revision for the following reasons:

- It is not easy to convert CLEs and CLOs simply by using a deposition velocity, because this may not be well defined for some habitats and climatic conditions.
- New information from experiments and field surveys has become available that allows an update of the CLEs. Responses of higher plant species are now being detected at much lower concentrations than the current one-year CLE. Epiphytic lichens and bryophytes are affected at NH_3 concentrations much lower than the current CLE.

The WG decided that although experimental results and field surveys have shown responses to NH_3 in terms of intrinsic plant properties, such as tissue N content or soluble ammonium, there is no direct proven causal relation or sufficient understanding of the link between such changes and ecologically significant effects. Consequently, only measurements of changes in species composition, including loss or death of an individual species, have been used to assess the need for changing CLEs. Damage to individual species and changes in species composition have an immediate relevance for impacts on biodiversity and the integrity of ecosystems. Only studies with long-term (at least one year) monitoring data for NH_3 have been used, although data from studies with shorter periods of monitoring data have been cited as corroboration. The WG decided not to use results from laboratory or open-top chamber experiments with constant exposure levels because new information indicates that these treatments may underestimate the impact of NH_3 in the field situation, where exposure is more episodic.

The clearest evidence of an ecologically significant effect of NH$_3$ has come from:

- Changes in species composition observed in controlled field experiments and field surveys
- Damage to and death of individual species in controlled field experiments

Changes in species composition do not occur as short-term responses, but are the result of long-term exposure, often more than one year, so the conclusions drawn from this type of response advise on the setting of a long-term CLE.

The WG decided that there is now enough empirical evidence from which to derive an empirical long-term CLE, at least for some types of vegetation.

22.3 Methodology

Changes in species composition have been formulated in terms of the presence or absence of individual plant species that have been classified according to their preferences for nitrogen-rich or acidic conditions. This method has been shown to allow the detection of subtle NH$_3$-driven changes in species composition with great sensitivity, because it focuses on the species at the extremes of the response, for which exposure to NH$_3$ produces the greatest negative or positive effect. The difference between the proportions of 'acidophyte' lichens and 'nitrophytes' lichens at a site, or between 'nitrophobe' and 'nitrophile' higher plants, provides an index that can be compared against measured NH$_3$ concentrations, typically downwind of a point source (e.g. Pitcairn et al. 2009).

Conceptually, the response to NH$_3$ could be described as in Fig. 22.1, where there is no detectable response to increasing NH$_3$ concentrations until a threshold is reached, and then the response increases monotonically as NH$_3$ concentrations increase. At high enough concentrations this relationship will break down, but for the purpose of setting CLEs the focus is at low concentrations. At low NH$_3$ concentrations the relationship (based on field measurements) appears to be a linear response to a logarithmic increase in concentrations, but the method does not rely on the exact form of the relationship.

Observations of a response in the field may not include a concentration level representative for pristine environments, which are probably in the range of 0.01–0.3 µg m^{-3}. Background levels in the cleanest regions of Europe are in the range of 0.1–0.3 µg m^{-3}. The WG used the lowest measured concentration at the site where the experimental data were collected (in the range of 0.1–0.6 µg m^{-3}) as an estimate of the 'background' level, and then used the linear relationship between response and (log) concentration to calculate the concentration at which the predicted response was greater than the 95% Confidence Interval for the fitted linear relationship.

This approach takes account of the measurement uncertainties, and the uncertainty in the exact form of the relationship between response and concentration, but provides an overestimate of the 'no effect' concentration that can be detected in any one set of data. A clear linear relationship, with many data points, especially at or close to the true 'no effect' concentration, gives the most reliable estimate.

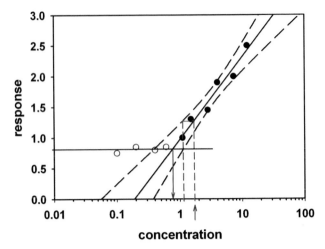

Fig. 22.1 Schematic diagram of system response to increasing concentrations. Symbols denote measurements across the whole range, including concentrations below the 'no effect' level (horizontal line, open symbols). The apparent 'no effect' concentration (left arrow) is around 0.7; for a set of measurements above the 'no effect' concentration (solid symbols) the method described here predicts an upper-bound no-effect concentration (right arrow) at a concentration of 1.6

22.4 The Evidence

The best evidence relates to studies on lichens and bryophytes, from long-term field experiments, from field measurements close to point sources and from large-scale field surveys. All the key data come from the UK. However, similar measurements in Portugal, Italy and Switzerland all show thresholds that are consistent with the UK results, although NH_3 concentration data are not available for a full year, and therefore we only used them as corroborative evidence.

The key results are summarised in Table 22.1, along with the corroborative results.

Based on these data, we recommend a long-term CLE which is appropriate for protecting epiphytic lichens and bryophytes, and ecosystems with significant abundance of ground dwelling lichens and/or bryophytes, such as bogs, fens, heaths and moor land. Exceeding this CLE would result in shifts in species composition, and increased potential for species extinctions. In this context, 'long-term' refers to a period of several years, sufficient for such changes to occur, but does not guarantee protection beyond 20–30 years.

The recommended long-term CLE for these systems is $1\,\mu g\ m^{-3}$ NH_3

For higher plants, there is much less information, but two studies indicate that the current annual CLE of $8\,\mu g\ m^{-3}$ is too high in the longer term:

1. Changes in woodland ground flora downwind of an intensive animal unit in SW England suggest a threshold of $4\,\mu g\ m^{-3}$.
2. Comparison of the increasing rate of death of Calluna at a field fumigation experiment in Scotland (Whim bog) with the death rate of the lichen *Cladonia*

Table 22.1 Summary of experimental data showing the effects of long-term exposure to NH$_3$ on species composition

Location	Vegetation type	Lowest measured NH$_3$ concentration (μg m^{-3})	Estimated 'no effect' concentration (μg m^{-3})	Reference
SW England	Epiphytic lichens	1.5	ca. 2	Leith et al. (2005) Pitcairn et al. (2009)
SE Scotland, poultry farm	Epiphytic lichens	0.6	0.7 (twigs) 1.8 (trunks)	Sutton et al. (2004a, b; 2009)
Devon, SW England	Epiphytic lichens diversity (twig)	0.8	1.6	Wolseley et al. (2006)
United Kingdom, national NH$_3$ network	Epiphytic lichens	0.1	1.0	Leith et al. (2005) Sutton et al. (2009a) Wolseley et al. (2009)
Switzerland	Lichen population index	Modelled concentrations	2.4	Rihm et al. (2009)
SE Scotland, field NH$_3$ experiment, Whim bog	Lichens and bryophytes – damage and death	0.5	<4	Sheppard et al. (2009)
Corroborative evidence (limited NH$_3$ data)				
Portugal, cattle farm	Epiphytic lichens	–	1	Pinho et al.. (2009)
Italy, pig farm	Epiphytic lichens	0.7	2.5	Frati et al. (2006)

indicates consistently that *Calluna* death occurs at a concentration 2.2 times that at which *Cladonia* is killed; this implies a 'no effect' concentration for *Calluna* of around $2 \mu g \ m^{-3}$.

Based on this evidence, the long-term CLE for higher plants is probably in the range $2–4 \mu g \ m^{-3}$, so on the basis of expert judgement we set the long-term CLE for higher plants as $3 \mu g \ NH_3 \ m^{-3}$.

We assume that this long-term CLE (defined as above) will protect heathland, woodland ground flora and probably also oligotrophic grassland, from NH_3-driven shifts in species composition, and the potential for species extinctions.

The proposed long-term CLEs should be taken as replacing the current annual CLE for all vegetation and habitat types of $8 \mu g \ m^{-3}$.

There is no additional information on short-term exposures, but we would provisionally retain the existing monthly CLE of $23 \mu g \ m^{-3}$, to ensure appropriate protection against short-term peak exposures in regions where the ratio of peak concentrations to mean concentrations is relatively high. Changes in agricultural practice (e.g. manure spreading in spring) may lead to higher peak/mean ratios and short term concentrations than have been measured to date. Moreover, to our knowledge no information exists on the peak/mean ratio in Southern Europe. The existing monthly CLE is related to the newly proposed long-term CLEs as it was based on the sensitivity of heathland species and bryophytes. But, because it was derived from fumigation experiments with constant exposure levels, and because another methodology was used for data evaluation, the WG decided that the monthly CLE has expert judgement as its basis.

The statistical distribution of NH_3 concentrations with time from several monitoring sites in northern and central Europe shows that the annual CLE will be a stricter constraint on exposure than the existing hourly or daily CLEs (Sutton et al. 2009b). However, there are too few monitoring data from southern Europe to generalise on the relationship between peak concentrations and annual averages.

22.5   Evaluation

Because the proposed CLEs are based on empirical evidence from measured exposure to NH_3, no assumptions have been made on the pathways or mechanisms of action. This is a similar approach to the estimation of empirical CLOs.

No 'safety factor' has been applied in estimating the long-term CLEs; rather, they have been chosen as the concentration above which significant effects of NH_3 exposure are measurable.

The proposed long-term CLEs do not apply to all ecosystems or habitats because of a lack of relevant information. However, the proposed long-term CLE for higher plants is likely to be no more restrictive than the existing CLO for most habitats, based on estimating the contribution of $3 \mu g \ NH_3 \ m^{-3}$ to dry deposition of N in most ecosystems. For example, typical values for the UK would be $15–20 kg \ N \ ha^{-1} \ y^{-1}$ for short vegetation, and up to $30 kg \ N \ ha^{-1} \ y^{-1}$ for tall vegetation, in addition to the deposition of other N species (wet and dry).

22.6   Recommendations

The basis for deriving CLEs would be greatly strengthened by the following:

- Standardising methodologies on
  - Lichen and vegetation mapping
  - Classification of acidic/nitrogen preferences of individual species
- Standardising NH₃ measurement protocols, including measurement height, and sampling for at least 1 year
- Defining priority habitats for future work, because of current lack of data/knowledge, such as southern European (Mediterranean) ecosystems, and continental Eastern Europe ecosystems
- More experimental data from long-term experiments such as the Whim bog study (Sheppard et al. 2009)
- Deriving methods for linking physiological and biochemical measurements on plants to observed shifts in species composition
- Better understanding of the effects of temporal variation in exposure concentration in relation to temporal variation in sensitivity
- Better understanding of the acceleration of growth (resulting in shorter life span) and of N cycles
- Development of a mechanistic model (based on N fluxes and physiological relations) to underpin and refine the proposed CLEs. The model input would probably be NH₃ concentrations and site characteristics, and the output an estimate of ecosystem response

References

Frati L., Santoni S,. Nicolardi V., Gaggi C., Brunialti G., Guttova A., Gaudino S., Pati A., Pirintsos S.A., Loppi. S. (2006) Lichen biomonitoring of ammonia emission and nitrogen deposition around a pig stockfarm. *Environmental Pollution* 146 (2), 311–316.

Leith I.D., van Dijk N., Pitcairn C.E.R., Wolseley P.A., Whitfield C.P., Sutton M.A. (2005) Biomonitoring methods for assessing the impacts of nitrogen pollution: Refinement and testing, 290, JNCC Report No. 386, Peterborough.

Pinho P., Branquinho C., Cruz C. Tang Y.S., Dias T., Rosa A.P., Máguas C., Loução M.A.M., Sutton M.A. (2009) Assessment of critical levels of atmospheric ammonia for lichen diversity in cork-oak woodland, Portugal. In: Sutton M.A., Reis S., Baker S.M.H. (eds.), *Atmospheric Ammonia - Detecting Emission Changes and Environmental Impacts. Results of an Expert Workshop Under the Convention on Long-range Transboundary Air Pollution*. Springer (Chapter 10).

Pitcairn C.E.R., Leith I.D., Sheppard L.J., Sutton M.A. (2006) Development of a nitrophobe/nitrophile classification for woodlands, grasslands and upland vegetation in Scotland, 21, Centre for Ecology & Hydrology, Penicuik, UK.

Pitcairn C.E.R., Leith I.D., van Dijk N., Sheppard L.J., Sutton M.A., Fowler D. (2009) The application of transects to assess the effects of ammonia on woodland groundflora. In: Sutton M.A., Reis S., Baker S.M.H. (eds.), *Atmospheric Ammonia – Detecting Emission Changes and Environmental Impacts. Results of an Expert Workshop Under the Convention on Long-range Transboundary Air Pollution*. Springer (Chapter 5).

Rihm B., Urech M., Peter K. (2009) Mapping ammonia emissions and concentrations for Switzerland – effects on lichen vegetation. In: Sutton M.A., Reis S., Baker S.M.H. (eds.), *Atmospheric Ammonia – Detecting Emission Changes and Environmental Impacts. Results of an Expert Workshop Under the Convention on Long-range Transboundary Air Pollution.* Springer (Chapter 7).

Sheppard L.J., Leith I.D., Crossley A., van Dijk N., Cape J.N., Fowler D., Sutton M.A. (2009) Long-term cumulative exposure exacerbates the effects of atmospheric ammonia on an ombrotrophic bog: Implications for Critical Levels. In: Sutton M.A., Reis S., Baker S.M.H. (eds.), *Atmospheric Ammonia – Detecting Emission Changes and Environmental Impacts. Results of an Expert Workshop Under the Convention on Long-range Transboundary Air Pollution.* Springer (Chapter 4).

Sutton M.A., Leith I.D., Pitcairn C.E.R., van Dijk N., Tang Y.S., Sheppard L., Dragosits U., Fowler D., James P.W., Wolseley P.A. (2004a) Exposure of ecosystems to atmospheric ammonia in the UK and the development of practical bioindicator methods. In: Lambley, P. & Wolseley, P.A. (eds.), *Lichens in a Changing Environment.* English Nature Research Report No. 525. English Nature, Peterborough.

Sutton M.A., Pitcairn C.E.R., Leith I.D., van Dijk N., Tang Y.S., Skiba U., Smart S., Mitchell R., Wolseley P., James P., Purvis W., Fowler D. (2004b) Bioindicator and biomonitoring methods for assessing the effects of atmospheric nitrogen on statutory nature conservation sites. Edited by M.A. Sutton, C.E.R. Pitcairn & C.P. Whitfield. JNCC Report No. 356, Peterborough.

Sutton M.A., Wolseley P.A., Leith I.D., van Dijk N., Tang Y.S., James P.W., Theobald M.R., Whitfield C. (2009a) Estimation of the ammonia critical level for epiphytic lichens based on observations at farm, landscape and national scales. In: Sutton M.A., Reis S., Baker S.M.H. (eds.), *Atmospheric Ammonia – Detecting Emission Changes and Environmental Impacts. Results of an Expert Workshop Under the Convention on Long-range Transboundary Air Pollution.* Springer (Chapter 6).

Sutton M.A., van Pul W.A.J., Sauter F., Tang Y.S., Horvath L. (2009b) Over which averaging period is the ammonia critical level most precautionary? In: Sutton M.A., Reis S., Baker S.M.H. (eds.), *Atmospheric Ammonia – Detecting Emission Changes and Environmental Impacts. Results of an Expert Workshop Under the Convention on Long-range Transboundary Air Pollution.* Springer (Chapter 8).

Wolseley P.A., James P.W., Theobald M. R., Sutton M.A. (2006) Detecting changes in epiphytic lichen communities at sites affected by atmospheric ammonia from agricultural sources. *The Lichenologist* 38: 161–176.

Wolseley P.A., Leith I.D., van Dijk N., Sutton M.A. (2009) Machrolichens on twigs and trunks as indicators of ammonia concentrations across the UK – a practical method. In: Sutton M.A., Reis S., Baker S.M.H. (eds.), *Atmospheric Ammonia – Detecting Emission Changes and Environmental Impacts. Results of an Expert Workshop under the Convention on Long-range Transboundary Air Pollution.* Springer (Chapter 9).

Chapter 23
Detecting Change in Atmospheric Ammonia Following Emission Changes

Jan Willem Erisman, Albert Bleeker, Albrecht Neftel, Viney Aneja, Nick Hutchings, Liam Kinsella, Y. Sim Tang, J. Webb, Michel Sponar, Caroline Raes, Marta Mitosinkova, Sonja Vidic, Helle Vibeke Andersen, Zbigniew Klimont, Rob Pinder, Samantha Baker, Beat Reidy, Chris Flechard, Laszlo Horvath, Anita Lewandowska, Colin Gillespie, Marcus Wallasch, Robert Gehrig, and Thomas Ellerman

23.1 Summary

The Working Group discussed the progress on the state of knowledge on deriving trends from measurements and their use to verify abatement measures or other causes for decrease in emissions of ammonia to the atmosphere. The conclusions from the 2000 Berne meeting (Menzi and Achermann 2001), the background review (Bleeker et al. 2009) and presentations during the session (Horvath et al. 2009; Tang et al. 2009; Webb et al. 2009), as well as the discussions served as input for the conclusions of this report.

We have seen some clear advancement in closing the gap between the observed and expected values for reduced nitrogen, where we do get a better understanding of the reasons behind it. The long-term measurements that are available follow the emission trend. Current measurements make it possible to evaluate policy progress on ammonia emission abatement. Especially in those countries where there were big (>25%) changes in emissions, such as in the Netherlands and Denmark the trend is followed quite closely, especially when meteorology is well taken into account. In order countries, such as the UK, the trend was much smaller, but there was no gap between measurements and model estimates. In the Netherlands there still is an ammonia gap: a significant (30%) difference between emissions based ammonia concentrations and measurements. The trend is the same. The difference might be due to either an underestimation of the emission or an overestimation of the dry deposition. It is recommended to further explore this gap, especially by investigating the high temporal resolution measurements, improving the emission/deposition modeling, by having a model intercomparison with countries that use models that do not show a gap and finally by doing a thorough uncertainty analysis.

On the European scale it is difficult to follow the emission changes, both because of lack of measurements, especially in the Eastern part of Europe and because of the confounding factor of the SO_2 emission reductions, affecting the ammonium

J.W. Erisman,
Energy Research Centre of the Netherlands (ECN), Petten, The Netherlands

concentrations in aerosol and in rain water. It is recommended to fully implement the EMEP monitoring strategy and to improve the models in order to quantify the influence of a changing chemical climate.

The EMEP monitoring strategy can be a good starting point for development of a strategy that is focused on the right questions. Therefore first it is necessary to evaluate policies and the indicators derived from them that need to be assessed (time and space). Using existing models a pre-modeling study should be done to select the monitoring sites that eventually will give you the answer to the basic (policy) question using improved models and assessment tools. The best and economic feasible instrumentation should be selected with an extensive QA/QC program to make the measurements comparable. After implementation, especially for trend evaluation, the monitors used should not be changed.

23.2   Introduction

The Working Group addressed the issues involved in making the link between estimated national NH_3 emissions and measurements of NH_x concentrations. Monitoring NH_3 emissions directly is not possible and therefore there is a need to quantify the effectiveness of NH_3 emission abatement in a more independent way by using measurements of NH_x.

The Working Group addressed the following objectives:

- To quantify the extend to which estimated regional changes in ammonia emissions have been reflected in measurements of ammonia and ammonium in the atmosphere
- To distinguish cases where the estimated changes in ammonia emission are due to altered sectoral activity or the implementation of abatement policies and thereby assess the extent to which atmospheric measurements verify the effectiveness of ammonia abatement policies
- To make recommendations for future air monitoring and systems for assessing the national implementation of ammonia abatement policies

During the discussions within the working group the following items were addressed, which will be described in more detail in the following sections:

- Update the current scientific understanding based on new datasets and assessments
- Is there still an 'ammonia gap' in the Netherlands?
- Does such a gap exist in other countries of Europe?
- Are we confident about the effectiveness of ammonia mitigation policies?
- How can we best address the relationships between emission and deposition using atmospheric modelling and improved monitoring activities?

Update the current scientific understanding based on new datasets and assessments

The update of the scientific understanding is presented in the Background Document prepared for this workshop. The document builds on the Berne Background Document (Sutton et al. 2001), which was used to facilitate the discussion about following trends by means of measurement data at the UNECE Ammonia Expert Group meeting in Berne (Switzerland) in 2000. It is now several years since the Berne Workshop and major new datasets on European NH_3 and NH_4^+ monitoring and their relationship to estimated NH_3 emissions have become available for the following countries: UK, Germany, Hungary, Switzerland, Denmark, The Netherlands, North Carolina (USA), Slovak Republic, Norway and Croatia. Bases on these datasets the findings of the previous workshop were evaluated, updating our current scientific understanding about the different issues that were addressed in the previous document. In particular, input is given to questions like: is there still an 'Ammonia Gap' in the Netherlands, does such a gap exist in other countries, can we be confident of the effectiveness of ammonia mitigation policies and how can we best address the relationships between emission and deposition, using atmospheric modelling and improved monitoring activities?

Next to country-specific case studies, also an overview of the European situation is given with respect to the link between emissions of NH_3 and the modelled and measured concentration/deposition of reduced nitrogen. Changes in chemistry, especially due to decreasing SO_2 emission, are evaluated in terms of its effect on the levels of reduced nitrogen over European both temporal and spatial, providing a better understanding of the observed levels in NH_3 and NH_4^+ over Europe.

In general, the conclusions from the Berne Background document (Sutton et al. 2001) are (to a large extent) still valid. Any further elaboration on the update of our current scientific understanding of the problems addressed in the background document and the conclusions made is not given here, but will be worked out in more detail when addressing the other items in the following sections. However, looking at the differences between Berne Background document and the current review (Bleeker et al. 2009), we can distinguish some major advances in our understanding of the different issues we mentioned before:

- We have seen some clear advancement in closing the gap between the observed and expected values for reduced nitrogen, where we do get a better understanding of the reasons behind it. This especially builds on case studies from the UK (study on Foot and Mouth Disease) and the longer term datasets that became available for the Netherlands and Denmark.
- The long-term measurements for reduced nitrogen follow the emission trend. The extended datasets presented in the background document made it clear that following emission trends by means of measurements can only adequately be done when long-term measurements are available (longer than, e.g. 10 years).
- Current measurements make it possible to evaluate policy progress on ammonia emission abatement. This is especially true for the situation in Denmark and the Netherlands, where the NH_3 ammonia reduction is followed by the monitoring results from the national monitoring networks. However, this is only possible since both these networks were designed to follow the expected changes in

emissions, based on extensive pre-studies used for developing the measuring networks. The UK also performed such a pre-study for developing their national monitoring network. Clear trends from this network could not be found there, but this is mainly due to only small changes in emission levels. Another case in the UK, where it was possible to detect changing emissions in measured concentrations of NH_3, was related to a study following an event of Foot and Mouth Disease in different regions in the UK. Also here a modelling pre-study was used for designing the layout of the measurement network, in order to detect effectively the expected emission changes within the different regions.

- In recent years the instrumentation (i.e. models, monitoring equipment) to evaluate the link between ammonia emission and the concentration and/or deposition of reduced nitrogen has improved.

23.3   Reflection of Emissions and Changes Therein in Monitoring Data

23.3.1   Conclusions

23.3.1.1   Is There Still an Ammonia Gap in the Netherlands?

Originally, the ammonia gap existed in two parts: (i) an absolute difference between measurements and concentrations based on modelling and (ii) a difference in trend in measurements and modelled data. There have been several studies done focusing on explaining these two gaps (see Bleeker et al. 2009). There is no difference between the measured and modelled trend. By extending the monitoring period, improving the emission estimates and taking the meteorological conditions into account this is solved. However, the absolute systematic difference is still in the order of 30%. The explanation of the difference can be by two factors, probably contributing both: (i) underestimation of certain emissions and (ii) parameterisation (overestimating) the dry deposition in agricultural areas. In source areas advection might play a significant role, but this is of more importance for the very small scale. However, when selecting monitoring sites this aspect has to be taken into account.

23.3.1.2   Does this Ammonia Gap Exist in Other Countries in Europe?

Apart from the Netherlands and Denmark, where emissions decreased by about 30%, in most countries where monitoring takes place the concentrations did not change much (UK, Switzerland). In these countries the emission reductions were only very limited. However, no systematic gap such as in the Netherlands is signalled (UK, DK) or clearly detected by deposition or concentration measurements (Switzerland).

For the whole of Europe it is questionable if the decrease in emissions can be detected that took place in the early nineties, during the transition period (to market economy) when significant structural changes occurred leading to vast improvements in efficiency of production in the Eastern part of Europe. This is troubled by the fact that at the same time the SO_2 emissions strongly decreased affecting the lifetime and transport distance of ammonia (Horvath et al. 2009).

The EMEP monitoring sites that have been in place are not aimed to detect ammonia from agriculture and therefore the signal is not detectable. Furthermore, for the new EMEP monitoring strategy there is lack of implementation especially in Eastern Europe. It is therefore concluded that the evaluation of the absolute emissions in Europe and the changes therein is difficult because of lack of monitoring data covering the whole of Europe.

23.3.2   Recommendations

Even though several studies and sensitivity analysis on the Dutch ammonia gap have been done, we suggest that the high temporal resolution site data might be used to interpret difference between model and measurements. Furthermore, emission modeling and the effect of meteorology and/or the net surface exchange need further improvement. Gradually the emission-deposition modeling should be integrated because they are strongly related and influenced by the same factors (meteorology). A quantification of the uncertainty in model input (emissions), parameters and output, together with a quantification of the uncertainty in the measurements is necessary. Because Denmark and the UK could not detect a significant gap the Working Group felt that it would be beneficial to exchange models and apply the Dutch model on the Danish and UK data and vice versa. Furthermore, modeling experience from the US could be taken on board.

23.4   Are We Confident About the Effectiveness of Ammonia Mitigation Policies?

23.4.1   Conclusions

Abatement options are usually tested on their effectiveness in the lab or under controlled conditions. The controlled conditions should reflect the practice in the field. These experiments form the basis of the abatement effectiveness. In practice the efficiency can differ because the practical applicability might be different. This is seen in the Netherlands for example in the case of application techniques of manure. The effectiveness of abatement options is also addressed in the Cross-Cutting Group A on the reliability of ammonia emission data and abatement efficiencies (Section 6.5).

Current monitoring is not focused on evaluating individual abatement options. The monitoring is focused on evaluating the changes in concentration/deposition to see if exposure of ecosystems improves. There are however, special case studies which focus on the abatement efficiency, such as the STOP program and the Veld study in the Netherlands, the food and mouth disease (FMD) measurement campaign in the UK, etc. Sectoral changes or individual abatement options can therefore currently not be detected with existing monitoring data. Local studies in this case are relevant to find sectoral changes.

Big changes in emissions have been detected using monitoring data. These changes either result from implementation of abatement options, such as in the Netherlands or Denmark, or as the result of reducing animal numbers due to the economic situation (Eastern Europe). The trends in emissions (including abatement measures) are in agreement with trends in measurements.

In Europe, where there is only limited monitoring of wet deposition and aerosol concentrations, there are confounding factors that result in deviating trends, such as the effect of the reduction of SO_2 emissions in Europe. This resulted in changes in ammonium deposition and concentrations. Furthermore, year to year changes in meteorology can have effects on observed trends and need to be filtered out with models.

23.4.2   Recommendations

It is recommended to initiate local studies when large changes are expected (e.g. FMD). Furthermore, we need to quantify the chemical effects better in models. Also the meteorology effects on emissions and depositions have to be quantified better. For the monitoring it is useful to follow a monitoring strategy aimed at asking the question and resulting indicators.

23.5   Recommendations on Improvement of Modelling and Monitoring Activities

23.5.1   Conclusions

From the background document and the Working Group discussions it was concluded that evaluation of emissions and the changes therein need a modelling and monitoring strategy. Currently, the monitoring activities are not suitable to evaluate emissions because they have not started aiming for evaluating agricultural data. Two exceptions are the Netherlands and the UK. However, EMEP modelling and monitoring activities were started in a time when sulphur and oxidized nitrogen emissions caused impacts and protocols were agreed upon to reduce its emissions.

The sites that have been established at that time not necessarily are suitable for evaluating agricultural emissions, such as ammonia.

23.5.2 *Recommendations*

A good modelling/monitoring strategy therefore starts with the basic question we want to address. This starting point is driven by the current policy items based on worries about impacts, such as impacts on biodiversity, human health and climate change. These impacts lead to questions about the contribution of ammonia, its emissions, abatement options to reduce emissions and validation and verification options. These form the starting point for the modelling/monitoring strategy. The first activity would therefore to evaluate CLRTAP/EU/national policy objectives, based on impacts (biodiversity, PM and human health, climate change) and to determine the expected changes at the different scales (emission, concentration, deposition). The policies might, e.g. be focused on the protection of ecosystems (biodiversity) through decreasing critical load exceedances. This means that deposition needs to be monitored for different ecosystems. However, since deposition monitoring, especially of dry deposition, is not possible, the concentration monitoring provides a good alternative, provided issues like the changes in surface affinities are taken into account.

We therefore have to take into account what is already in place; a monitoring strategy builds on current facilities. Therefore it is strongly advised to implement the EMEP monitoring strategy and improve and extend it by focusing on:

- Monitoring of the spatial variations in ammonia emissions
- Detect the expected changes in emissions
- Focusing on all output parameters of the model relevant for changes in ammonia (e.g. N-balance)

The way to move forward with this is to do a pre-study based on current knowledge. The aim of such a study would be to optimise the spatial and temporal resolution of monitoring data, given the current monitoring sites and policy questions, the current emission, transport and deposition modelling, the impacts and the level of integration (e.g. pollutant swapping, climate change, etc.). Once the pre-study has been done the locations are selected and the temporal resolution, the required precision and accuracy and the representativeness are known. It is essential to use monitoring equipment with enough and known quality for specific applications; harmonised, intercompared, QA/QC, etc., not changing over the years. The past years instrumentation has been improved and new methods might be considered for monitoring depending on their application (MARGA, photoacoustic, lasers, passive samplers, DELTA, etc.). Finally, resources should be reserved for improvement and application of models to do good assessment of monitoring results (emissions, dry deposition, atmospheric chemistry, dispersion, transport) and answer the policy questions. Additional to this local studies and impact assessment for special issues might be initiated, e.g. in cases where large emission reductions are expected (e.g. the FMD study in the UK).

References

Bleeker A., Sutton M.A., Acherman B., Alebic-Juretic A., Aneja V.P., Ellermann T., Erisman J.W., Fowler D., Fagerli H., Gauger T., Harlen K.S., Hole L.R., Horvath L., Mitosinkova M., Smith R.I., Tang Y.S., van Pul W.A.J. (2009) Linking ammonia emission trends to measured concentrations and deposition of reduced nitrogen at different scales. In: Sutton M.A., Reis S., Baker S.M.H. (eds.), *Atmospheric Ammonia – Detecting Emission Changes and Environmental Impacts. Results of an Expert Workshop Under the Convention on Long-range Transboundary Air Pollution.* Springer (Section 3.1).

Horvath L., Fagerli H., Sutton M.A. (2009) Long-term record (1981–2005) of ammonia and ammonium concentrations at K-puszta Hungary and the effect of SO2 emission change on measured and modelled concentrations. In: Sutton M.A., Reis S., Baker S.M.H. (eds.), *Atmospheric Ammonia – Detecting Emission Changes and Environmental Impacts. Results of an Expert Workshop Under the Convention on Long-range Transboundary Air Pollution.* Springer (Section 3.2).

Menzi H. and Achermann B. (2001) UNECE Ammonia Expert Group (Berne 18–20 Sept 2000) Proceedings (eds.) 157 Swiss Agency for Environment, Forest and Landscape (SAEFL), Berne. (Copy of Working Group 2 report - http://www.nitroeurope.eu/ammonia_ws/documents/AEG_bern_wg2_report.pdf)

Sutton M.A., Asman W.A.H., Ellerman T., van Jaarsveld J.A., Acker K., Aneja V., Duyzer J.H., Horvath L., Paramonov S., Mitosinkova M., Tang Y.S., Achermann B., Gauger T., Bartnicki J., Neftel A., Erisman J.W. (2001) Establishing the link between ammonia emission con-trol and measurements of reduced nitrogen concentrations and deposition. In: UNECE Am-monia Expert Group (Berne 18–20 Sept 2000) Proceedings (Edited by Menzi H. and Achermann B.) pp. 57–84. Swiss Agency for Environment, Forest and Landscape (SAEFL), Berne. (Revised version published in *Environmental Monitoring & Assessment* (2003) 82: 149–185.)

Tang Y.S., Dragosits U., van Dijk N., Love L., Simmons I., Sutton M.A. (2009) Assessment of NH_3 and NH_4^+ trends and relationship to critical levels in the UK National Ammonia Monitoring Network (NAMN). In: Sutton M.A., Reis S., Baker S.M.H. (eds.), *Atmospheric Ammonia – Detecting Emission Changes and Environmental Impacts. Results of an Expert Workshop Under the Convention on Long-range Transboundary Air Pollution.* Springer (Section 3.3).

Webb J., Eurich-Menden B., Dämmgen U., Agostini F. (2009) Review of published studies estimating the abatement efficacy of reduced-emission slurry spreading techniques. In: Sutton M.A., Reis S., Baker S.M.H. (eds.), *Atmospheric Ammonia – Detecting Emission Changes and Environmental Impacts. Results of an Expert Workshop Under the Convention on Long-range Transboundary Air Pollution.* Springer (Section 3.4).

Chapter 24
Assessment Methods for Ammonia Hot-Spots

Pierre Cellier, Mark R. Theobald, Willem Asman, William Bealey, Shabtai Bittman, Ulrike Dragosits, Janina Fudala, Matthew Jones, Per Løfstrøm, Benjamin Loubet, Tom Misselbrook, Beat Rihm, Ken Smith, Michal Strizik, Klaas van der Hoek, Hans van Jaarsveld, John Walker, and Zdenek Zelinger

24.1 Summary

To date, most attention in modelling NH_3 dispersion and deposition has focused on the regional and European scale, and little attention has been given to dealing with NH_3 in hot-spot areas. The Working Group addressed the issue of hot-spots with four main objectives:

- Identify what are the main issues of hot-spots and their consequences at different scales.
- Review current modelling methods for accounting for NH_3 dispersion and deposition in hot-spots.
- Examine the status of methods for effect assessment and air monitoring in NH_3 hot-spots.
- Recommend broad principles for assessment approaches in ammonia hot-spots, including spatial approaches and the interactions between transboundary ammonia emission reduction targets and other policy measures.

24.1.1 Key Findings

Accounting for hot-spots for either up-scaling of fluxes or risk assessment for nearby ecosystems requires a precise description of all the processes involved. Hot-spot assessment should also account for a more general context, e.g. background concentrations and the deposition history.

The key uncertainties in the models are on the emission and the dry deposition. Sufficient local input data are required for making effects assessments and landscape analysis. For dry deposition, this requires better knowledge of compensation points and surface resistances for different ecosystems, their dependence on climatic variables and the deposition history for ammonia and other pollutants.

P. Cellier
Institut National de la Recherche Agronomique (INRA), Unité Mixte de Recherche Environnement et Grandes Cultures, 78850 Thiverval-Grignon, France

Using different models allows the analysis of landscape interactions between sources and receptors with sufficient accuracy for a range of conditions to consider real cases and scenarios. It also allows the assessment of local, tailored abatement measures.

Scenarios from local-scale modelling can be used in a statistical way to provide estimates of within-grid cell recapture for national- and regional-scale models, linked with global descriptors of the spatial variability in land cover.

24.1.2 Recommendations

- To obtain maximum results from local modelling on hot-spots, it is necessary to gather local values instead of national standard values. This holds for ammonia emission inventories and for dispersion and deposition modelling.
- To further develop dynamical models to estimate the diurnal and seasonal changes in emission strengths from point sources (animal houses) and area sources (land spreading of animal manure); for area sources in detailed plot studies, this should include the effects of meteorological and soil variables
- To make a synthesis of available databases to build up some reference cases against which the different models could be tested and compared
- An intercomparison of regional-scale and sub-grid models would be highly beneficial for highlighting the differences between the modelling approaches and the abilities of regional models to simulate local-scale interactions
- To define scenarios to investigate the possible effect of in-grid fragmentation of land use on net ammonia fluxes; a sensitivity study would allow to investigate the range of local recapture and possible effects on air quality
- To promote the development of deposition measurement methods that could apply to advective conditions (e.g. one measurement height) in order to provide means of direct validation of dispersion/deposition models

24.2 Introduction

To date, most attention in modelling NH_3 dispersion and deposition has focused on the regional and European scale. Certainly, within the UNECE Convention on Long-Range Transboundary Air Pollution, little attention has been given to dealing with NH_3 in hot-spot areas. In the past, it was considered that this was a local problem and not relevant for the transboundary interest of the Convention. However, the role of hot-spots is increasingly recognized by the Task Force on Measurement and Modelling, for example, when modelling the so called city-delta, which is the urban enhancement of particulate matter concentrations above background.

In the case of NH_3, developing assessment approaches in hot-spot areas is similarly important since hot-spots are the areas of acutest environmental impact and regulatory focus, e.g. for the protection of designated nature conservation areas near farm sources. Moreover, the properties of the surface (vegetation, soil, water) very close to hotspots determines the local dry deposition and thereby also the fraction that is left for long-range transport which was shown to vary from about 0.4 to 0.9, demonstrating the influence of the dry deposition near hotspots. This shows that deposition assessments need to consider the fraction that derives from local sources versus that which is of more distant national or transboundary origin.

The analysis of ammonia hot-spots by this Working Group is thus the first time that this issue has been treated specifically in a UNECE Expert Workshop. As a result, a large part of the work has been to identify the main issues of this topic at different scales, from very local scale (1–100 m) to the regional model grid scale (1–100 km) and to extract the relevant results and approaches from the experiences in different countries. The discussions of the Working Group have considered whether similar framework approaches are being taken by the research community and the comparability and reliability of the detailed models and methods being implemented. The modelling tools that have been considered range from detailed models for assessment of landscape structure, such as the effect of tree belts round farms through general dispersion models at the landscape and sub-grid scale (i.e. within a grid cell of a national or regional-scale model) to the development and application of screening tools. Most of the work to date has focused on land-atmosphere exchange of NH_3 and the assessment of the impacts of NH_3 dry deposition. However, guidelines also need to be considered regarding the contribution of locally enhanced wet and aerosol NH_4^+ deposition as well as the interaction with other policy issues (e.g. *Habitats Directive*, other forms of N pollution). Practical methods to assess ammonia interactions in hot-spots have also been considered. This work covered the available modelling and monitoring tools and how these can be combined to provide an integrated approach to assessment.

The Working Group had four main objectives:

- Identify what are the main issues of hot-spots and their consequences at local and regional scales
- Review current emission/dispersion/deposition modelling methods for accounting for NH_3 dispersion and deposition in hot-spots
- Examine the status of methods for effect assessment and air monitoring in NH_3 hot-spots
- Recommend broad principles for assessment approaches in ammonia hot-spots, including spatial approaches and the interactions between transboundary ammonia emission reduction targets and other policy measures

The work done to achieve these objectives has been divided into three sections:

- The objectives for assessing ammonia hot-spots
- The modelling approaches available and their uncertainties
- Tools for local ammonia management

24.3   Objectives for Assessing Ammonia Hot-Spots

Ammonia hot-spots refer to strong sources (or groups of sources) of ammonia that may have effects at a range of scales on either nearby sensitive ecosystems or on the atmosphere at a regional scale. These sources might be highly variable in space (e.g. livestock housing) or in time (e.g. land-spreading of manures).

For assessment of NH_3 hot-spots, one should consider processes and integration at:

- Spatial scales ranging from local scale (farm sources and nearby receptors) to regional model sub-grid scale (typically $10 \times 10\,km^2 - 50 \times 50\,km^2$)
- Time scales of:
  - Diurnal cycles of emissions, chemical reactions, atmosphere-vegetation exchange fand meteorology
  - Seasonal cycles of crop and vegetation growth, agricultural practices (particularly fertiliser and slurry application), livestock lifecycles and meteorology

The optimal spatial and temporal resolution of the assessment will depend on the objectives of the study, the accuracy required in outputs and the accessibility of the input data.

The objectives for assessing hot-spots can be broadly divided into two classes: (i) local effects and (ii) effects on transboundary air pollution.

24.3.1   Local Deposition and Other Local Effects

Local deposition of NH_3 to nearby sensitive ecosystems can be much larger than critical loads and can have serious consequences for the protection of sensitive ecosystems. The potential impact of a source (or sources) on one or more sensitive ecosystems is often the main objective of any assessment but there are many other objectives that may be important. For example, it is often important to investigate individual processes such as recapture by vegetation or the temporal dynamics of exchange processes that are not well understood and need to be investigated in detail. Another relevant objective can be the understanding of the interactions between sources and receptors within a landscape. Modifying the characteristics of the environment close to sources and receptors (e.g. tree belts or buffer zones) has been shown to potentially mitigate impacts on sensitive ecosystems. Spatial planning provides a useful tool for reducing impacts to sensitive ecosystems by assessing these landscape interactions. Also of importance, however, is the air quality of nearby urban environments, particularly in relation to the contribution of ammonia to fine particle formation ($PM_{2.5}$) and the associated impacts on human health. This latter issue is particularly of importance for very intensive agricultural areas that can modify the chemical environment at the regional scale. Moreover, possible linked positive or negative issues such as N_2O emission from locally deposited NH_3, odours or landscape quality should be taken into account when considering management near hot-spots.

24.3.2   Link with Transboundary Air Pollution

National and regional chemistry-transport models like the EMEP Unified Model run on a grid with cells of several tens of km. Within these grid cells, the properties are the same across the cell (emission rate, dry deposition velocity, wet deposition rate, etc.). In most models, this results in a single value of NH_3 concentration or dry deposition for the entire grid cell. In reality concentrations and dry deposition have a high spatial variability and these types of models will overestimate the values in some locations and underestimate in others.

The question arises whether a better description of sub-grid variability could help improve the national and regional models and understand unexplained differences between observations and modelling (e.g. the so called "ammonia gap"). This would also allow a better analysis of the impacts on individual sensitive ecosystems within the grid cell. Some of the NH_3 emitted within a grid cell of a national or regional model will be deposited within the same grid cell. This results in a significant fraction of the emitted NH_3 not being available for long-range transport. The concept of "net emission" corresponds to the total emission in a grid cell minus the fraction of the emissions deposited in the same cell. Models such as the EMEP Unified Model do simulate the deposition of ammonia emitted from the same grid cell, but the interactions between individual source and sink areas within the grid cell cannot be taken into account. Local-scale modelling has shown that these interactions can be significant, at either the very local scale (effect of features surrounding the sources, e.g. tree belts) or at larger scale (interactions within a network of sources and sinks in a landscape). One of the principal goals is to estimate the net emission of NH_3 from the grid cell of a national or regional model, which may paradoxically require a good representation of processes at a very small scale (e.g. to take account of horizontal change in land use such as tree belts). An improved estimate of net NH_3 emission from a grid cell could also lead to improved estimates of chemical reactions and particle formation and therefore improved estimates of air quality in national and regional models.

24.4   Local Emission/Dispersion/Deposition Modelling Approaches and Uncertainties for Ammonia

For effect assessment a quantitative approach must be conducted that should account for the main processes from emission to deposition at the relevant time and space scale. This requires special attention not only to dispersion but also to the emission source that must be described with great attention both in space (location relative to the sink area) and time. Calculating deposition requires a good understanding of atmosphere-vegetation exchange as well as vegetation characteristics for a range of ecosystems and vegetation types. Atmospheric dispersion and transport processes are reasonably well understood and the Working

Group identified that the largest current uncertainties lie in the emission and deposition components of the models.

24.4.1 Emissions Modelling

- Assessing ammonia emission at local to regional scale require to consider the nitrogen flow principle meaning that the ammonia emission in the four compartments (animal housing, manure storage outside the animal housing, fertilizer and manure application in the field, and animal grazing) is calculated as a percentage loss from the incoming nitrogen flow into each compartment. This has the advantage that for example covering of manure storage facilities outside the animal house automatically will lead to higher emissions during the subsequent land application of the animal manure. This should especially be considered when assessing the effects of abatement options linked to e.g. animal housing or manure manipulation.

- For local emission modelling, the local characteristics of the sources should be considered more accurately than national or regional averages which are generally used in inventories.

  - The information used for a local inventory should be based on actual local data, as far as possible. This could include animal type and numbers, the type of animal houses and techniques of manure and fertilizer application within the area under concern.
  - The location of the animal houses and agricultural fields for grazing or animal manure application should be considered as precisely as possible with respect to the sensitive area in case of quantitative analysis. Care should be taken that they are not always located at the farmer's house as is usually given in the National Agricultural Census data.
  - Local environmental conditions might also influence ammonia emissions: soil type, soil pH, vegetation, etc.
  - The vegetation compensation point contributes to determining the air ammonia background concentration. It might be determined by local conditions like events of fertilizer application or the proximity of a hot-spot by the way of local deposition.

- Much attention should be given to the time dependence of ammonia emissions from the sources and hence the relation to variables that make emission change:

  - Animal or manure management is a major cause of change in emission at weekly to seasonal time scales. When cattle are in the meadow in summertime leads to less ammonia emission from the animal house and more in the fields. For area sources like animal manure application, the emission period coincides with events of manure application which might change with locations and local measures. The same applies to mineral fertilizer application, but with lower fluxes. For poultry or pig farming, the duration and timing of the peaks should be known.

– Climatic factors: the ammonia volatilization rate often exhibits a diurnal cycle as emissions depend on temperature and atmospheric diffusion. Emission from mechanically ventilated animal housing is not directly influence by meteorological conditions, but indirectly because the ventilation rate during part of the year depends on the temperature in the housing and thereby on the solar radiation. All other emissions (from naturally ventilated animal housings, storage facilities, after application of manure) depend on the wind speed/turbulence, the air and surface temperatures and sometimes also on the occurrence of rainfall (open storage facilities, emission after application). Solar radiation is also important, as it increases surface temperature.

- Uncertainties in national ammonia emission inventories can be better than 20%, for instance the Netherlands uses 17%. (Van Gijlswijk et al. 2004). Using the national shares of housing types and manure application techniques in the area under study, the uncertainty in the total local emissions will certainly be higher. Carefully registration of available housing types and manure application techniques is therefore necessary.
- Moreover, ammonia emissions from area sources are sensitive to the same meteorological parameters as dispersion and deposition processes. Dry deposition depends also on the wind speed/turbulence and atmospheric diffusion does as well. If the wind speed, e.g., increases the emission rate will increase and the emitted amount will be transported over longer distances, due to increased diffusion. It is therefore important that the temporal resolution of these processes is the same, otherwise such interactions cannot be modelled. Fine tuning between emission modelling and atmospheric chemistry and dispersion/transport modelling is therefore necessary.
- Other N-containing compounds are emitted to the atmosphere such as amines and dusts (from animal housings or crop harvest). Moreover, volatile fatty acids, that might contain N, are emitted. The emission of those compounds is, however, considered to be much less important than the emission of ammonia.
- Finally, for a correct estimate of the effects of local emissions and local measures it is necessary to have an estimate of the contribution of the local emissions to the local ammonia concentrations. If the background ammonia concentration is high and the local contribution is low, then the impact of the local emissions also will be low and potential measures will likely have little effect.

24.4.2 Exchange Processes

- Exchange of ammonia with vegetation can be bidirectional, depending on the relative values of the atmospheric concentration and the canopy compensation point. Downwind and close to sources, atmospheric concentrations are generally higher than compensation points resulting in deposition of NH_3. This is not the case for vegetation located upwind of sources where the concentrations can be lower than the compensation points resulting in emission of NH_3 from the vegetation. Emission is also likely after exposure of vegetation to high concentrations for a short period, e.g. during land spreading and subsequent exposure to small air concentrations. Thus, the stomatal emission potential (Γ_s) may exhibit large temporal variability downwind of ammonia sources. Additionally, the feedback between stomatal uptake and Γ_s may produce compensation point "gradients" in the vicinity of sources, producing large spatial variability in Γ_s.

- For assessment studies that require quantitative estimates of ammonia deposition to sensitive ecosystems, there is a clear need to have a better estimate of plant compensation points for both estimating deposition to these ecosystems and the vegetation-atmosphere exchange processes for the area between the source and the ecosystem. Improved quantitative knowledge of compensation point for different natural and agricultural species, including the relationship with the fertilisation level of crops, would be necessary. There is a clear need for the development of a database containing Γ_s for primary vegetation types along with fertilization level (agricultural systems) or atmospheric nitrogen input (semi-natural systems). By necessity, the development of such a database will rely heavily on direct measurements of leaf apoplast chemistry. However, given the observed disagreement between direct measurements of Γ_s and estimates derived from canopy-level flux measurements, the representativeness of direct methods is currently unknown. More work is urgently needed to resolve this issue.

- Cuticular deposition (R_w) is the dominant pathway for NH_3 deposition for most atmospheric concentrations. However, when NH_3 concentrations increase R_w also increases due to saturation of the plant cuticle. Therefore, at higher concentrations – for example in the vicinity of a source – there will be relatively less NH_3 deposited to the cuticle than the stomata. Currently, in modelling NH_3 deposition, a fixed value of R_w is applied irrespective of the atmospheric concentration. Implementing a variable R_w accounting for atmospheric concentrations will dramatically decrease the predicted amounts of NH_3 deposited near to sources, and conversely increase the amount deposited in areas of low atmospheric NH_3 further downwind. However, cuticular deposition is also strongly influenced by other factors such as surface wetness, as after rainfall the R_w will be dramatically reduced due to surface water, and by the duration of previous exposure – as the greater the amount the higher the R_w. Therefore, to improve the surface resistance parameterisation – R_w should be a function of vegetation type, atmospheric concentrations of NH_3, surface wetness and the deposition history. Currently, concentration dependent R_w has only been calculated for moorland and sand dune communities and further work is needed in applying these values to other habitat types. A special focus should also be given to changes in surface resistance during and after rainfall events. The diurnal cycle of canopy wetness should also be examined in greater detail, as drying periods after sunrise often coincide with large emission fluxes. In some cases, a large fraction of ammonia deposited to the cuticle at night may be re-emitted during morning hours, thus changing R_w for further deposition.

- Within-canopy transfer is a key-process for the recapture of NH_3, especially when a plume from a point or area source passes through a plant canopy, which is the case when a tree belt is located around the farm. Depending on the relative heights of the source and the leaves of the canopy downwind, the canopy recapture can change significantly, both due to changes in turbulent diffusion and exposure to a varying area of leaves. To account for this process, a model that simulates canopy recapture close to a source should be able to simulate within-canopy exchange at canopy or even leaf level.

- For simulating deposition to sensitive ecosystems over a range of canopies and periods of the year, the deposition model should be able to separate exchange with the soil, understorey and the foliage. A two-layer bi-directional exchange model, in which the foliage and soil components are separated (e.g. Nemitz et al. 2000), is likely sufficient for most applications. The two-layer resistance modelling approach is relatively easy to implement and represents a compromise between the mechanistically simple single-layer model and more complex multi-layer approaches that separate the foliage or ground into multiple compartments.

24.4.3   Modelling Tools: Range of Application and Uncertainties

24.4.3.1   Range of Application

- Modelling is necessary when an assessment of the impact of an existing or new installation (animal housing, manure storage) on the environment is to be made. The spatial resolution required for these assessments include the very local scale (for assessing the effect of e.g. a tree belt), and a larger scale of tens of kilometres where the impact is felt. The temporal resolution should be hourly (to account for coupled emission/dispersion/deposition processes), and the results should be integrated over at least one year to account for seasonal variability in emissions, vegetation growth and meteorology.
- Modelling could also be used to help understand the main environmental variables and processes (meteorological conditions, description of the environment and its functioning in relation to the atmosphere) that can influence local deposition and impacts.
- Due to the range of different objectives of ammonia hot-spot assessment from estimating deposition close to sources to investigating within-grid cell interactions in regional models, a range of modelling tools have been developed. These range from models that give a detailed description of the air flow and the exchange with vegetation to models that describe transfer at landscape or sub-grid scales (typically $10\,m - 10\,km$) and can be used for both impact assessment and landscape planning. Model used as screening tools were also developed to provide rapid, approximate estimates of concentrations and deposition. To date, no common modelling approach has been agreed and different models are used in different countries (and even within the same country) and for different purposes. The choice of the model is driven by the objectives of the study and by the available models/expertise. These models can be categorised in three groups:

- Detailed, short-range models for very local application such as ammonia recapture by a tree belt downwind of an animal house. These comprise e.g. Lagrangian Stochastic Models (LSM), Eulerian second-order models and Computational Fluid Dynamics (CFD).
- Models used for describing the emission, dispersion and deposition at a scale of several km^2. These are generally Gaussian Plume (GPM) or Eulerian first-order models (EM) which were developed for simulating transfer at landscape or larger scales and might be applied for hot-spot assessment and interactions within a grid cell of a national or regional model. They often account for chemical reactions in the atmosphere with several degrees of complexity.
- Simpler Models (SM) such as analytical relationships describing the decrease of ammonia concentration with distance from the source were developed from experiments or outputs of more complex models. They can cover a range of conditions and their application range from screening to regional estimates of concentration and deposition by coupling with an emission inventory. They generally only require basic input data and they can be fully implemented within a GIS.

- These different groups should be seen as complementary since they provide tools that operate at different spatial and temporal scales with differing complexities. The choice of model to use should be based on the objectives of the assessment. The following points were discussed in the Working Group and related to the different types of models:

  - **Background concentrations.** As the plant compensation point of sensitive areas is generally not far from the air concentration at distance from the hot spot, an assessment model needs to consider the background concentration from local measurements or from a national- or regional-scale model (i.e. a nested approach).
  - **Deposition velocities.** Using a constant deposition velocity, which was often made for calculating ammonia deposition from atmospheric concentrations in the past, is not adequate as the dry deposition velocity shows temporal variations that are partly a function of the meteorological conditions. Dynamic models accounting for the dependence of deposition velocity on meteorological and vegetation characteristics should be used if an accurate representation of the exchange processes is required. Moreover, two-pathways and two-layer models should be considered when possible, although these will increase the uncertainty on deposition estimates.
  - **Verifying deposition.** estimates Most models are verified using concentration measurements rather than direct deposition measurements. This can lead to large uncertainties in the deposition estimates. Measuring deposition fluxes close to a source, thus in advective conditions is still a challenge. Developing direct (Relaxed Eddy Accumulation, Disjunct Eddy Correlation, Eddy Covariance) or indirect (balance, tracer-ratios, labelling) flux measurement methods would help to improve short-range deposition models.
  - **Temporal dynamics.** Assessments made over long time periods or under variable meteorological conditions should be able to take into account a large

range of conditions. For example, they should be able to simulate such processes as re-emission of ammonia subsequent to deposition when air concentration decreases rapidly with time e.g. following rain events, changes in source strengths or changes in wind speed and direction.

- For making operational assessments, the results of atmospheric transport and deposition models in the form of emission-deposition matrices can be incorporated in a GIS. Such a system would facilitate the study of the effects of changes in emission strength or positions of sources.
- **Wet deposition from nearby sources.** Dry deposition caused by a source is more important than wet deposition up to a distance of the order of 50 km from the source. This distance varies somehow with wind direction, the conversion rate of ammonia and other conditions. So very near the source it is not important, but within a current EMEP grid cell ($50 \times 50 km^2$) it is important.

24.4.3.2 Priority Uncertainties in Modelling

The priority uncertainties for the three different types of models are:

- Short range models (e.g. LS, Eulerian 2nd order, CFD) are generally used for ideal case studies and scenario analyses, their main requirements are improving the exchanges parameter description (compensation point, surface resistance) and the turbulence field especially in transition zones and in the canopy.
- For GPM and EM which are used for estimating possible critical load exceedance, the priority will be on the spatial location of sources and receptors and on accounting for the background concentration. The exchange parameters (compensation point, surface resistance etc.) and the strength of the sources are also important.
- For simpler models, the relative location of source and receptor (i.e. distance and direction) is the most important as well as a good estimate of the source strength. A special focus should be put on the ability to apply these models over a range of conditions (source types, land cover, climates etc.).

All model types should be improved in order to better account for stable conditions (such as night time) for it is a situation that may lead to large deposition rates due to low dispersion of the plume and decreased surface resistance due to surface wetness by dew.

24.4.4 Links Between Hot-Spot Assessment and National/Regional Modelling

There are two important links between assessments at local scale and modelling at larger scales:

- There is a need to know the fraction of ammonia emitted that is dry deposited locally in order to know the amount that is left for long range transport. This was discussed above. In the past it was assumed in long-range transport models that a certain fraction of the emission was dry deposited within a grid cell. This was often the same fraction for all grid cells. In reality, however, the surface properties and meteorology differ between grid cells and for that reason the fraction of the emission that is dry deposited should be a function of these factors. As we have seen before, a substantial fraction of the emission is deposited very locally (<1 km from the source). For that reason it would be worthwhile to take into account sub-grid information on the vegetation (type, distance from source, surface area, fragmentation of land use, etc.) near sources of NH_3 (not for other components). If sub-grid information is not available, one possible way to incorporate local source-sink interactions into larger-scale modelling is to simulate a large number of 'within-grid' scenarios with a local-scale model. A large range of source and sink configurations could be simulated to provide a probability distribution of the proportion of NH_3 that is recaptured 'within-grid'. These statistics could then be integrated into the larger-scale model (e.g. the EMEP unified model).

- Information is needed on the deposition to sensitive ecosystems within the grid cells of a national or regional model (e.g. the EMEP Unified Model). At the moment the deposition to these ecosystems is based on an average concentration within the grid cell and an ecosystem specific dry deposition velocity. This can lead to an underestimate of dry deposition near to sources and an overestimate far from sources. It is important, therefore, to redistribute the concentration fields within a grid cell in a realistic way. Such a redistribution can only be made if a spatially detailed emission and land cover inventory is available for the grid cell. The spatial resolution should at least be that of the nature areas. The concentration redistribution can then be made by local atmospheric dispersion and deposition models or by simpler models. It would be useful if differences in dry deposition velocity between different vegetation types could be taken into account, but this is less important than a high spatial resolution of concentrations.

24.5   Tools for Local Ammonia Assessment and Management

There are many different tools that can be used to evaluate and manage the effects of NH_3 in hot-spot areas. These include field monitoring, numerical models and national and international policies and regulations. An integrated assessment approach will combine several of these tools into an assessment strategy.

24.5.1   Elements for Assessment and Management Strategies

Bearing in mind the range of objectives described above, strategies for assessment and management of NH_3 hot-spots will vary. They should consider the main objectives, which is generally to quantify the risk to ecosystems near sources, but also linked issues:

- A first step is to consider the definition of "sensitive ecosystem". In many cases, this should be considered in the frame of national policies or the EU Habitats Directive. The characteristics of the habitat that is to be protected should be considered, since all ecosystems do not exhibit the same sensitivities. Synergies with the findings of Working Group 1 (development of new Critical Levels for NH_3) must be borne in mind with regard to the use of critical levels and critical loads.
- In case studies (e.g. for authorising installation of new animal housing in a specific area), the NH_3 concentration and deposition before and after the changes should be considered to study the effect of the proposal on the exceedance of thresholds such as critical levels and critical loads of nearby nature areas.
- Risks of N swapping (i.e. swapping one N pollutant for another) should be evaluated and avoided where possible. Large N deposition rates close to sources could lead to enhanced nitrate leaching or nitrous oxide emissions which may be in contradiction to the objectives of the EU Nitrates Directive and the greenhouse gas balance of agricultural activities.
- Measures to reduce the impacts of NH_3 should be assessed in conjunction with other potential benefits, such as odour or noise reduction and general benefits to landscape design (aesthetics, etc.), which are often given more weight in planning decisions.
- Assessments of the impacts to sensitive ecosystems may include an economic dimension since a cost/benefit analysis is useful and sometimes necessary to evaluate what could best be protected for a given cost of abatement options. Economic considerations were not one of the objectives of this working group although it was highlighted that the (legal) protective status of the ecosystem and the ecological value should be considered in such an analysis.

24.5.2 Monitoring

Monitoring NH_3 concentrations or deposition provides a method of assessing the impact of local sources on sensitive ecosystems by comparison of the measured values with the relevant critical levels or critical loads. The data obtained are also very valuable for the verification of numerical models. Concentration measurements are often made using low cost technologies such as passive samplers or diffusion tubes. Low cost methods are often used because the monitoring is frequently done at several locations simultaneously and over time periods of a year or more. The following guidelines should be followed when using low cost monitoring methods:

- Measurements should be made close to the sources and receptors as well as in areas far from sources (to provide an estimate of background concentrations). The use of a dispersion model (even a simple one), or expert knowledge should be used to decide where it is best to place the measurements.

- Measurements should be made over a period of at least a year to take into account the seasonal variability of emissions, dispersion and deposition.
- The sampling method should be compared with a continuous measurement (e.g. by a denuder, laser or photo-acoustic method), to assess the accuracy of the measurement and whether the measured concentration is a realistic average of the temporally varying value.

When using more expensive methods (denuders, lasers, photoacoustic, etc.), it may not be possible to monitor in more than one location or over long time periods. In cases such as these it is recommended that the measurement is made where an assessment of the impact is needed (e.g. the edge of a nature reserve) and that the measurements are made when the impact is likely to be greatest (e.g. when the emissions are largest). If one of the objectives of the assessment is the quantification of N deposition to a sensitive ecosystem then the best approach is to directly measure the deposition using a technique such as eddy covariance, relaxed eddy accumulation or a gradient method. However, these methods are expensive and require frequent maintenance and are, therefore, often unsuitable for long-term measurements at more than one location.

As mentioned above, monitoring of deposition could also be useful for assessment.

24.5.3 Spatial Planning

As a consequence of the high spatial variability of NH_3 emissions and rapid decreases in atmospheric concentration away from sources, dependence of deposition on habitat characteristics etc, effects are very variable at a landscape scale. This variability points to the potential to include locally tailored abatement measures as part of strategies to protect sensitive vegetation from NH_3 deposition.

Currently, national abatement policies typically include uniform recommendations for technical abatement measures, such as ploughing in manures after land spreading. In general, ammonia abatement strategies may be implemented in the following ways:

a. All sources and source sectors reduce emissions by the same percentage (blanket approach, "common misery")
b. Certain source sectors reduce more, because they are either cheaper/easier to abate or fall within specific regulations (e.g. Integrated Pollution Prevention and Control (IPPC) directive; this is also "common misery", but for the selected sectors)
c. Certain sources and source sectors reduce more or less depending on their source/receptor interactions, so that abatement measures are targeted to the geographic areas where they have the most significant impact.

The advantage of this last approach is that abatement is targeted precisely and only where necessary. The disadvantage is that it is less equal for farmers, and depends on the spatial location of sources relative to sensitive sink areas, which requires

individual assessment. If this is followed through at a local level, the logical consequence is to look at individual farmsteads and fields near sensitive ecosystems, and
prescribe local solutions (bearing in mind unequal costs for farmers and political
solutions required for this).

Spatial Planning may be defined in this context as measures designed to decrease
impacts of emissions from nearby sources on sensitive sink areas by taking account
of relative spatial relationships. Examples are:

- Introducing buffer zones of low emission agriculture around NH_3 sources or sinks
- Planting tree belts around NH_3 sources or sinks
- Consideration of alternative locations for new sources at the planning stage

Results of a recent study using modelling at landscape scale in central England
(Dragosits et al. 2006) showed that:

- Tree belts can reduce deposition to sensitive areas, with trees surrounding the
  sensitive habitats being more effective than trees around the sources.
- Low emission buffer zones around sink areas also result in useful reductions in
  NH_3 deposition.
- Smaller nature reserve sites benefit to a greater degree from such spatial
  planning measures, as large reserves can provide their own buffer zone to
  some degree.
- Relocating point sources or using planning policies to ensure the location of
  large NH_3 point sources is at least 2–3 km from the sensitive habitats results in
  substantial reductions in NH_3 dry deposition

Overall, tree belts appear more successful at decreasing NH_3 dry deposition than low-
emission buffer zones, and spatial planning measures around sinks seem to be more
effective than around sources in the study area. The latter is partially due to measures
directly at the reserves decreasing incoming deposition from all surrounding sources,
not just nearby farmsteads. An additional benefit of tree belts is the increased dispersion of atmospheric NH_3 due to higher surface roughness of trees, and thus turbulence
created by tree belts, compared with low-emission buffer zones.

Spatial planning measures have been shown to provide a viable complementary
approach to technical abatement measures. However wider aspects and practical
advantages and disadvantages for both farms and nature reserves need to be considered in detail on a case-by-case basis.

24.5.4   Integrated Assessment Approaches

An integrated assessment approach is one that combines monitoring, numerical modelling and the relevant policies and regulations. A typical approach will consist of a network of concentration measurements and the application of a local-scale dispersion and
deposition model. The measured concentrations can be used to verify the concentration
field predicted by the model as well as to assess the likely impact on sensitive ecosystems by comparison with the critical level. The model can be used to assess scenarios

such as a reduction in the NH$_3$ emissions or the effect of spatial abatement measures. In addition the measured concentrations can also be used to estimate NH$_3$ dry deposition. This is often done by multiplying the concentration by a land cover-specific deposition velocity. However, this is a very simplistic approach since it does not take into account the meteorology, the N status of the vegetation or the atmospheric NH$_3$ concentrations. It is recommended that an inferential model such as that proposed by Nemitz et al. (2001) is used to estimate deposition rates from concentration measurements. Biomonitoring also has potential for use within integrated assessment approaches but the Working Group had no sufficient expertise to investigate this option.

24.6   Key Findings

The discussions showed that accounting for hot-spots for either up-scaling of fluxes or risk assessment for nearby ecosystems requires a precise description of all the processes involved. The degree of detail depends on the objectives, the scale and the model that is used for assessment. These findings will apply to all cases, with specific adaptations.

- The Working Group was in agreement that the key uncertainties in the models are the emission and the dry deposition. We need good emission and dry deposition models and sufficient input data when making effects assessments and landscape analysis. These should be adapted to the objective and the scale of the study.
- As deposition rates may often be close to critical loads and there are complex interactions and feedbacks between processes and local variables, the dry deposition processes should be described sufficiently well. This requires better knowledge of compensation points and surface resistances for different ecosystems, their dependence on climatic variables and the deposition history for ammonia and other pollutants.
- Using different models allows the analysis of landscape interactions between sources and receptors with sufficient accuracy for a range of conditions to consider real cases and scenarios. It also allows the assessment of local, tailored abatement measures.
- Scenarios from local-scale modelling can be used in a statistical way to provide estimates of within-grid cell recapture for national- and regional-scale models, linked with global descriptors of the spatial variability in land cover.
- Hot-spot assessment should account for a more general context, e.g. background concentrations and the deposition history.

24.7   Recommendations

- To obtain maximum results from local modelling on hot-spots, it is necessary to gather local values for all parameters instead of national standard values. This holds for ammonia emission inventories and for dispersion and deposition modelling.

- There is a need to further develop dynamical models to estimate the diurnal and seasonal changes in emission strengths from point sources (animal houses) and area sources (land spreading of animal manure). For area sources in detailed plot studies, this should include the effects of meteorological and soil variables.
- Simple dynamical schemes should be improved to calculate canopy resistances and stomatal compensation points based on meteorology, fertilisation rates, vegetation type and exposure to ammonia and other pollutants.
- A synthesis should be made of available databases to build up some reference cases against which the different models could be tested and compared,
- Model intercomparison: the findings of Working Group dealing with regional modelling of ammonia suggest that an intercomparison of regional-scale models is necessary. A parallel intercomparison of sub-grid models would be highly beneficial, which would also provide an insight into differences between the regional-scale models, their abilities to simulate local-scale interactions and the differences between regional-scale models and measured concentrations.
- It is necessary to define scenarios to investigate the possible effect of in-grid fragmentation of land use on net ammonia fluxes. A sensitivity study would allow investigation of the range of local recapture and possible effects on air quality.
- The development of deposition measurement methods that could apply to advective conditions (e.g. one measurement height) should be promoted, in order to provide means of direct validation of dispersion/deposition models.

References

Dragosits U., Theobald M.R., Place C.J., ApSimon H.M., and Sutton M.A. (2006) The potential for spatial planning at the landscape level to mitigate the effects of atmospheric ammonia deposition. *Environmental Science & Policy* 9, 626–638.

Nemitz E., Milford C., and Sutton M.A. (2001) A two-layer canopy compensation point model for describing bi-directional biosphere-atmosphere exchange of ammonia. *Quarterly Journal of Royal Meteorological Society* 127, 815–833.

Van Gijlswijk R., Coenen P., Pulles T., and van der Sluijs J. (2004) Uncertainty assessment of NO_x, SO_2 and NH_3 emissions in the Netherlands. Rapport nr. R 2004/100, TNO Environment, Energy and Process Innovation, Apeldoorn, The Netherlands.

Chapter 25
Modelling the National and Regional Transport and Deposition of Ammonia

Addo van Pul, Stefan Reis, Tony Dore, Liu Xuejun, Hilde Fagerli, Camilla Geels, Ole Hertel, Roy Wichink Kruijt, Maciej Kryza, Robert Bergström, Massimo Vieno, Ron Smith, and Eiko Nemitz

25.1 Summary

- A range of chemical transport models are used across the Convention to model the emission, transport and deposition of atmospheric ammonia at the national and regional scale. These models have been developed from a range of historical backgrounds and with different purposes in mind.
- Six models ranging from describing the national scale up to the full European scale were considered in the Working Group. The models differ in concepts particularly in their chemical scheme and in scale, ranging from Lagrangian models at a national scale, via Eulerian models at the European scale to nested models coupling the European scale to the local scale.
- Key uncertainties in the modeling of atmospheric ammonia were found to be linked to the emissions (absolute level and spatial and temporal allocation), dry deposition parameterization, spatial resolution of the model and the description vertical diffusion.
- All models on a European scale currently underestimate the measured ammonia concentration whereas the national models in general find a better agreement with the ammonia measurements.
- The concentration of ammonium aerosol was fairly well described by all models. However, both under- and overestimates of measured concentrations were found. The magnitude of the wet deposition of ammonium was in general reproduced well by all models.
- The main reasons for the observed differences between measured and the modeled ammonia concentrations were the spatial resolution of the models and the (parameterization of the) dry deposition process.
- All of the models do not routinely use the compensation point (bi-directional exchange scheme) as a parameterization in the dry deposition process of ammonia. This is thought to be one of the reasons why some models tend to underestimate concentrations particularly in summer. The main reason for not taking

A. van Pul

RIVM, LVM, PO Box 1, 3720 BA Bilthoven, The Netherlands

M. Sutton, S. Reis and S.M.H. Baker (eds), *Atmospheric Ammonia,*
© Springer Science+Business Media B.V. 2009

this process into account is the lack of a generalized scheme for the compensation point with respect to the main land cover types used in the models. The models should carry out sensitivity tests with implementations of the compensation point (bi-directional exchange) schemes, to estimate the magnitude of the effect on the dry deposition in the model.

- The siting of the measurements plays an important role in the comparison with modeled concentrations; some stations that are situated in agricultural areas should not be used for validation of the Eulerian models with large grid size (50 km), because of the significant contribution of sources close to the measurement stations that cannot be picked up by the models on this spatial scale.
- Currently, differences in the model performance between countries are not fully understood as they may reflect differences in (i) the quality of the emissions inventory, (ii) differences in the model parameterization schemes, (iii) geographical differences (climate, terrain) or (iv) differences in measurement datasets. Hence, a co-ordinated model intercomparison, using a common model domain, input database and measurement database, is urgently needed to assess relative model performance.
- In many countries, better data (emissions and monitoring data) are available for national modeling efforts than are submitted to EMEP and thus available for other countries. The reporting to EMEP has to be made more flexible to improve data availability.

25.2   Introduction

The Working Group discussed the current status of the regional modeling of ammonia. The group consisted mainly of researchers from institutes and organizations that own and operate the chemical transport models (CTMs) that run on national to regional scale. The discussion was based on the information in the background document and the presentations of the modelers. The overall goal of the Working Group was:

> *To review the current status of the regional modeling of the transport and deposition of ammonia and to review the performance in relation to the model formulations.*

25.3   Objectives for Modeling Atmospheric Ammonia

The main reasons for modeling the emission, transport and deposition of atmospheric ammonia are:

1. To assess the local contributions vs. the long-range transport
2. To provide spatial coverage (beyond what can be achieved with measurements)
3. To establish the link between emissions and concentrations and deposition

4. To provide deposition estimates for comparison with critical loads
5. To provide a prediction of the concentrations and depositions in response to emission abatement measures and scenarios
6. To improve and assess our understanding of the processes that determine the concentrations and depositions of ammonia;
7. To quantify the role of ammonia in the formation of particulate matter;
8. To establish source/receptor relationships
9. To establish national budgets and country-to-country transport matrices

No model currently exists that is ideally suited to address all of these objectives at once. Instead, the exact objective of a modeling study and the availability of input data determine the model that needs to be used, including its spatial scale.

25.4 Current Status of the Regional Modeling of Ammonia

A number of models are currently in use in research as well as operational tools to address environmental issues related to ammonia. In Table 25.1 the models that were considered in this Working Group are listed along with the main features of the models (see the background document for more detail). The models can be divided into three types: (a) Langrangian models working at a national scale (FRAME, OPS), (b) Eulerian models working at a European scale (MATCH, LOTOS/EUROS, EMEP) and (c) a nested model (DAMOS) coupling a Eulerian model (DEHM) to a Gaussian model (OML). The models differ in purpose; the FRAME, OPS and DAMOS systems are constructed to describe concentrations and depositions at a high spatial resolution whereas the other models are (initially) built for describing long range transport of pollution in general. At a later stage in the preparation of the Workshop, the EMEP4UK model was added to the cluster of models and is briefly described in the background document. EMEP4UK is a refinement of the EMEP model for the UK domain and is still under development. No validation has been carried out yet. It is therefore not evaluated here for its performance.

The models differ in concepts particularly on (1) the spatial scale, i.e. the horizontal and vertical resolution, with both the consequences on the dispersion close to sources and the spatial representation of the emissions, and (2) the chemical conversion mechanism of ammonia to ammonium.

The performance of the models in terms of the validation with measurements and the suggested reasons for observed deviations from measurements were inventoried (Table 25.2).

In general it was concluded that:

1. The concentration of ammonia was largely underestimated by all Eulerian models, particularly in source areas, whereas the models with a better spatial resolution at the national scale show good agreement with the ammonia measurements or a moderate underestimate (30%).

Table 25.1 Overview of ammonia models at a national or regional scale

Name	Type	Vertical structure; lowest troposphere	Emissions Spatial scale = horiz. resolution Eulerian models	Dry deposition description; all models use resistance model	Detail deposition; ammonium aerosol	Wet deposition description	Chemical conversion description
FRAME CEH, UK	Lagrangian	Variable; starting first layer from 1 m	5 × 5 km	Applied to five land use classes	Fixed deposition velocity	One coefficient for scavenging process	Pseudo first order reaction rate?
OPS MNP/ RIVM, NL	Lagrangian	Variable; calculation of profile	1 × 1 km (500 × 500 m)	Applied to nine land use classes	Surface resistance; function of u^* and for forests also wind speed and rel. hum.	Scavenging coefficients for in-cloud, below cloud process	Pseudo first order reaction rate?
MATCH SMHI, SE	Eulerian	Variable, usually first layer 60m, 6–11 layers below 1,500m (1999–2003)	44 × 44km Europe 11 × 11 km² Sweden	Applied to ten land use classes (for Sweden)	Usually a simplified parameterised resistance used	One coefficient for scavenging process (vertical variation of scavenging coeff sometimes used)	reactions with sulphate and nitrate explicitly solved
DAMOS/ DEHM NERI, DK	Nested grid 3-D Eulerian	20 vertical layers with highest resolution in the PBL.	Europe on 50 × 50km², and for Denmark and nearby areas 16.67 × 16.67km²	Applied to 14 land use classes, water and ice	Assume size of ammonium containing aerosols of 0.8 ∝cm.	Scavenging coefficient for in-cloud, below cloud scavenging	Explicit chemical mechanism with 63 compounds and 120 chemical reactions.
LOTOS/ EUROS TNO/ MNP/ RIVM, NL	Eulerian	First layer 25 m		Applied to nine land use classes	Surface resistance; function of u^* and for forests also wind speed and rel. hum.	One coefficient for scavenging process	reactions with sulphate and nitrate explicitly solved
EMEP-unified UN-ECE	Eulerian	First layer 90m	50 × 50km	Applied to 16 land use classes	Landuse and meteorology dependent	Scavenging coefficient for in-cloud, below cloud scavenging	reactions with sulphate and nitrate explicitly solved (EQSAM module)

Table 25.2 Summary of model performance (-- /-/o/ + / + +: underestimation/correct/overestimation, compared with measurements)

Model name	European scale				National scale			
	NH$_3$ conc	NH$_4^+$ conc	NH$_4^+$ wet dep	Precip	NH$_3$ conc	NH$_4^+$ conc	NH$_4^+$ in precip	Precip
EMEP unified	-- (scale, dry dep)	O	o	Location?				
EMEP4UK (UK)								
DAMOS (DK)	-- (scale, dry dep)	+ (conversion rate)	O	–	+ (siting)			
FRAME (UK, PL)					O	-- (chemistry, import?)	o	prescribed
LOTUS/EUROS (NL)	– (scale) (polluted sites: --)	o	o	prescribed	– (polluted sites: --)	o	o	prescribed
MATCH (SE)	-- (dry dep, scale)	– (N Europe) + (Spain) (dep)	o		?	–	o	
OPS (NL)					– (emission, dry dep)	– (emission, dry dep)	– (emission, dry dep)	prescribed

2. The concentration of the ammonium aerosol was reasonably well reproduced by all models. However, both under- and overestimates of the concentrations were found.
3. The wet deposition of ammonium was represented well by all models with the exception of OPS.

The OPS model was the only model showing underestimates of all three measured components of ammonia. However, it should be noted that this assessment is only indicative of the true model performance as the measurement database differs between model domains. In addition, some measurements suffer sampling artifacts of the gas/aerosol partitioning.

The main reasons for the observed differences between the measured and the modeled concentrations were the spatial scale of the models and the (parameterization of the) dry deposition process. Another aspect in the comparison is the siting of the measurements used for the comparison. It appeared that some sites are not suited to represent a background concentration for ammonia (as they had originally been chosen for measurements of SO_x and NO_x) and should rather be considered as a site influenced by sources within a few kilometers. This obviously can not be captured by the Eulerian models working on a typical scale of 50 km. On the other hand there are stations that represent a rather remote level but are treated in the modeling as located in a grid with high emissions. In those cases higher concentrations are calculated than measured (Danish experience).

In the next sections more detail on specific modeling issues that describe the differences between the models and the above model performance are given; the horizontal and vertical scale of the models, the dry deposition process and the chemical conversion (mechanisms).

25.5 Model Issues Particular to Ammonia

It is noted that the chemical transport models (CTMs) to predict concentration and deposition fields of ammonia are subject to the same constraints and uncertainties as the transport modeling of any atmospheric pollutant. This includes uncertainties in the meteorological field, circulation patterns and boundary layer height. By contrast, ammonia poses additional demands on the models, due to (i) the dominance of ground level emissions, (ii) the small scale at which emission and receptor areas are interspersed in the landscape and (iii) its relatively high reactivity that results in a rapid atmospheric chemical conversion and a high deposition rate.

There are few systematic sensitivity studies which have assessed the relative uncertainties of the different elements, as the spread of the parameter spaces of most parameters (inputs and model parameters) are poorly defined. In the Working Group therefore only a qualitative ranking of the uncertainties was estimated. For national modeling, the uncertainties tend to follow the order:

$$\text{emissions} > \text{dry deposition} > \text{diffusion,}$$

although the quality of the emissions differs greatly between countries. Similar, for regional (e.g. European) modeling, the uncertainties follow the order:

emission > dry deposition/vertical model resolution (coupled problem)
> diffusion/transport.

25.5.1 Emissions

It has long been recognized that uncertainties in the emissions inventories are a key uncertainty in ammonia modeling. The quality of ammonia emission inventories varies greatly across the member states of the Convention. For example, highly resolved emission inventories are available in countries with a recognized ammonia problem, for example, in Denmark (field level), in the UK (1 km) and in the Netherlands (1 km). Issues with data confidentiality and restricted disclosure limit the availability of data at higher spatial resolution in many countries. This contrasts with industrial emissions from large point sources, which have to be reported to the IPPC national inventories and are generally publicly available. The rationale for this distinction should be revisited.

Across the Convention, the uncertainty of the ammonia emission data (annual country total) is deemed to be in the region of 60–70%, while significant better results can be achieved in detailed local studies. This is lower than the accuracy of emission inventories of CO, CO_2, NO_x and SO_x, similar to the situation for primary PM, but compares favorably to emissions of heavy metals, persistent organic pollutants (POPs) and N_2O.

It has been noted that better national, spatially disaggregated emission data are often available at national level, compiled by the same process as the official disaggregated emissions that are submitted to EMEP only every 5 years. It is suggested that a mechanism be developed to increase submission to EMEP for those countries which compile disaggregated emissions at higher frequency than currently required by EMEP.

For some areas of the Convention, independent emission inventories compiled within research projects on various spatial scales are expected to provide more accurate emission estimates than the official national emission inventories. EMEP should be further encouraged to explore these alternative emission inventories in scenario studies to investigate the reasons for disagreement between modeled and measured concentrations in some countries.

While all transport models use annual average emissions as an input, the treatment of these emissions differs greatly between models. Sensitivity studies (e.g. of the Danish model) have shown that pre-processing these emissions to provide temporally explicit emissions in response to management and meteorological conditions achieves better temporal correlation between modeled and measured concentrations.

25.5.2 Spatial Scales and Model Resolution

A key challenge in modeling atmospheric ammonia is the small spatial scale at which (agricultural) ammonia sources are interspersed with semi-natural receptor areas. Accordingly, there is a strong push for chemical transport models (CTMs) to predict fields of NH_3 concentrations and deposition at an increasingly finer spatial resolution. The spatial scale that is currently reasonably achievable in national and regional CTMs is limited by a combination of (a) the resolution of the input data (in particular of the emissions), (b) computing power (particularly true for regional modeling) and (c) the resolution for which the model parameterizations are formulated. For example, while the EMEP unified model is expected to work reasonably down to a scale of 5×5 km, there is a general limitation for the Eulerian modeling approach, probably in the region of 1×1 km. To achieve higher spatial resolution, a combination of Eulerian, Lagrangian and Gaussian modeling approaches to account for near source effects may be required.

A large spatial resolution fails to resolve processes, resulting in a general underprediction of air concentrations in the regional Eulerian models (cf. Table 25.2). By contrast, high spatial resolution increases the demand on the input parameters. Hence the certainty in the model prediction for a given grid cell peaks at a spatial resolution which depends on data availability. Thus, whether the move to higher resolution modeling is necessary or even desirable depends on the model application. For example, the improvement of the spatial resolution clearly improved the ability of the EMEP unified model to predict Critical Loads exceedances to forest, but it is realized that even high resolution models still underestimate Critical Loads exceedances by spatially averaging concentrations and deposition. However, since CLs are statistically defined, the exact location of the exceedance is of secondary importance and sub grid parameterizations may be sufficient to deal with the effect of heterogeneity. Similarly, for future scenario calculations, meteorological conditions and emissions are not known at high spatial and temporal resolution.

By contrast, explicit high resolution modeling is desirable (a) for comparison of model performance with measurement data, (b) to demonstrate that processes are correctly represented and (c) to provide plot-scale inputs of N and acidity to specific ecosystems, also in relation to the needs of the Habitat Directive. However, at present, in most cases the latter still needs to be achieved in targeted local scale modeling studies. The development of nesting approaches of local scale dispersion models into larger-range transport model appears to be the most promising way forward.

A particular challenge for the modeling of atmospheric transport model remains the large thickness of the surface layer in Eulerian models (typically 50–100 m), which tends to lead to an under-estimation of surface concentrations, with implications for dry deposition, chemistry and comparisons with measurements. Most models include a sub grid parameterization based on K-theory. However, more effort should go into the development of such sub-grid parameterizations. Here the local scale modeling efforts should be used to derive operational statistical relationships.

To minimise the problem of vertical resolution on model assessment, the validation of long-range transport models should concentrate on measurements in remote sites and exclude those in source areas.

25.5.3 Parameterisation of Surface/Atmosphere Exchange

Sensitivity studies have repeatedly identified dry deposition schemes as a second key uncertainty in estimating deposition. Conclusions on the accuracy of NH_3 dry deposition are currently drawn very indirectly from the comparison of modeled and measured air concentrations, with confounding effects from uncertainties related to emissions and the parameterizations of the model behavior. There is a clear lack of dry deposition measurements against which to validate models.

There is a suspicion that most current models tend to overpredict deposition (and thus underestimate the transport distance), especially in summer, although the US CMAQ modeling approach is suspected to underpredict dry deposition. Elevated compensation points during warm periods are likely to suppress deposition in the real environment, an effect which would partly close the ammonia gap in several models, including the EMEP model. It is anticipated that the implementation of bi-directional exchange schemes based on compensation point models will increase the transport distance of atmospheric ammonia, with policy implications related to long range transport. None of the models currently applied within the Convention uses a bi-directional exchange parameterization on a routine basis. The implementation of bi-directional exchange schemes has been hampered by three problems:

1. Currently, robust generalized parameterizations of bi-directional exchange for application across Europe are not available, but meta-analysis of the existing field measurements has started and such a parameterization is likely to become available in the next year or two, once parameterizations have been expanded for the two parameters governing the exchange of ammonia with vegetation: it is anticipated that the cuticular resistance is a function of surface humidity and the ratio of SO_2/NH_3 in the air, while the compensation point is a function of surface temperature and agricultural management (for agricultural surfaces) and atmospheric N deposition (semi-natural vegetation).
2. To deal with bi-directional exchange schemes, sub-grid parameterizations need to deal with land-cover specific surface concentrations, as the use of surface concentrations that is averaged over various land cover types with different compensation points would result in an over-prediction of emission from fertilized vegetation and over-prediction of deposition to semi-natural vegetation.
3. The inclusion of compensation points is likely to improve the capability of the models to correctly predict ammonia concentrations and dry deposition inputs. However, these inputs may be inconsistent with the deposition that was estimated for the effects studies that were used to derive the dose/effect relationship that underpin current Critical Loads.

It is suggested, that, once generalized bi-directional exchange schemes become available, these be tested in the atmospheric transport models to provide sensitivity studies. In the bi-directional exchange model, the emission from agricultural crops is predicted within the CTM, potentially resulting in inconsistencies with the official national emission inventories. In the long term, emission from vegetation (agricultural and semi-natural) should be calculated in an atmospheric transport model as they are intrinsically dependent on the air concentrations. It is realized that more measurements of dry deposition are required to validate the dry deposition schemes, and to improve the robustness of the parameterizations, for example by providing more data on compensation points. New international initiatives, such as the establishment of a nitrogen flux network with the European NitroEurope IP, will improve the availability of ammonia dry deposition measurements across Europe, but will be insufficient to validate dry deposition schemes over the full range of climate, pollution climate and vegetation types.

25.5.4   Chemical Schemes

The chemical schemes are not considered to be the major uncertainty for modeling NH_3 deposition. Because the different models have different histories and different objectives, the complexity of chemical schemes differs greatly between models. In general, currently implemented schemes are deemed fit for the purpose for which the individual models were developed, with models targeted at predicting the contribution of ammonia to the formation of secondary PM using more advanced schemes. It should be noted, however, that modeling of the full aerosol mass (including secondary organic aerosol production) is still in its infancy, and requires much advanced schemes than are currently contained in most models that target ammonia, describing the total aerosol size-distributions. Some existing models use empirical chemistry schemes which have been derived from current air concentrations and these have limited applicability for future scenario calculations under modified chemical climates.

More measurements of gas/aerosol partitioning are required to assess model performance. These will gradually become available as the new EMEP monitoring strategy with Level-2 and Level-3 supersites is implemented across the area of the Convention.

25.5.5   Effect of Uncertainties on the Integrated Assessment Modelling

Integrated assessment modeling results cannot be better than the source/receptor relationships that are provided by the CTMs. Thus the current uncertainties in the CTMs directly affect the data that form the basis on which policy decision are made.

In addition, the source/receptor relationships and response curves that are fed into the CTM process only account for a small number of non-linear responses. However, a whole range of non-linear interactions are expected to affect the concentrations and behaviour of atmospheric ammonia in the future. These include:

- Changes in the oxidation capacity and oxidation rate with time, affecting the future applicability of chemical bulk reaction schemes not coupled to a photo-chemical scheme
- Changes in the NO_x/SO_x ratio, increasing the relative importance of NO_3^- over SO_4^{2-} and resulting in a different chemical regime, driven by equilibrium reactions
- Changes in the SO_2/NH_3 ratio, resulting in changes in the deposition rate of NH_3
- Increase in free available NH_3 (resulting in lower acid concentrations means less incorporation into aerosol)
- Increase in the relative importance of import from outside the EMEP domain and shipping
- Increase of compensation points in response to increases in surface temperature

The use of simplified source/receptor relationships and responses, in general, does not fully account for these non-linearities in the system.

25.6 Recommendations

25.6.1 Model Inter-Comparison Study

The agreement between modeled outputs with measurements is used as indirect assessment of the accuracy of the emissions, the model processes and the deposition scheme. In many cases the model performance can be improved by a number of modifications and models can get the correct results for the wrong reasons. While sensitivity studies can shed light on the relative contributions of the individual parameters to the overall uncertainty, they do not provide a means to identify erroneous input data or process parameterization.

An attempt was made by the Working Group to compare national modeling studies with the EMEP model as a common reference. For example, the application of EMEP4UK at the UK scale predicts a very different spatial precipitation field in Scotland than the regional EMEP model, which fails to predict the fine-scale orographic effects. In addition, comparison of the dry deposition maps generated with the FRAME model for the UK and Poland shows similar country totals, but spatial patterns that are different enough to suspect that the emission database is not identical. In general, national modeling studies tend to have access to a more detailed emission database than currently used by EMEP. In addition, it was realized that the performance of different models was often assessed against different measurement databases. Even where EMEP data were used for model assessment, these were filtered in different ways for different studies.

A strong recommendation therefore has to be that a model inter-comparison based on the same input database is urgently required to fully understand the uncertainties in relation to emissions, model parameterizations, deposition schemes and spatial scales of the different models. The individual objectives of such inter-comparison would be:

1. To explore which differences in the models causes which differences in the results
2. To assess model performance against a common and robust measurement database
3. To validate the model response to emission changes
4. To assess the impact of model resolution on the results
5. To provide an uncertainty assessment through a model ensemble approach

Such an intercomparison study should focus on a small number of countries of contrasting conditions (ammonia emission strength, meteorology) as well as the regional scale. The countries should be selected to have (a) a high quality emission inventory and (b) a robust and long-term measurement database. It is recognized that such a model intercomparison would require resources from national funding agencies. In addition, it should be explored whether the interaction of the groups (meetings, compilation of results) could be further supported through the European Network of Excellence 'ACCENT', ESF-NINE or COST Action 729.

25.6.2   Measurement Database

It needs to be emphasized that the EMEP database contains few measurements of NH_3 and NH_4^+, while more measurements are held in individual national databases. This is partly due to the relatively poor links between the EMEP reporting process and the scientific institutes that conduct the measurements. In addition, many measurements did not fulfill the historical EMEP requirements related to siting, instrumentation and temporal resolution.

It is expected that the new EMEP Monitoring Strategy, which is only beginning to be implemented, will improve the situation through (a) a more flexible approach that allows measurements made at a lower temporal resolution to be submitted to EMEP and (b) the establishment of Level-2 and Level-3 monitoring sites across Europe, which will monitor concentrations of inorganic gases and aerosols at a daily resolution (minimum). Despite these advances, harmonization of instrumentation is not satisfactorily addressed in the new Monitoring Strategy. It should also be emphasized that the EMEP monitoring network was originally designed to monitor background concentrations of SO_2 and NO_x and therefore sites may not be ideally suited to the measurement of compounds of agricultural origin. Although the Measurement Strategy encourages dry deposition measurements of gases and aerosols at selected sites as part of the Level-3 methodology, these are not currently compulsory and, in this area, not much improvement is expected through the EMEP activities over the next few years.

It is recommended that EMEP/TFMM develops an approach to maximize the international availability of existing national data, at a minimum, by compiling a meta-database. In addition, it should foster the submission of historical data.

Chapter 26
Reliability of Ammonia Emission Estimates and Abatement Efficiencies

J. Webb, Nicholas J. Hutchings, Shabtai Bittman, Samantha M.H. Baker, Beat Reidy, Caroline Raes, Ken Smith, John Ayres, and Tom Misselbrook

26.1 Summary

This group addressed the key concerns relating to ammonia (NH_3) emission estimates:

- The current status of uncertainty analysis of national NH_3 inventories
- The discrepancies between atmospheric measurements and models, and the most likely explanations
- The extent to which the uncertainty in regional NH_3 emissions can be distinguished between the absolute magnitude and the trends in emissions
- The agreement on the assessment of key NH_3 mitigation methods as good, promising and unsuitable in relation to current UNECE guidance
- The potential for 'soft' approaches to NH_3 abatement
- A forward look for mitigation measures and the extent to which costs are expected to reduce in the future as methods become more widely adopted

Few countries have considered uncertainty in detail. Results available indicate national estimates may be accurate to within ±20%. For those countries which have created inventories using emission factors (EFs) measured elsewhere, the uncertainty may be much greater, perhaps c. 100%. The greatest uncertainty is likely to be for emission estimates for regions within countries. Sensitivity analysis of the UK Inventory showed that activity data, information on a range of relevant farming practices, were the inputs to which the system was most dependent. Cattle diets, especially those which are grass-based, were considered particularly uncertain.

The UK and DK reported good agreement between modelled and measured NH_3 concentrations. However, these countries use a finer grid size in their models than the NL. A detailed discussion of the Dutch 'ammonia gap' suggested that the EFs used in the inventory were accurate. The discrepancy was considered to be due to either

J.Webb

AEA, Gemini Building, Harwell Business Centre, Didcot, Oxfordshire 0X11 OQR, United Kindgom

overestimation of abatement efficiencies or overestimation of dry deposition velocities. Adjustment of either could eliminate the gap but it was not yet known which was
responsible.

The abatement efficiencies quoted in the UNECE Guidance document were considered robust. While means will not reflect the variability in the data, quoting ranges
may create uncertainty over which point in the range is most appropriate to use. Since
the data were obtained almost exclusively from Northern and North-western Europe,
the efficiencies should not be assumed to be applicable across the whole UNECE area.
Only a brief statement is given in the Guidance on the impacts of reducing emissions of
NH_3 following spreading on losses of other N pollutants, because nitrate leaching and
nitrous oxide emissions are very specific to the site and season of manure application.

'Soft' approaches to NH_3 abatement are those implemented using existing facilities and equipment (e.g. applying manure during weather conditions associated
with little emission). While these offer an economically attractive method of reducing NH_3 emissions, it is often difficult to know their uptake by farmers and their
efficiency, and therefore to convince environmental authorities of their efficacy.

Experience from the adoption of abatement technologies in other areas, suggests that ex ante cost assessments tend to over-estimate the cost of implementation.
However, taking emerging technologies into the industry can lead to a reduction in
abatement efficiency. A number of emerging abatement options are discussed in the
full report together with a summary of other developments that may have an impact
on NH_3 emissions.

26.2   Introduction

This Cross-cutting Group addressed the key concerns relating to ammonia (NH_3)
emission estimates.

The group addressed the following questions:

- What is the current status of uncertainty analysis of European regional ammonia
  emission estimates?
- Where there are remaining discrepancies between atmospheric measurements
  and models, what are the most likely factors that might explain over or under
  estimation of ammonia emission estimates?
- To what extent can the uncertainty in regional ammonia emissions be distinguished between the absolute magnitude and the trends in emissions (e.g. due to
  sector activity and abatement measures)?
- Is there agreement on the key ammonia mitigation methods that should be considered as good (tier 1), promising (tier 2) and unsuitable (tier 3) in relation to current
  UNECE guidance? That is, have some countries now shown success in measures
  formerly thought to be tier 3 (either on grounds of cost or effectiveness)?
- What is the potential for 'soft' approaches to ammonia emission reduction, such
  as good management practices (e.g. cleaning surfaces, reducing dirty area) and
  altering timing of management practices (e.g. delayed fertilizer and manure
  application after cutting)?

- Provide a forward look for mitigation measures. To what extent are costs expected to reduce in the future as methods become more widely adopted?

26.3 Regional Ammonia Emission Uncertainties

26.3.1 What is the Current Status of Uncertainty Analysis of European Regional Ammonia Emission Estimates?

Few countries have considered uncertainty in detail, but the number doing so is increasing rapidly. The future requirements of the CLRTP for Inventory submissions to be accompanied by an Informative Inventory Report (IIR) will make an uncertainty analysis a requirement.

For those national Inventories for which uncertainty analyses are available, results indicate that the estimates may be accurate to within ±20%. The group considered that, for those countries which have created inventories using emission factors (EFs) or approaches developed elsewhere, the uncertainty could be much greater, perhaps in the region of 100%.

The group agreed that uncertainty is likely to be greater for emission estimates for regions within a country than at the country scale. It was considered appropriate to base the calculation of national annual emissions on EFs that are the average measured values, together with national average activity data, i.e. information relating to a range of relevant farming practices (e.g. livestock numbers, N excretion, manure management, length of the housing period for grazing animals). It may also be possible to estimate emissions from small discrete areas (i.e. sub-regional) quite accurately, provided that it is possible to obtain activity data for the specific sources and process-based models are used to take account of local environmental conditions. However, when estimating emissions from large areas at the sub-national scale, significant bias may be introduced by using national average data, as farming practice or weather patterns may differ greatly across a country. This difficulty can be partially overcome with respect to EFs by using a more process-based approach to calculating national emissions. A geospatial sensitivity analysis of the UK Inventory demonstrated that adoption of a more process-based approach could lead to better spatial and temporal estimates of emissions.

An analysis of the sensitivity of the UK ammonia (NH_3) emission inventory identified farm activity data as the inputs to which the output was generally most sensitive. With respect to activity data, information on cattle diets, especially those that are predominantly grass-fed, was considered particularly uncertain. The analysis of uncertainty of EFs identified emissions from buildings in which livestock were bedded on straw and emissions during grazing by beef cattle and sheep to be the most uncertain.

The Working Group recommended that an renewed approach be made to the Commission to fund a collaborative project, in concert with EUROSTAT, to harmonize approaches to collecting key activity data from the livestock sector for use not only with NH_3 abatement but also to assist with determining more accurately other impacts of livestock production on the environment.

26.4   Exploring Model-Measurement Discrepencies

Where there are remaining discrepancies between atmospheric measurements and models, what are the most likely factors that might explain over or under estimation of ammonia emission estimates?

To what extent can the uncertainty in regional ammonia emissions be distinguished between the absolute magnitude and the trends in emissions (e.g. due to sector activity and abatement measures)?

In the background document on trends in ammonia concentrations (Bleeker et al. 2009), it was noted that for the Netherlands, the modelled atmospheric NH_3 concentrations were smaller than those measured and there appeared to be less effect from the abatement measures previously imposed (the 'ammonia gap'). In the intervening period, modifications to the extent to which abatement measures were adopted and a better definition of the different abatement measures, means that in 2006, the modelled trend in NH_3 concentrations now reflects the estimated reductions in emissions. However, the modelled NH_3 concentrations remain about 30% below the measured concentrations. In Switzerland an emission reduction of 17% from 1990 to 2000 has been calculated while no clear reductions in either deposition or concentration measurements could so far be detected. Other countries (UK, DK) have found good agreement between the modelled and measured NH_3 concentrations. A detailed discussion of the Dutch 'ammonia gap' (Erisman et al. 2009) did not suggest that inaccurate EFs were responsible for creating an inaccurate inventory. Two possibilities exist to explain the discrepancy. One was an overestimation of abatement efficiencies. The other was an overestimate of dry deposition velocities. Adjustment of either could eliminate the gap but it was yet to be determined which was in fact responsible.

26.5   Progress in Abatement Technologies

Is there Agreement on the Key Ammonia Mitigation Methods that Should be Considered as Good (Tier 1), Promising (Tier 2) and Unsuitable (Tier 3) in Relation to Current UNECE Guidance? that is, have some Countries Now Shown Success in Measures Formerly Thought to be Tier 3 (Either on Grounds of Cost or Effectiveness)?

The efficiency, applicability and reliability of abatement methods have been extensively discussed within the Expert Group on Ammonia Abatement, which was established by the Working Group on Strategies and Review (see http://www.unece.org/env/lrtap/welcome.html). Results were presented of a recent review of techniques to reduce NH_3 emissions following slurry spreading (Erisman et al. 2009). These were in good agreement with those reported with the abatement efficiencies quoted in the UNECE Guidance document. However, it was also agreed that since the data were obtained almost exclusively from Northern and North-western Europe, those efficiencies could not be assumed to be applicable across the whole UNECE area. A programme of work quantifying a range of abatement options is nearing completion in Spain. When available those results will be a particularly useful contribution to the information on abatement efficiencies.

It should be noted that the Guidance document is intended to provide background information and the abatement efficiencies have to be adapted to specific countries/regions. The abatement efficiencies can be climate dependent. For example, incorporation of cattle slurry within 6 hours in a cool climate (7°C) will lead to an emission of NH_3 equivalent to about 20% of the total ammoniacal N, whereas about 40% would be emitted in the absence of incorporation. Hence the abatement efficiency will be 50%. In contrast, at 20°C, the corresponding figures are about 30% and 50% respectively, giving an abatement efficiency of 40% (using the ALFAM model).

There has been discussion of the most appropriate way to present the abatement efficiencies in the Guidance Document (UNECE 1999, 2007). Overall means may be misleading as these will not reflect the variability in the data and consequent uncertainty. However, quoting ranges may create uncertainty over which point in the range is most appropriate to use. At present, some brief information is given on the factors that may influence the abatement to be achieved in practice together with an estimate of the range where this can be reliably assessed.

Only a brief qualifying statement is given in the Guidance Document (UNECE 1999, 2007) with respect to the impact of reducing emissions of NH_3 following spreading on losses of other N pollutants. This is because the impacts of NH_3 abatement on nitrate (NO_3) leaching and nitrous oxide (N_2O) emissions are very specific to the site and season of manure application. For example, published results have indicated that the proportion of NH_4-N conserved following immediate incorporation of solid manures or application of slurry by injection or band spreading that is subsequently lost by NO_3 leaching may range from 0% to 100% depending on the soil type and subsequent hydrologically effective rainfall at the site of application. Published results have also shown a wide range of impacts of slurry injection and manure incorporation on subsequent emissions of N_2O. In some cases immediate incorporation of litter-based manures has reduced N_2O emissions compared with surface applications, while N_2O emissions may increase if the soil remains wet and crop growth, and hence N uptake, is limited.

26.6 Soft Approaches for Ammonia Abatement

What is the potential for 'soft' approaches to ammonia emission reduction, such as good management practices (e.g. cleaning surfaces, reducing dirty area) and altering timing of management practices (e.g. delayed fertilizer and manure application after cutting)?

'Soft' approaches to NH_3 abatement are defined as those that the farmer can implement using the existing facilities and equipment on the farm.

Producing hay instead of silage has the potential to reduce N intake and N excretion at a given energy intake. However, with only one crop per year from hay as opposed to four or more from silage, the capacity to produce on-farm feed is reduced and so overall costs increase. Moreover, rumen methane emissions are likely to increase because the feed is less digestible. Non-protein N, which is high in grass silage, could be reduced by using forages low in proteases such as red clover; maize which has high N use efficiency and energy and low protein with high bypass value, high sugar ryegrass. Moreover, maize requires less water than cool season grasses and legumes so may stabilize yield in warm dry summers.

Storing FYM for 6 months prior to field application can lead to a reduction in the ammonium content of the manure and reduced NH_3 emissions after application. However, little is gained if composting occurs and the reduction is achieved via NH_3 emission. If the reduction is achieved via denitrification, there is a risk of increased N_2O emissions. Good management of the composting process can reduce NH_3 emissions and deliver other benefits (e.g. reduction in the viability of weed seeds); composting is likely to increase in some regions.

Slurry dilution has been adopted in Switzerland but the greater volumes will substantially increase spreading costs if slurry is not applied with drag hose systems. Moreover, slurry dilution may increase storage requirement, and if the same amount of slurry N is applied, there is a greater risk of runoff.

The emission of NH_3 after field application is related to air temperature, solar radiation and wind speed, so timing the application to correspond with times when these factors are low (e.g. in the evening and at night) is likely to reduce emissions of NH_3 but may increase formation of PM This option is available to farmers who are under relatively few time constraints. In many areas, agronomic considerations or regulations such as the EU Nitrates Directive impose restrictions on the timing of manure applications, so the window of opportunity for manure application is limited and farmers have little flexibility in the timing. An additional consideration is that the conditions likely to reduce NH_3 emissions are in conflict with those recommended in some countries to reduce the odour nuisance.

Soft approaches offer an economically attractive method of reducing NH_3 emissions, and can therefore be a valuable alternative for small-scale farming or for farmers in regions with limited access to efficient abatement techniques. However, it is often difficult to know their uptake by farmers and their efficiency, and therefore to convince environmental authorities of their efficacy. Restricting use of N to below the economic optimum, provides a strong incentive to minimize NH_3 loss. However, such restrictions are usually only applied in situations where N management is already tightly controlled, so there is little opportunity to apply soft approaches. If restricting N-use reduces yield, total costs need to be assessed, including increasing production acreages.

26.7   Forward Look for Mitigation Measures

To What Extent are Costs Expected to Reduce
in the Future as Methods Become More Widely Adopted?

There are a number of existing technologies currently employed to reduce emissions of NH_3 in Europe. In Canada abatement is a side-effect of odour and runoff control. Based on assessments of cost, together with the observation that emission reductions achieved following spreading will not be lost during subsequent manure handling (in contrast to reductions achieved during housing and storage), the greatest focus has been on abatement of field emissions. The technologies most commonly adopted are rapid incorporation (for all manure types) and trailing hose, trailing shoe and injection (slurry only). These technologies are being adopted to meet current emission targets, although the use of slurry injection is often driven as much by the need to reduce odours as NH_3. For manure storage, a variety of covers are available, ranging from natural crusts (mainly cattle slurry) to man-made covers. Relatively few abatement technologies have been adopted in animal housing, partly due to the cost. However, in the Netherlands, new animal housing must include low emission technologies, leading to a wider occurrence in this country.

Experience from the adoption of technologies in the past, suggests that ex ante cost assessments tend to over-estimate the cost of implementation of new technologies. However, experience also suggests that taking emerging technologies into the industry can lead to a reduction in abatement efficiency. The growing use of commercial manure application services offers opportunity for reducing costs and faster adoption and more rigorous implementation.

Looking to the future, the following abatement technologies are under consideration:

- **Feeding practices**. Optimising feeding to avoid surplus protein in the animal diet is one of the most effective methods of reducing NH_3 emission from all stages of manure management. By reducing both the total- and ammoniacal-N excreted, the method is also effective in reducing emissions of NO_3 and N_2O as well as avoiding any subsequent increase in N_2O emissions and NO_3 leaching as a consequence of reducing losses of NH_3. The technology is well tested in pigs and poultry and in principle can also be applied to ruminants. The main difficulty is the large variation in protein content of forage.
- **Woodchip corals**. Since NH_3 emission from excreta deposited by animals kept outdoors is nearly always less than if it were deposited in buildings, the option of keeping animals outdoors for a greater part or all of the year has promise. Woodchips can form a cheap, absorbent material for outdoor enclosures and NH_3 emission from such enclosures is expected to be significantly less, although no robust measurements have yet been completed. However, the eventual fate of any ammonium-N retained in the woodchips has to be considered, though most is known to percolate through and should be collected in a tank or lagoon. If the

"spent woodchip" is composted, there is the possibility that subsequent NH_3 emissions may outweigh the reductions achieved during the overwintering period. Alternatively, if denitrification is encouraged, there may be increased N_2O emissions.

- **Manure incineration**. The incineration of poultry manure is practiced in the UK and the pressure to increase non-fossil fuel energy production may present opportunities for co-firing of manure with other fuel sources.
- **Air scrubbers**. Scrubbing the exhaust air from forced ventilated animal housing can reduce the housing NH_3 emissions by up to 90%. There are two main types of scrubbers; chemical scrubbers and biofilters. The chemical scrubbers capture and retain the NH_3 as a salt in solution. This is either stored separately or added to the slurry tank, prior to field application, so storage and field application losses will increase accordingly. Biofilters first capture the NH_3 then nitrify and denitrify the N. It is unclear to what extent the denitrification leads to N_2O emissions.
- **Slurry acidification**. This is a technique well-tested in research facilities but which has not gained widespread acceptance in the industry, due to the cost and the difficulties associated with the handling of strong acids on farms and the risk of hydrogen sulphide emissions. A commercial system is in use on a number of farms in Denmark. The technology is attractive if large emission reductions are to be achieved since it is effective in housing, storage and after field application. A disadvantage is that emissions of N_2O may be increased. However, it is too early to know if the technique will gain widespread acceptance, as health and safety legislation may limit adoption. Acidification of litter may be done in poultry housing. It is also unclear how repeated field applications of acidified slurry (usually pH < 6) may affect soil pH and fertility on the long term.

The task of assessing the cost effectiveness of both existing and emerging technologies is made more difficult by parallel developments in the industry. The structural development of the industry (fewer, larger farms) means that there is an increase in the replacement rate of farm facilities (e.g. animal housing, manure storage) and the cost of introducing new technology tends to be less if included in new facilities rather than retro-fitted. Increased farm size and pressure to reduce NH_3 emission creates a market for contracting out the entire manure management operation.

The potential use of the commonly used N fertiliser ammonium nitrate (AN) for illegal explosives means that there had been a proposal to ban this compound. If this were to occur, the use of urea may increase, since it is often the cheapest alternative. The NH_3 emission from urea is considerably greater than from AN unless it is drilled into the soil at the time of application, so an increase in its use would lead to a significant increase in NH_3 emissions. Subsurface application of entire dose in early season may not be practical and urease inhibitors do not reduce NH_3 emissions to those of AN. Polymer-coated urea products release N according to moisture and allow application of relatively large rates even near seed. These products are gaining acceptance in US and there are promising research results from Canada (especially reduction of N_2O). The potential for this approach needs further investigation.

It appears that bedding cattle on sand can lead to a reduction in mastitis infection. This technology is has become significant on dairy farms in Canada and has been adopted on some farms in Europe. It is unclear what implications this technology has for manure, although it is known to cause serious problems with sludge formation in storage tanks and abrasion of pumps and slurry handling equipment, without active management to separate the sand from the slurry after removal from the animal house. Similarly implications for NH_3 emissions are uncertain, although there is some recent US data reporting reduced emissions.

The demands for improved animal welfare may increase the use of straw-based housing systems in the future. There is evidence that increased straw use may lead to a reduction in NH_3 emissions from systems in which animals are already bedded on straw. Hence any trend to increase bedding of animals on straw may reduce NH_3 emissions.

The demands to reduce the surplus of phosphate on farms in connection with the EU Water Framework Directive are likely to cause problems for farmers in areas with a high livestock density. One solution for these farmers may be to adopt slurry separation, since the phosphate is mainly associated with the solid fraction. After separation, this fraction might be transported to areas where the phosphate can be used as a fertiliser or to an incineration plant. Applying the separated liquid may result in reduced emissions due to better soil infiltration.

As part of the drive to produce power from renewable energy sources, there is increasing interest in anaerobic digestion of livestock slurries. Studies suggest the impact on NH_3 emissions is likely to be small, the increased potential for NH_3 emissions due to the greater ammoniacal-N content and pH being counter-balanced by the reduced slurry viscosity and hence greater infiltration rate following spreading.

References

Bleeker A., Sutton M.A., Acherman B., Alebic-Juretic A., Aneja V.P., Ellermann T., Erisman J.W., Fowler D., Fagerli H., Gauger T., Harlen K.S., Hole L.R., Horvath L., Mitosinkova M., Smith R.I., Tang Y.S., van Pul W.A.J. (2009) Linking ammonia emission trends to measured concentrations and deposition of reduced nitrogen at different scales. In: Sutton M.A., Reis S., Baker S.M.H. (eds.), *Atmospheric Ammonia – Detecting Emission Changes and Environmental Impacts. Results of an Expert Workshop Under the Convention on Long-range Transboundary Air Pollution.* Springer (Chapter 11).

Erisman J.W., Bleeker A., Neftel A., Aneja V.P., Hutchings N., Kinsella L., Tang Y.S., Webb J., Sponar M., Raes C., Mitosinkova M., Vidic S., Andersen H.V., Klimont Z., Pinder R., Baker S.M.H., Reidy B., Flechard C.R., Horvath L., Lewandowska A., Gillespie C., Wallasch M., Gehrig R., Ellermann T. (2009) Detecting change in atmospheric ammonia following emission changes. In: Sutton M.A., Reis S., Baker S.M.H. (eds.), *Atmospheric Ammonia – Detecting Emission Changes and Environmental Impacts. Results of an Expert Workshop Under the Convention on Long-range Transboundary Air Pollution.* Springer (Chapter 23).

UNECE (1999) Guidance document on control techniques for preventing and abating emissions of ammonia. EB.AIR/1999/2 (www.unece.org/env/documents/1999/eb/eb.air.1999.2.e.pdf)

UNECE (2007) Guidance document on control techniques for preventing and abating emissions of ammonia. ECE/EB.AIR/WG.5/2007/13. (www.unece.org/env/documents/2007/ eb/wg5/ WGSR40/ece.eb.air.wg.5.2007.13.e.pdf)

Chapter 27
Ammonia Policy Context and Future Challenges

Till Spranger, Zbigniew Klimont, Michel Sponar, Caroline Raes, Samantha M.H. Baker, Mark A. Sutton, Collin Gillespie, Y. Sim Tang, Helle Vibeke Andersen, Thomas Ellerman, Chris Flechard, and Nick J. Hutchings

27.1 Summary

Ammonia emissions are major contributors to eutrophication and acidification of ecosystems and secondary $PM_{2.5}$ concentrations in Europe. Reduction of the ammonia emissions in Europe has been on the agenda for more than a decade, first on a national scale, e.g. in Denmark and the Netherlands, followed by international efforts. The latter include the UNECE CLRTAP Gothenburg Protocol and EU directives and strategies.

This Cross Cutting Group considered the policy context of the ammonia problem, including socio-economic, environmental, institutional and technological aspects. Drawing on the scientific findings and recommendations from the other Working Groups and independent contributions from the participants, the Cross Cutting Group addressed the potential role of different policy options to help mitigate ecosystem and health impacts of ammonia emissions. It also discussed a need to adapt tools used in policy analysis (integrated assessment models, IAMs) and consequently evaluate policies in view of new findings.

Ammonia policies are becoming strongly interlinked with a number of other environmental and agricultural policies. In order to avoid the problem of pollution swapping, future policies need to consider these interactions. This in turn calls for extensions the tools currently used, verification of specific elements of the models, adaptation of monitoring networks, targeted measurement programs, but also possible revision of legislation in order to close existing loopholes and increase synergies in addressing nitrogen pollution at large. In that sense, priority should be given to measures aiming at reducing all kinds of nitrogen losses at farm level. Ammonia emission reduction policies must be analysed in a multi-effect (human health, acidification and eutrophication of the ecosystems and related biodiversity loss), multi-media (air, water, soil), multi-scale (hot spots, regional, European, global) framework.

T. Spranger
German Federal Environment Agency (UBA), Wörlitzer Platz 1, 06844 Dessau, Germany

Z. Klimont
Institute for Applied Systems Analysis (IIASA), APD, Schlossplatz 1, Laxenburg 2361, Austria

M. Sutton, S. Reis and S.M.H. Baker (eds), *Atmospheric Ammonia,*   433
© Springer Science + Business Media B.V. 2009

Following the recommendation of Working Group on critical levels for ammonia, of lowing the values set (Cape et al. 2009), there is a need for careful evaluation of the representativeness of EMEP model results for ammonia concentration. Another important implication is a need for discussion whether, and if so how the new critical levels will be used in addition to critical loads in formulating air pollution targets. This is particularly true at local or regional levels in regions with spatially variable ammonia emissions and concentrations.

Responding to some of the policies like the EU Nitrate Directive or Biodiversity related directives, farmers in certain areas adjusted agricultural practices, e.g., by shifting application of manures from autumn to spring. This leads to changes in seasonal patterns of ammonia concentrations. More attention is needed how to monitor and incorporate impacts of such policies in modelling tools.

It is necessary to explore possibilities of considering local biodiversity action plans in larger scale modelling. Strategies exist to integrate them into the European scale, e.g., via the FFH Directive and the *Natura2000* network. However, the role of air pollution effects is often not explicitly taken into account even though nitrogen inputs have a large effect on biodiversity: there is room for improvement on local, national and European levels.

27.2   Introduction

This Cross Cutting Group considered the policy context of the ammonia problem, including socio-economic, environmental, institutional and technological aspects. It addressed the links between different Working Group (WG) themes to summarize the current challenges faced in relation to the different environment and health effects of ammonia emissions. Drawing on the scientific findings and recommendations from the WGs and independent contributions from the participants, the Cross Cutting Group addressed the potential role of different policy options to help mitigate these impacts. It also discussed a need to adapt tools used in policy analysis (IAMs) and consequently evaluate policies in view of new findings.

The scope of the Cross Cutting Group can be summarized in two key questions:

1. What is the agricultural and environmental policy context of ammonia abatement and effects?
2. How can scientific understanding help address the future challenges to reduce the negative effects of ammonia?

The discussion was guided by a number of more specific points:

- What are the different policy issues to which ammonia emissions and their effects relate and where are the benefits and conflicts?
- How adequate is our scientific knowledge of ammonia to respond to the different policy concerns?
- How adequate are models for the integrated assessment of ammonia impacts and their relationship with other environmental and health concerns?

- What are the major challenges for reducing ammonia, from the perspective of key threats, scientific uncertainty and difficulties to agree upon measures that are acceptable to different stakeholders?
- How can problems of spatial scale be turned into a benefit that can be exploited, e.g., by identifying spatial priorities for assessment and abatement?
- What are the scientific tools and indicators that we need to inform the setting of priorities (both policy and spatial priorities) as well as to measure successes?
- What lessons can be learned from previous mitigation policies for different air and water pollutants which can inform the science-policy interaction for ammonia?

27.3 Policy Background

Ammonia emissions are major or even dominating contributors to eutrophication and acidification of ecosystems and secondary $PM_{2.5}$ concentrations in Europe. Reduction of the ammonia emissions in Europe has been on the agenda for more than a decade, first on a national scale, e.g. in Denmark and the Netherlands, followed by international efforts. The latter include the UNECE Convention on Long-range Transboundary Air Pollution (Gothenburg Protocol, UNECE 1999) and the EU directives and strategies specifically the National Emission Ceiling Directive (NECD), the Clean Air For Europe (CAFE) and the Thematic Strategy on Air Pollution (TSAP, see CEC 2005).

In the TSAP, the European Commission expressed the environmental objectives for 2020 in terms of relative improvements compared to the situation as it has been assessed for the year 2000 (Amann et al. 2006b) aiming at inter alia reducing area of forest ecosystems where acid deposition exceeds the critical loads for acidification by 74%, and reducing ecosystems area where nitrogen deposition exceeds the critical loads for eutrophication by 43%. Furthermore, TSAP set the target to reduce live years lost from particulate matter (YOLLs) at 47% compared to the assessment for 2000. Achieving these goals requires significant further emissions reductions of various air pollutants. It has been estimated that emissions of ammonia 80–95% typically originates from agricultural sources) would need to be reduced by 27% in the EU25 by 2020 compared to 2000 (CEC 2005). Since existing legislation and projections of livestock changes accounted for only 4% decline, remaining 23% would have to be met by introduction of specific abatement measures in agriculture.

However, there are many other environmental and agricultural policies driving, constraining or otherwise influencing ammonia emissions and abatement policies in Europe:

- EU Common Agricultural Policy (CAP)
- EU Biofuel Action Plan
- EU Directive on Integrated Pollution Prevention and Control (IPPC) for pigs and poultry
- EU Directives on animal welfare for pigs, Calves, and laying hens
- Other environmental policies related to nitrogen inputs, e.g., EU Water Framework, Nitrate and Ground Water Directives, Marine Conventions (OSPAR, and the EU Thematic Strategy on Marine Pollution)

- UN Framework Convention on Climate Change (UNFCCC)
- Biodiversity policy including UN Convention on Biological Diversity (CBD), the EU Directive on Fauna, Flora and Habitat (FFH) with Natura2000 and Streamlining European Biodiversity Indicators (SEBI2010) networks

These policies and their implications for ammonia abatement need to be considered in the development of long-term European strategies to reduce effects from ammonia emissions. More specific discussion is provided below.

27.4   Towards an Ammonia Abatement Strategy

As indicated in previous section, there exist ambitious goals to reduce the impact of ammonia on ecosystems and secondary PM on human health. The more recent work on the revision of the NEC Directive follows the same path and does integrate some of the elements mentioned in the background discussion above, namely CAP reform and IPPC Directive. They are estimated to have important impact on emissions of ammonia from agriculture reducing total EU25 emissions by up to 11% by 2020 compared to 2000 (Amann et al. 2006a; Klimont et al. 2007). This is a significantly larger reduction than previously estimated within CAFE. However, reaching TSAP targets calls for much larger cuts, i.e., about 16% more, leading to total reduction of 27% compared to the level of 2000 (Amann et al. 2006b, Klimont et al. 2007). This again is associated with a need to extend introduction of abatement in agriculture or finding other ways of minimizing nitrogen losses to the environment.

It was concluded during the discussion that this calls for

- extensions of currently used tools,
- verifications of specific elements of the models,
- improved information on impacts originating from a mixture of pollutants,
- new knowledge on the sensitivity of ecosystems as well as their recovery,
- adaptation of monitoring networks and targeted measurement programmes, but also
- possibly the revision of legislation in order to close existing loopholes in addressing nitrogen pollution at large and improve synergies between existing legislation.

The following paragraphs expand on specific elements of such a long-term strategy.

27.4.1   Important Interactions and Additional Information

Designing ammonia abatement strategies necessitates consideration of current knowledge concerning a number of interactions, results of the latest monitoring and measurement programmes and policy developments, and may lead to proposals for more research to tackle issues on the policy agenda. Some of the key elements are listed below and are addressed in a greater detail in the reports from the other Working Groups and Cross Cutting Group (Chapters 7.1–7.5).

However, some of the issues were either addressed in this Cross Cutting Group B (CCGB) discussions from a different perspective, while other issues were not addressed. The latter includes scope for potential abatement and information on the state and evolution of agricultural systems. Both are vital when determining the scope for analysis when preparing the strategy. Specifically, constraints on the application of measures are key to the development of realistic strategies, while information on production systems, farm size distribution and their evolution is important to assure meaningful cost estimates and allows assessing the affordability of measures to farmers. Furthermore, this information is also needed to consider issues like, e.g., animal welfare, trends in organic farming, and the application of the IPPC directive, to name a few.

In summary, information is needed on the following:

- The biogeochemical nitrogen cycle and its effects (including on biodiversity), and its link to the carbon cycle and climate change
- Secondary particulate formation and effects with particular focus on the relative toxicity of the various secondary PM compounds
- Emissions and abatement efficiencies
- Scope and potential for abatement
- Agricultural production systems and their development
- Interaction of ammonia emissions with/pollutant swapping between different air pollutants (SO_2, NO_x, NMVOC, CH_4,...)
- Synergies and conflicts concerning pollutant swapping between various forms of reactive nitrogen.

It is generally agreed that with current scientific knowledge and available data, air pollution strategies can be designed for ammonia. However, these do not take full account of the interactions discussed above. More research, data and consequently further development of tools is required to respond to the different policy concerns mentioned in previous sections. They should allow for comprehensive inclusion of synergies and interactions between various processes and multi-media and multi-scale (smoothly integrating across local and regional scales) policies.

Some of the interactions listed above are included in current analysis while others need certainly more work to allow for quantitative assessments at different scales, specifically to improve representation of mentioned processes in IAMs used in policy work. For example, a number of measurement campaigns have demonstrated strongly non-linear responses to abatement of one form of N vs other N species, e.g., Harrison and Webb (2001). The processes responsible are not parameterised well in models (see discussion in the CCGA on NH_3 abatement vs N_2O emissions) (see Chapter 26).

But systematic approaches where there above is taken into account (at least to some extent) are being developed for both regional and pan-European applications, e.g., NARSES (Webb and Misselbrook 2004), GAS-EM (Dämmgen et al. 2002) and for the latter GAINS model (Klaassen et al. 2004; Amann 2004; Klimont and Brink 2004).

27.4.2   Policy Process, Strategy Approaches, Tools, and Science Interactions

The process of developing an ammonia strategy involves a number of steps and depending on the scope and geographical coverage may involve several stakeholders. The drivers for controlling ammonia emissions in Europe and USA have been different, i.e., fine PM in the USA and acidification and euthrophication in Europe. Only recently the connection between secondary PM formation and ammonia emissions has entered European policy agenda, and both are important drivers now.

In Europe, the approach has been to first evaluate knowledge on effects and atmospheric transport and set targets, and in the next step develop strategies for emission abatement. Science plays an important role in both of these steps and models that support the process play a very important role. They can provide objective information on feasibility and costs of strategies while policy makers have to value the risks and choose final targets. Integrated assessment tools are available and have been widely used for policy support in Europe.

Drawing on the information and knowledge discussed in previous section, several strategy elements and approaches can be evaluated and proposed, e.g.:

1. **Effects:**

– Evaluation of present and future effects of ammonia emissions, including pollutant interactions,
– Target setting (air quality limit values) for
  • ecosystem and human health endpoints
  • critical loads (compared to deposition rates) and critical levels (compared to concentrations)
  • ecosystem sensitivity and recovery on various temporal (seasonal to decades) and spatial (European scale integrated assessment models, local scale approaches) scales

2. **Emissions:**

  – Reducing nitrogen fluxes at large, especially in agricultural systems
  – Reducing intensities (animal numbers, fertiliser application rates)
  – Technical abatement measures in the whole agricultural nitrogen chain
  – Economic instruments

The new knowledge and the policy objective to develop an even more comprehensive ammonia strategy that has been discussed at this meeting presents new challenges not only for IAMs but also for several other models that address a particular area and often operate on a much finer resolution. In particular the following issues were discussed:

• Following the recommendation (Cape et al. 2009, see Section 2.1) on lowering critical levels for ammonia there is a need for careful evaluation of representativeness of EMEP model results for ammonia concentration (see also discussion in the report of van Pul et al. (2009, Chapter 7.4) where need for systematic model intercomparison is addressed). Another important conclusion is the need for a discussion, whether and if so, how the new critical levels should be used in addition to critical loads in formulating air pollution targets. This is particularly true at the local or regional scale in regions with spatially variable ammonia emissions and concentrations.

- Responding to some of the policies like the EU Nitrate Directive or Biodiversity related directives farmers in certain areas adjusted practises. For instance, shifting application of manures from autumn to spring leading to changes in seasonal patterns of ammonia concentrations. More attention is needed how to monitor and incorporate impacts of such policies in modelling tools.
- The above points (as well as discussion within other WGs) indicate a need for a review of EMEP and national monitoring strategies.
- Monitoring is also needed to document and explain differences between measured and modelled ammonia concentrations and their trends (trends are seen as indicator of the effectiveness of economic and pollution abatement policies). The findings then have to be used to revise estimates of emissions, abatement efficiencies, and atmospheric transport. Based on this, transfer matrices might have to be modified to provide an improved basis for negotiations of emission ceilings.
- Ammonia emission reduction policies must be analysed in a multi-effect, multi-media, multi-scale (hot spots, regional, European, global) framework. Conflicts and trade-offs between different policies (e.g., climate change policies aiming at N_2O and CH_4 reduction vs. air pollution policies aiming at ammonia reduction) need to be assessed and solved. Specifically, IAMs need to link to or include swapping reactive nitrogen emission; in this respect they should in particular better integrate the nitrate vulnerable zones (NVZ) and the measures already adopted in the frame of the Nitrates Directive. This would ensure that ammonia measures, especially those implemented in the NVZ, do not lead to increased nitrate leaching; an example approach used for the Netherlands is the so called N-cascade study (Erisman et al. 2005).
- It is necessary to explore possibilities of considering local biodiversity action plans in larger scale modelling. Strategies exist to integrate them into the European scale, e.g., via the FFH Directive and the Natura 2000 network. However, the role of air pollution effects are often not explicitly taken into account even though nitrogen inputs have a large effect on biodiversity; there is room for improvement on local, national and European levels. In several countries (e.g., Netherlands, Denmark, Germany) assessments and strategies were developed that include, e.g., buffer zones for agricultural activities in the vicinity of Natura 2000 areas and/or the use of critical loads as deposition targets for abatement (e.g. Ambelas Skjoth 2004, LAI 2006). On the European level, the programme "Streamlining European Biodiversity Indicators (SEBI2010) has chosen the exceedance of critical loads of nutrient nitrogen as headline indicator for risk of biodiversity loss.
- Similarly to the above ("bottom-up"), concepts on how to translate landscape scale measures and planning rules (e.g. in nitrate vulnerable zones) into the European scale (e.g., integrated assessment modelling) are being developed, e.g., in the NitroEurope project but a far from completion.

27.4.3   Discussion of Existing Legislation and Challenges for Stakeholders

The key conclusions on existing legislation were:

- Work is needed to eliminate loopholes and improve synergies of different legislation – the example used here was related to IPPC farms: following implementation of, often expensive, reduction measures in the building and storage to preserve nitrogen in the manure, it is sometimes allowed in certain situations to apply the manure to land using broadcast spreading, rather than to use a technique that would reduce the loss of ammonia. This is not a sustainable strategy from the perspective of nitrogen loss prevention. More explicit and active help from the legislator on how to achieve set goals would be beneficial – this comment originated from local regulators who are often confronted with difficulties in the interpretation of laws and lack examples of measures and ways of achieving the aims.
- There was no discussion on possible lessons to be learned from previous mitigation policies addressing other air pollution issues. However, the approach adopted for developing an ammonia strategy is guided by the same principles as for other air pollutants, especially when being expanded to link to greenhouse gases emissions, water, soil, landscape, and biodiversity analysis. Some of the key new elements introduced in the last years were a more active involvement of stakeholders and opening of the access to the scientific findings, model assumptions and results to parties participating in the process. One of the major benefits of this cooperation is higher acceptance of the proposed targets and resulting abatement strategy that is further translated into legislation.
- In spite of all the effort to improve process understanding and increase the reliability of measurements and modelling, important uncertainties remain and policy makers have to cope with it. The treatment of uncertainties in translating model results of a proposed European strategy into national laws is one of the big challenges stakeholders face. There are at least two types of uncertainties that need to be addressed in this context; one concerns limitations associated with incomplete knowledge on any scale, for example the 'Dutch Ammonia gap'; the other is linked to capability of used tools to reflect on specific national issues and important local scale phenomena or sensitivities. Both of them need to be dealt with when agreeing on a European strategy and when designing national/local strategies compatible with the European legislation. Science can play an important role in supporting these processes both at the European and local level.
- Legislation development and implementation is often a lengthy process confronted with an ever changing background in terms of other legislation or unexpected economic or political developments. For instance, in the future we might expect more urea fertilizer to be produced and used in Europe and slow decline in ammonium nitrate production because of security concerns. In contrast, the current strategies, aiming at ammonia emission reduction, recommend a significant reduction in the use of urea and its replacement by with ammonium nitrate as a cost-effective measure – exactly the opposite strategy resulting in a very

different 'ammonia emission balance'. The increase in urea use might greatly increase the costs of achieving emission and effects targets currently proposed within the EU *Thematic Strategy on Air Pollution*.

- One of the further challenges in designing future policies in response to the need for more integrated approach is how to cope with reduced abatement potential for technical measures. This is especially the case for countries where extensive ammonia reduction programs have been in place for more than a decade, e.g., the Netherlands, Denmark. This question will be more widely posed when next emission reduction targets will be discussed since during the current debate technical measures received most of the attention and proposed strategies rely on their limited potential. This stresses the importance of the long-term integrated N approach where efficient use of N throughout the farming system is promoted.

The group felt a strong need for futher research and data collection necessary to assist policies to achieve ambitious targets and a comprehensive and consistent strategy development. Specifically mentioned areas included:

- N-cycle and effects, e.g., on biodiversity
- PM health impact assessment, role of PM speciation
- Data on management practices (possible use of the remote sensing results from MARS project)
- Measurements and modelling at different scales
- Development of multi-media emission reduction strategies

27.5 Conclusions

The Group concluded that integrated assessment should continue to rely on a gap closure approach, i.e., linking national emission reduction needs to the spatial pattern of all critical loads exceedances via country-to-grid transfer matrices. A possible complementation of this approach by emission reduction aims for specific protection of areas with high conservation value could be tested.

There is a need for careful evaluation of representativeness of EMEP model results for ammonia concentration and for discussion how the new critical levels may be used in addition to critical loads in formulating air pollution targets.

Ammonia emission reduction policies must be analysed in a multi-pollutant, multi-effect, multi-media, multi-scale framework. Specifically, IAMs need to either link to or include swapping reactive nitrogen emission in order to advise future policies. Besides these extensions, verification of specific elements of the models, adaptation of monitoring networks, targeted measurement programs are needed, as well as a possible revision of legislation to close existing loopholes and increase synergies in addressing nitrogen pollution at large. In that sense, priority should be given to measures aiming at reducing all kinds of nitrogen losses at farm level.

It is necessary to explore possibilities of considering local biodiversity action plans and landscape scale measures and planning rules into the European scale, i.e., larger scale modelling.

More attention is needed how to monitor and incorporate impacts of certain policies in modelling tools, e.g., policies which lead to different seasonal patterns of ammonia concentrations.

A number of issues addressed in the discussion within the Cross Cutting Group and discussion within other Working Groups indicate a need for a review of EMEP and national monitoring strategies.

References

Amann M. (2004) The RAINS model: Documentation of the model approach prepared for the RAINS peer review 2004. http://www.iiasa.ac.at/rains/review/review-full.pdf (see also http://www.iiasa.ac.at/rains/gains-methodology.html for several background reports).

Amann M. Asman W.A.H., Bertok I., Cofala J., Heyes C., Klimont Z., Posch M., Schöpp W., Wagner F., Hettellingh J-P. (2006a) NEC Scenario Analysis Report Nr. 2: Emission control scenarios that meet the environmental objectives of the Thematic Strategy on Air Pollution. Part 1: Methodology and assumptions. International Institute for Applied Systems Analysis (IIASA), Laxenburg, Austria, December 2006.

Amann M. Asman W., Bertok I., Cofala J., Heyes C., Klimont Z., Posch M., Schöpp W., Wagner F., Hettellingh J-P. (2006b) NEC Scenario Analysis Report Nr. 2: Emission control scenarios that meet the environmental objectives of the Thematic Strategy on Air Pollution Part 2: Scenario analysis. International Institute for Applied Systems Analysis (IIASA), Laxenburg, Austria, December 2006.

Ambelas Skjoth C., Hertel O., Gyldenkaerne S., Ellermann T. (2004) Implementing a dynamical ammonia emission parameterization in the large-scale air pollution model ACDEP. Journal of Geophysical Research – Atmospheres 109(D6), 1–13.

Cape J.N., van der Eerden L., Fangmeier A., Ayres J., Bareham S. Bobbing R., Branguinho C., Crittenden P., Cruz C., Dias T., Leith I.D., Martins-Loução M.A.M., Pitcairn C.E.R., Sheppard L.Jl, Spranger T., Sutton M.A., van Dijk N., Wolseley P.A. (2009) Critical Levels for NH3. In: Sutton M.A., Reis S., Baker S.M.H. (eds.), *Atmospheric Ammonia – Detecting Emission Changes and Environmental Impacts. Results of an Expert Workshop Under the Convention on Long-range Transboundary Air Pollution*. Springer (Chapter 22).

CEC (2005) Communication from the Commission to the Council and the European Parliament on a Thematic Strategy on Air Pollution. SEC(2005) 1132. Commission of the European Communities, Brussels, http://eurlex.europa.eu/LexUriServ/site/en/com/2005/com2005_0446en01.pdf.

Dämmgen U., Lüttich M., Döhler H., Eurich-Menden B., Osterburg B. (2004) GAS-EM – ein Kalkulationsprogramm für Emissionen aus der Landwirtschaft. *Landbauforschung Völkenrode* 52, 19–42.

Erisman J.W., Domburg N., De Vries W., Kros H., De Haan B., Sanders K. (2005) The Dutch N-cascade in the European perspective. Science in China: Series C – Life Sciences Vol. 48 Special Issue, 827–842.

Harrison R., Webb J. (2001) A review of the effect of N fertilizer type on gaseous emission. *Advanced Agronomy* 73, 65–108.

Klaassen G., Amann M., Berglund C., Cofala J., Höglund-Isaksson L., Heyes C., Mechler R., Tohka A., Schöpp W., Winiwarter W. (2004) The Extension of the RAINS Model to Greenhouse Gases. IR-04-015. International Institute for Applied Systems Analysis (IIASA), Laxenburg, Austria.

Klimont Z., Brink C. (2004) Modelling of Emissions of Air Pollutants and Greenhouse Gases from Agricultural Sources in Europe. IR 04-048, IIASA, www.iiasa.ac.at/rains/reports/ir-04-048.pdf

Klimont Z., Asman W.A.H., Bertok I., Gyarfas F., Heyes C., Wagner F., Winiwarter W., Höglund-Isaksson L., Sandler R. (2007) Measures in agriculture to reduce ammonia emissions. Final report to the European Commission under Service Contract No. 070501/2006/433072/FRA/C1, www.iiasa.ac.at/rains/reports/agri-report.pdf

LAI (2006) Ermittlung und Bewertung von Schadstoffeintraegen. Abschlussbericht des Arbeitskreises (in German)

UNECE (1999) Protocol to the 1979 Convention on long-range transboundary air pollution to abate acidification, eutrophication and ground-level ozone. United Nations Economic Commission for Europe (UNECE), Geneva; www.unece.org/env/lrtap/welcome.htm

van Pul W.A.J., Reis S., Dore A.J., Xuejun L., Fagerli H., Geels C., Hertel O., Wichink Kruijt R., Kryza M., Bergström R., Vieno R., Smith R.I., Nemitz E. (2009) Modelling the national and regional transport and deposition of ammonia. In: Sutton M.A., Reis S., Baker S.M.H. (eds.), *Atmospheric Ammonia – Detecting Emission Changes and Environmental Impacts. Results of an Expert Workshop Under the Convention on Long-range Transboundary Air Pollution.* Springer (Chapter 25).

Webb J., Misselbrook T.H. (2004) A mass-flow model of ammonia emissions from UK livestock production. *Atmospheric Environment*, 38(14), 2163–2176.

Chapter 28
Synthesis and Summary for Policy Makers

Mark A. Sutton, Stefan Reis, and Samantha M.H. Baker

28.1   Preface

In accordance with the workplan of the Working Group on Strategies and Review (WGSR) of the UNECE Convention on Long-Range Transboundary Air Pollution (CLRTAP) (ECE/EB.AIR/WG.5/2006/9, item 1.8(c)), the Workshop on Atmospheric Ammonia: Detecting Emission Changes and Environmental Impacts was held on 4–6 December 2006 in Edinburgh (United Kingdom). It was organized and supported by the Centre for Ecology & Hydrology (CEH), the Department for Environment, Food and Rural Affairs (DEFRA), the Scottish Executive Rural Affairs Department (SEERAD), COST 729 and the NitroEurope Integrated Project (NEU). Background documents and presentations are available at www.ammonia-ws.ceh.ac.uk/documents.html.

The workshop was attended by 80 experts from the following 19 Parties to the Convention: Austria, Canada, Croatia, the Czech Republic, Denmark, France, Germany, Hungary, Ireland, Italy, the Netherlands, Norway, Poland, Portugal, Slovakia, Sweden, Switzerland, the United Kingdom and the United States. The European Commission, the EMEP *Meteorological Synthesizing Centre – West* (MSC-W), the International Institute for Applied Systems Analysis and the secretariat were represented.

Several bodies under the Convention provided input to the organization of the workshop: the *Expert Group on Ammonia Abatement*, the *Task Force on Emission Inventories and Projections* and the *Task Force on Mapping and Modelling* under EMEP, and the *International Cooperative Programme on Mapping and Modelling* under the *Working Group on Effects*.

The findings of this workshop as summarised in the previous chapters were reported to the WGSR at the 39th Session in Geneva (UNECE 2007)

Mark A. Sutton
Centre for Ecology & Hydrology, Bush Estate,
Penicuik, Midlothian, EH26 0QB, United Kingdom

M. Sutton, S. Reis and S.M.H. Baker (eds), *Atmospheric Ammonia,*
© Springer Science + Business Media B.V. 2009

28.2 Aims and Approach of the Workshop

The objectives of the workshop were to:

- Assess the extent to which the existing critical thresholds for ammonia reflect current scientific understanding by:

  - Examining the case for setting new ammonia critical threshold(s) based on current evidence of direct impacts of ammonia on different receptors
  - Discussing the extent to which vegetation and sensitive ecosystems appeared to be differentially sensitive to ammonia versus other forms of reactive nitrogen (N)
  - Debating the case for establishing indicative air concentration limits for indirect effects of ammonia which would be consistent with current critical loads for N

- Assess the extent to which independent atmospheric measurements can verify where regional changes in ammonia (NH$_3$) emissions have and have not occurred by:

  - Quantifying the extent to which estimated regional changes in NH$_3$ emissions have been reflected in measurements of atmospheric NH$_3$ and ammonium
  - Distinguishing cases where the estimated changes in NH$_3$ emission are due to altered sectoral activity or the implementation of abatement policies, and thereby assess the extent to which atmospheric measurements verify the effectiveness of NH$_3$ abatement policies
  - Making recommendations for future air monitoring and systems for assessing the national implementation of NH$_3$ abatement policies and considering the implications of any non-linearities for integrated assessment models

- Review approaches for downscaling transboundary assessments to deal with ammonia hot spots in relation to operational modelling and monitoring by:

  - Reviewing current emission and atmospheric dispersion modelling methods for downscaling NH$_3$ dispersion and deposition in hot spots
  - Examining the status of methods for effect assessment and monitoring in hot spots
  - Recommending broad principles for assessment approaches in ammonia hot spots, including spatial approaches and interactions between transboundary NH$_3$ emission reduction targets and other policy measures

- Review mesoscale atmospheric transport and chemistry models in relation to their formulation and results for NH$_3$ by:

  - Reviewing emission parameterizations used in the models, establishing comparability, spatial and temporal resolution and uncertainties
  - Reviewing dispersion, air chemistry and deposition formulations identifying key differences and uncertainties
  - Assessing the overall performance of the models against measurements and against a common reference, and thereupon making recommendations for improving mesoscale models of NH$_3$ transport and deposition, including the implications of any non-linearities for source-receptor matrices and integrated assessment models

Ms. R. Brankin, Deputy Minister for Environment and Rural Development of Scotland, opened the workshop. She introduced the background and needs of the workshop, highlighting the dependence of future NH_3 strategies on sound scientific evidence. Mr. K. Bull (secretariat) presented the history and development of the Convention and its Protocols, noting that the workshop was the first one specifically devoted to NH_3 that linked expertise across relevant subsidiary bodies of the Convention. Mr. M. Sponar (European Commission) described the European Union (EU) Thematic Strategy on Air Pollution, noting the increasing importance of NH_3 and highlighting the need for long-term development of an integrated approach to mitigate NH_3 in relation to other policies and other forms of N pollution.

On behalf of the organizers, Mr. M. Sutton (United Kingdom) explained that the workshop would consist of four separate working groups and two cross-cutting groups. Each group would produce conclusions and recommendations (Chapter 22 to 27 of this volume) that were agreed by the workshop. A full report, including working group reports, background documents, posters and a list of participants, is available on the website given in the first paragraph of this chapter, with this volume reporting the final synthesis.

28.3   Conclusions

28.3.1   Critical Levels for Gaseous Ammonia

The current NH_3 critical levels (CLEs) for vegetation under the Convention, agreed in Egham (United Kingdom) in 1992, were based on measurements and observations from the 1980s, mostly from the Netherlands, and were set at $3,300\,\mu g\ m^{-3}$ (hourly), $270\,\mu g\ m^{-3}$ (daily), $23\,\mu g\ m^{-3}$ (monthly) and $8\,\mu g\ m^{-3}$ (annual). The workshop concluded that these levels required revision in light of new evidence from field-based experiments and surveys.

The existing annual CLE ($8\,\mu g\ NH_3\ m^{-3}$), when expressed as an equivalent deposition of N to an ecosystem, was less protective than the current critical load for most, if not all, European ecosystems and habitats. Field-based evidence relating effects on vegetation to NH_3 concentrations measured over one year or longer showed that the current annual CLE was too high.

A new long-term CLE for the most sensitive vegetation types (lichens and bryophytes) and the associated habitats was proposed, based on observed changes to species composition in the field. Most of the evidence came from studies in the United Kingdom, but there was corroborative evidence from Italy, Portugal and Switzerland. The proposed long-term CLE for NH_3 for (a) sensitive lichen communities and bryophytes, and (b) ecosystems where sensitive lichens and bryophytes were an important part of the ecosystem integrity was set at $1\,\mu g\ NH_3\ m^{-3}$.

There was less evidence available to quantify the concentrations at which long-term effects of NH_3 caused species changes in communities of higher plants. The workshop proposed a long-term CLE for higher plants of $3\,\mu g\ NH_3\ m^{-3}$. This value was set for higher plants in general, but was particularly based on data from heathlands and forest ground flora. Given the larger uncertainties in this estimate, an uncertainty range was proposed of 2–4 µg m^{-3}, depending on the degree of precaution appropriate to different regulatory contexts.

On the basis of current knowledge, it could not be assumed that each of these new long-term CLE values would be protective for periods longer than 20–30 years. No assumptions had been made on the mechanism by which NH$_3$ exposure led to changes in species composition. By emphasizing long-term rather than daily NH$_3$ concentrations, the NH$_3$ critical level was concluded to have the advantage of providing a practical tool complementing the critical loads approach which was simple to apply for cost-effective regulation and monitoring of NH$_3$ specific measures.

28.3.2 Detecting Changes in Atmospheric Ammonia

The workshop discussed progress in the state of knowledge in deriving trends from measurements and their use to verify abatement measures or other causes for decrease in emissions of NH$_3$ to the atmosphere. The workshop identified clear progress in closing the gap between the observed and expected values for reduced N, as well as a better understanding of the reasons behind this.

The long-term measurements available followed the emission trend. Current measurements made it possible to evaluate policy progress on NH$_3$ emission abatement. In those countries where there were large (>25%) changes in emissions, such as in the Netherlands and Denmark, the trend followed closely, especially when meteorology was taken into account. In other countries, such as the United Kingdom, the trend was much smaller, but there was no significant gap between measurements and model estimates. In the Netherlands, there was still an NH$_3$ gap – a significant (30%) difference between emissions-based NH$_3$ concentrations and measurements – but the temporal trend was the same. The difference might be due to either an underestimation of the emission or an overestimation of the dry deposition.

On the European scale it was difficult to match the emission change, both because of lack of measurements, especially in the eastern part of Europe, and because of the confounding factor of the SO$_2$ emission reductions, which affect the ammonium concentrations in aerosol and rain water.

28.3.3 Assessment Methods for Ammonia Hot Spots

The workshop agreed that accounting for hot spots for either upscaling of fluxes or risk assessment for nearby ecosystems required a precise description of all processes involved. Hot-spot assessment should also account for background concentrations and deposition history.

The key uncertainties in the models were emissions and dry deposition. Sufficient local input data were required for making effects assessments and landscape analyses. For dry deposition, this required better knowledge of NH_3 compensation points and surface resistances for different ecosystems, their dependence on climatic variables and the deposition history for NH_3 and other pollutants.

The workshop concluded that using different models allowed the analysis of landscape interactions between sources and receptors with sufficient accuracy for a range of conditions to consider real cases and scenarios. It also allowed the assessment of local, tailored abatement measures.

The workshop agreed that scenarios from local-scale modelling could be used in a statistical way to provide estimates of within-grid cell recapture for national- and regional-scale models, linked with global descriptors of the spatial variability in land cover.

28.3.4 Regional Modelling of Atmospheric NH_3 Transport and Deposition

A range of chemical transport models was used across the Convention to model the emission, transport and deposition of atmospheric NH_3 on the national and regional scales. These models had been developed from a range of historical backgrounds and with different purposes. Six models were considered, ranging from the national scale up to the full European scale. The models differed in concept, particularly in their chemical scheme and in scale, ranging from Lagrangian models on the national scale via Eulerian models on the European scale to nested models coupling the European scale with the local scale.

Key uncertainties in the modelling of atmospheric NH_3 were linked to emissions (absolute level and spatial and temporal allocation) dry deposition parameterization, spatial resolution of the model and the description of vertical diffusion. All European-scale models (including the EMEP model) currently underestimated the measured NH_3 concentration. National models generally found better agreement with NH_3 measurements. The main reasons for the observed differences between the measured and modelled NH_3 concentrations were the spatial resolution of the models and the parameterization of the dry deposition process.

The concentration of ammonium aerosol was fairly well described by all models. However, both under- and overestimates of measured concentrations were found. The magnitude of the wet deposition of ammonium was in general reproduced well by all models.

None of the models routinely used the compensation point (bidirectional exchange scheme) as a parameterization in the dry deposition process of NH_3. This was thought to be one of the reasons why some models tended to underestimate concentrations, particularly in summer. The main reason for not taking this process into account was the lack of a generalized database for the compensation point with respect to the main land cover types used in the models.

The siting of measurements played an important role in the comparison with modelled concentrations. Some stations in agricultural areas should not be used for verification of the Eulerian models with large grid size (50 km), because of the significant contribution of sources close to the measurement stations that cannot be simulated by the models on this spatial scale.

28.3.5   *Reliability of NH₃ Emission Estimates and Abatement Efficiencies*

Few countries had considered uncertainty in NH_3 emissions in detail. Results indicated that national estimates may be accurate to within ±20%. For countries that had created inventories using emission factors (EFs) measured elsewhere, the uncertainty may be around 100%. The greatest uncertainty was likely to be for emission estimates for regions within countries. Sensitivity analysis of the United Kingdom inventory showed that activity data and other information on a range of relevant farming practices were the inputs for which the system was most sensitive. Cattle diets, especially grass-based ones, were considered particularly uncertain.

The United Kingdom and Denmark reported a high level of agreement between modelled and measured NH_3 concentrations, while models still underestimated measurements in the Netherlands. A detailed discussion of the Dutch "ammonia gap" suggested that the EFs used in the Dutch inventory were accurate. The discrepancy was considered to result either from overestimation of abatement efficiencies or from overestimation of dry deposition velocities. Adjustment of either could eliminate the gap, but it was not yet known which was responsible.

The abatement efficiencies in the guidance document on ammonia to the 1999 Gothenburg Protocol were considered robust. While averages did not reflect the variability in data, quoting ranges may create uncertainty regarding which point in the range is most appropriate to use. Since data were obtained almost exclusively from Northern and North-Western Europe, abatement efficiencies could not be assumed to be applicable across the whole UNECE region. Only a brief statement is given in the guidance document on the impacts of reducing emissions of NH_3 following spreading on losses of other N pollutants, because nitrate leaching and nitrous oxide emissions tend to be site- and season-specific.

"Soft" approaches to NH_3 abatement were those implemented using basic facilities and simple management approaches (e.g. applying manure during weather conditions associated with little emission). While these offered an economically attractive method of reducing NH_3 emissions, it was often difficult to know their uptake by farmers and their efficiency, and therefore to convince environmental authorities of their effectiveness or to measure the achievement in national reporting.

Experience from the adoption of abatement technologies in other areas, suggested that ex ante cost assessments tend to overestimate the cost of implementation. However, taking emerging technologies into the industry could lead to a reduction in abatement efficiency. A number of emerging abatement options were considered, together with a summary of other developments that may affect NH_3 emissions.

28.3.6 Ammonia Policy Context and Future Challenges

Ammonia emissions are major contributors to eutrophication and acidification of ecosystems and secondary $PM_{2.5}$ concentrations in Europe. Reduction of NH_3 emissions in Europe has been on the agenda for more than a decade, first on a national scale (e.g. in the Netherlands) and more recently through international efforts. The latter include the Convention's Protocols and EU directives and strategies.

The workshop considered the policy context of the NH_3 problem, including socio-economic, environmental, institutional and technological aspects, and the potential role of policy options in mitigating the ecosystem and health impacts of NH_3 emissions. The need to adapt tools used in policy analysis, such as integrated assessment models, and to consequently evaluate policies in view of new findings was also considered.

Ammonia policies were recognised as becoming interlinked with a number of other environmental and agricultural policies. In order to avoid the problem of "pollution swapping", future policies needed to consider these interactions.

The workshop noted that, in responding to some of the policies like the EU Nitrates Directive (91/676/EC) or biodiversity-related directives, farmers in certain areas adjusted agricultural practices (e.g. by shifting application of manures from autumn to spring). This led to different seasonal patterns of NH_3 concentrations, although there was little knowledge of the environmental consequences.

28.4 Recommendations

28.4.1 Critical Levels for Ammonia

The workshop recommended:

a. A revision of the currently set values of the ammonia critical level, since the data reviewed show that the existing CLE values of $3,300\,\mu g\ m^{-3}$ (hourly), $270\,\mu g\ m^{-3}$ (daily), $23\,\mu g\ m^{-3}$ (monthly) and $8\,\mu g\ m^{-3}$ (annually) are not sufficiently precautionary
b. A new long-term CLE for lichens and bryophytes, including for ecosystems where lichens and bryophytes are a key part of the ecosystem integrity, of $1\,\mu g\ m^{-3}$
c. A new long-term CLE for higher plants, including heathland, grassland and forest ground flora and their habitats, of $3\,\mu g\ m^{-3}$, with an uncertainty range of $2–4\,\mu g\ m^{-3}$; the workshop noted that these long-term CLE values could not be assumed to provide protection for longer than 20–30 years

d. Retaining the monthly critical level ($23\,\mu g\ m^{-3}$) as a provisional value in order to deal with the possibility of high peak emissions during periods of manure application (e.g. in spring)
e. Research to improve the future estimation of NH_3 critical levels. This included addressing uncertainties relating to the shortage of observational data and long-term NH_3 concentration measurements, particularly in Southern and Eastern Europe. Similarly, there was a need for a better understanding of the mechanisms whereby NH_3 affected plants, especially over decadal timescales, so that predictive models could be constructed for extrapolation to other types of vegetation and land use in different climatic zones.

28.4.2   Detecting Temporal Changes in Atmospheric Ammonia

The workshop recommended:

a. Exploring further the gap between measurements of NH_3 concentrations and model estimates, especially by: investigating high temporal resolution measurements; improving emission/deposition modeling; having a model inter-comparison with countries whose models do not show a gap and through uncertainty analysis
b. Fully implementing the EMEP monitoring strategy and improving the model treatments of NH_x in order to quantify the influence of a changing chemical climate. The EMEP monitoring strategy could be a good starting point for the development of a strategy focused on the appropriate questions. It was necessary to evaluate policies and indicators derived from them (in time and space). Using existing models, a pre-modelling study should select the monitoring sites that would eventually respond to the basic (policy) questions by using improved models and assessment tools. The best and most economically feasible instrumentation should be selected, and an extensive programme of quality assessment and quality control (QA/QC) should be used to make measurements comparable. After implementation, especially for trend evaluation, the monitors used should not be changed.

28.4.3   Assessment Methods for Ammonia Hot Spots

The workshop recommended:

a. Further developing dynamic NH_3 emission models to estimate the diurnal and seasonal changes in emission strengths from point sources (animal houses) and area sources (land spreading of animal manure). For area sources in detailed plot studies, this should include the effects of meteorological and soil variables.
b. Synthesizing information from available databases to identify reference cases against which different models could be tested and compared. It was agreed that an inter-comparison of regional-scale and subgrid models would greatly help in highlighting the differences between the modelling approaches and the abilities of regional models to simulate local-scale interactions.

c. Investigating scenarios of the possible effect of in-grid fragmentation of land use on net NH_3 fluxes. A sensitivity study would allow the investigation of the range of local recapture and possible effects on air quality. It was proposed that Parties should promote the development of deposition measurement methods that could apply to advective conditions (e.g. one measurement height) to allow verification of dispersion/deposition models.

28.4.4 Regional Modelling of Atmospheric Ammonia Transport and Deposition

The workshop concluded that more data were needed to provide a generalized scheme for ammonia compensation points with respect to the main land cover types in models. The models should carry out sensitivity tests with implementations of the compensation point (bi-directional exchange) schemes to estimate the magnitude of the effect on the dry deposition in the model.

Currently, differences in model performance between countries were not fully understood. These might reflect

a. Differences in the quality of the NH_3 emissions inventory
b. Differences in the model parameterization schemes
c. Geographical differences (climate, terrain)
d. Differences in measurement data sets

Hence a coordinated comparison of regional atmospheric ammonia models, using a common model domain, input database and measurement database, was urgently needed to assess relative model performance.

In many countries, better data (on emissions and monitoring) were available for national modelling efforts than those submitted under the Convention. EMEP reporting must be made more flexible to improve data availability.

28.4.5 Reliability of Ammonia Emission Estimates and Abatement Efficiencies

The workshop recommended devoting effort to estimating the uncertainty of regional and national NH_3 emission inventories. In particular, there was a need for international collaboration to obtain better activity data regarding agricultural management practices across Europe. This information was not typically available from statistical sources and was a key uncertainty in regional emissions.

Moreover, there was a need for further measurement data to underpin regional estimates of NH_3 emissions. In particular, data were needed from Southern and Eastern Europe. The workshop recommended devoting more effort to examining the quantitative synergies and trade-offs that occur in abating different forms of nitrogen emission (NH_3, nitrous oxide, nitrate leaching).

Noting that "soft" approaches to NH_3 abatement were an economically attractive method of reducing NH_3 emissions, the workshop recommended putting further research effort into methods to quantify the achievement of such approaches, so that the benefits could be considered within the Convention.

28.4.6   Ammonia Policy Context and Future Challenges

Considering the N trade-offs, the workshop recommended the extension of currently used tools, verification of specific elements of the models, adaptation of monitoring networks, targeted measurement programmes, and possible revision of legislation in order to close existing loopholes and increase synergies in addressing nitrogen pollution at large. Priority should be given to measures aiming at reducing all kinds of nitrogen losses at farm level. Ammonia emission reduction policies must be analysed in a multi-effect (human health, greenhouse balance, acidification and eutrophication and related biodiversity loss), multi-media (air, water, soil), multi-scale (hot spots, regional, European, global) framework.

Considering the recommendation to lower critical levels for NH_3, there was a need for careful evaluation of the representativeness of EMEP modelling results for NH_3 concentration. It was also recommended to give further consideration to whether, and if so, how, the new critical levels would be used in addition to critical loads in formulating air pollution targets, especially on local or regional scales in areas with spatially variable NH_3 emissions and concentrations.

Considering the increase in springtime NH_3 emissions that had occurred in implementing some policies such as the *EU Nitrates Directive*, further research was recommended to quantify the seasonal dependence of environmental impacts of NH_3. More attention was also needed on how to monitor and incorporate impacts of other N-related policies in modelling tools.

It was recommended to explore possibilities of considering local *Biodiversity Action Plans* in larger scale modelling. Strategies existed to integrate them into the European scale, e.g. via the *Flora Fauna and Habitats Directive* and *Natura 2000* network. However, the role of air pollution effects was often not explicitly taken into account even though N inputs had a large effect on biodiversity; there was room for improvement on local, national and European levels.

Reference

UNECE (2007) Review of the Gothenburg Protocol. Report on the workshop on atmospheric ammonia(ECE/EB.DIR/WG.5/2007/3),Geneva. http://www.unece.org/env/lrtap/WorkingGroups/wgs/documents.htm

Author Index

Subject Index

461